Defect and Fault Tolerance in VLSI Systems

Volume 1

Defect and Fault Tolerance in VLSI Systems

Volume 1

Edited by

Israel Koren

University of Massachusetts
Amherst, Massachusetts

Plenum Press • New York and London

Library of Congress Cataloging in Publication Data

Defect and fault tolerance in VLSI systems / edited by Israel Koren.

p. cm.

"Proceedings of the International Workshop on Defect and Fault Tolerance in VLSI Systems, held October 6-7, 1988, in Springfield, Massachusetts"—T.p. verso.

Includes bibliographies and index.

ISBN 978-1-4615-6801-8 ISBN 978-1-4615-6799-8 (eBook)

DOI 10.1007/978-1-4615-6799-8

1. Integrated circuits—Very large scale integration—Design and construction—Congresses. 2. Fault-tolerant computing—Congresses. I. Koren, Israel. II. International Workshop on Defect and Fault Tolerance in VLSI Systems (1988: Springfield, Mass.)

TK7874.D415 1989 89-8448

621.39′5—dc20 CIP

Proceedings of the International Workshop on
Defect and Fault Tolerance in VLSI Systems,
held October 6–7, 1988,
in Springfield, Massachusetts

Softcover reprint of the hardcover 1st edition 1989

A Division of Plenum Publishing Corporation
233 Spring Street, New York, N.Y. 10013

Program Committee

V.K. Agarwal, McGill, Canada

E. Fujiwara, NTT, Japan

I. Koren, UMass, USA

W. Maly, CMU, USA

E.J. McCluskey, Stanford, USA

J.F. McDonald, RPI, USA

W. Moore, Oxford, England

D.K. Pradhan, UMass, USA

G. Saucier, IMAG, France

A.D. Singh, UMass, USA

C.H. Stapper, IBM, USA

Organizing Committee

I. Koren, UMass (chairman)

A.D. Singh, UMass (vice-chairman)

Sponsors

IEEE Computer Society - Technical Committee on Fault-Tolerant Computing and Technical Committee on VLSI

University of Massachusetts at Amherst

Acknowledgment

The workshop and this book were supported in part by the National Science Foundation under grant MIP-8803418.

Preface

This book contains an edited selection of papers presented at the *International Workshop on Defect and Fault Tolerance in VLSI Systems* held October 6-7, 1988 in Springfield, Massachusetts. Our thanks go to all the contributors and especially the members of the program committee for the difficult and time-consuming work involved in selecting the papers that were presented in the workshop and reviewing the papers included in this book. Thanks are also due to the IEEE Computer Society (in particular, the Technical Committee on Fault-Tolerant Computing and the Technical Committee on VLSI) and the University of Massachusetts at Amherst for sponsoring the workshop, and to the National Science Foundation for supporting (under grant number MIP-8803418) the keynote address and the distribution of this book to all workshop attendees.

The objective of the workshop was to bring together researchers and practitioners from both industry and academia in the field of defect tolerance and yield enhancement in VLSI to discuss their mutual interests in defect-tolerant architectures and models for integrated circuit defects, faults, and yield. Progress in this area was slowed down by the proprietary nature of yield-related data, and by the lack of appropriate forums for disseminating such information. The goal of this workshop was therefore to provide a forum for a dialogue and exchange of views. A follow-up workshop in October 1989, with C.H. Stapper from IBM and V.K. Jain from the University of South Florida as general co-chairmen, is being organized.

This book summarizes the state-of-the-art contributions presented at the workshop. They are divided into ten chapters as outlined below.

Chapter 1 reviews yield models for defect tolerant VLSI circuits. I. Koren and C.H. Stapper show that several yield formulae that have been previously presented are equivalent and, consequently, result in the exact same value for the yield of defect-tolerant chips. In Chapter 2 the workshop keynote address is summarized. D.L. Peltzer reviews the Trilogy approach to wafer-scale integration and presents the design of a 112-Mbyte RAM on a six-inch wafer. This design may reduce the cost per bit by nearly a factor of two.

Chapter 3 includes three papers discussing models for defects and yield. A.V. Ferris-Prabhu reviews the relations between physical defects, device faults, and semiconductor yield. S.C. Seth and V.D. Agrawal study fault occurrence probabilities, which combined with detection probabilities provide the relation between the testing coverage and the chip quality, or reject ratio. The third paper, by C. Thibeault, Y. Savaria, and J-L. Houle, presents a new yield formula for fault-tolerant large-area chips that can be used to select the best fault tolerance strategy.

The three papers in Chapter 4 discuss defect-tolerant designs. M.C. Howells, R. Aitken, and V.K. Agarwal study defect-tolerant interconnects. They propose adding logic and spares into the interconnects to increase their testability and reconfigurability. R.J. Cosentino, B.L. Johnson, and J.J. Vaccaro present a defect-masking technique for designing fault-tolerant systolic architectures for signal processing. The added redundancy in this design is based on the residue number system. In the last paper, P. Mazumder discusses on-chip error correcting coding techniques for the design of fault-tolerant dynamic random-access memories.

Strategies for monitoring defects in VLSI chips and yield projection are presented in Chapter 5. R. Glang discusses monitoring of defects using defect-sensitive structures in test chips. He shows that the yield of product chips can be projected based on the defect data from these test chips. The second paper in this chapter, by J.J. Hammond, B. Boerman, and F.W. Voltmer, is concerned with the interaction between process development and circuit design in strategies for yield improvement. The authors present yield improvement techniques for the different phases of the manufacturing process from development to maturity. In the next paper, P. Schvan, D.Y. Montuno, and R. Hadaway describe a yield projection methodology that is based on fault distributions and critical structures statistics obtained from electrical measurements on simple test structures. Finally, S. Magdo presents several yield models for projecting the yield from test sites. The complexity and accuracy of these models are also compared.

Chapter 6 discusses testing and testable designs of VLSI chips. K. Yamashita *et al.* present the design of a wafer-scale FFT with built-in self-testing. D. Fussell, S. Rangarajan, and M. Malek describe a comparison-based algorithm for fault diagnosis of linear arrays. Fault diagnosis of processor arrays is also the subject of the third paper, by M. Cutler, M. Wang, and S.Y.H. Su. The authors present a built-in self-testing scheme for processor arrays with uniformly distributed faults.

Designs of fault-tolerant and high-yield processors are presented in Chapter 7. B. Maytal *et al.* describe the hierarchical methodology followed in the design of the recent 32-bit microprocessor NS32532 and its effect on the yield of this chip. The chosen method resulted in a significantly higher yield than would have been expected for a VLSI chip of this size. R. Leveugle, M. Soueidan, and N. Wehn present the detailed design of a high-yield defect-tolerant 16-bit microprocessor. The added redundancy occupied 30% of the chip area. S. Chau and D. Rennels describe a design methodology for a self-checking and self-exercising processor capable of concurrent detection of faults. The last paper in Chapter 7, by G.S. Sohi, discusses a defect-tolerant organization of an on-chip cache memory which can enhance the yield of the VLSI chip with a very small performance degradation.

Chapter 8 deals with defect- and fault-tolerant memory designs. W.K. Fuchs and M.-F. Chang review several algorithms for diagnosis and repair of large memories with spare rows and columns. B. Nasreddine *et al.* present the design of a 4.5-Mbit reconfigurable static RAM implemented on a 4-inch wafer. The configuration algorithm and the switches needed for replacing defective elements by redundant ones are described. C.H. Stapper presents a block alignment technique for steering blocks of defect-free data cells. This way, the number of faulty sections that can be tolerated is larger than with a straight partially good chip scheme, leading to a higher yield. Finally, C.A. Njinda, C.G. Guy, and W.R. Moore discuss the design of a fault-tolerant memory unit integrated in an application-specific processor. Error correction codes are used for defect tolerance.

The five papers in Chapter 9 discuss reconfigurable VLSI arrays. A. Jain and J. Rajski review several reconfiguration schemes for rectangular processor arrays and suggest a methodology for analyzing their yield and area utilization. S.Y. Kung, S.N. Jean, and C.W. Chang present a rectangular array grid based on single-track switches for tolerating defects (for yield enhancement) and run-time faults (for reliability improvement). A.D. Singh and H.Y. Youn present a restructuring method for wafer-scale processor arrays. They apply their scheme to rectangular arrays, binary trees, and hexagonal arrays. L. Jervis, F. Lombardi, and D. Sciuto describe a host-driven reconfiguration strategy for fault-tolerant two-dimensional arrays. The last paper, by N. Hasan, J. Cong, and C.L. Liu, introduces a general model for fault-covering problems in reconfigurable arrays representing the relationship between redundant elements and defective elements.

Fault-tolerant arrays are discussed in Chapter 10. G. Saucier *et al.* present the design of a defect-tolerant two-dimensional wafer-scale array for image processing. A hierarchical defect-tolerant strategy is implemented to maximize the yield enhancement. Next, F.J. Provost and R. Melhem describe fault-tolerance techniques for embedding binary trees and rings in hypercubes. Finally, Y.-X. Wang and J.A.B. Fortes address the problem of analytically estimating and optimizing the reliability of hierarchical fault-tolerant processor arrays.

Israel Koren

January 1989

Contents

1 YIELD MODELS FOR DEFECT-TOLERANT VLSI CIRCUITS: A REVIEW

YIELD MODELS FOR DEFECT-TOLERANT VLSI CIRCUITS: A REVIEW

Israel Koren and **Charles H. Stapper***

University of Massachusetts at Amherst
Department of Electrical and Computer Engineering
Amherst, MA 01003

* **IBM General Technology Division**
Essex Junction, VT 05452

ABSTRACT

The statistical models for estimating and predicting the manufacturing yields of VLSI circuits are reviewed. It is shown how defect clustering is taken into account, and how yield formulae for defect and fault tolerant VLSI circuits are developed. Different types of formulae for the yield of defect tolerant VLSI circuits have appeared in the literature. It is proven here for the first time that most of these approaches are equivalent.

INTRODUCTION

The designation *defect-tolerant* (or *fault-tolerant*) is often used in connection with integrated circuits that have some degree of tolerance to manufacturing flaws or defects. Such circuits are capable of functioning correctly when they contain certain types of faults. However, not all faults can be dealt with by the fault-tolerance schemes. As a result these circuits do not have 100% fabrication yields and the prediction of those yields is difficult. This difficulty is illustrated in at least three doctoral dissertations dealing with this subject, namely those of Mangir[1], Hedlund[2], and Harden[3].

The difficulty in modeling the yield of fault-tolerant integrated circuit chips is caused mainly by the clustering of manufacturing defects during chip fabrication. Several schemes for deriving expressions for the yield of fault-tolerant circuits in the presence of clustering have been proposed in different publications. We review these schemes and show that most of them are equivalent and result in the exact same value for the yield of such chips. We also present a simple way of handling the complexity of deriving yield expressions when defects are clustered.

In the next section we briefly review the most commonly used distributions for faults, namely, the Poisson distribution and the negative binomial distribution and show the resulting expressions for yield. We then present a method for determining the parameters of the negative binomial distribution and discuss the effect of the size of clusters relative to the size of chips. Next, the previously proposed models for the yield of fault tolerant chips are reviewed and the equivalence between them is proved. We also present a general method for deriving a yield expression for clustered faults from the simpler expression obtained when faults are assumed to be evenly distributed, i.e., follow the Poisson distribution. These results for chips with redundancy are then extended to partially good chips and multiple module-type chips. Final conclusions are presented in the last section.

YIELD MODELS

The earliest yield model published in the archival literature was the one used by Wallmark[4] in 1960. He expressed the probability of transistor failure as a ratio $S/100$, where S represents the number of failing transistors occurring in a batch of 100. For M transistors, the yield, denoted by Y_M, consequently becomes

$$Y_M = (1 - S/100)^M. \tag{1}$$

Wallmark used this result in a binomial distribution for estimating the yield of integrated circuits with redundant transistors. For obtaining exactly M good transistors out of a total of N this took the form

$$Y = \binom{N}{M} (S/100)^{N-M}(1 - S/100)^M. \tag{2}$$

This was the first step towards a yield model for integrated circuits with fault-tolerance. Several years later Hofstein and Heiman[5] also tackled the problem of manufacturing more circuits than were actually needed. Their chips contained field effect transistors, and they claimed that the transistor failures were predominately caused by pinholes in the oxide gates. Assuming a Poisson distribution for the number of defects per gate, they obtained a yield model for M transistors with the formula

$$Y_M = e^{-MA_GD} \tag{3}$$

where A_G is the gate area in each transistor and D an average defect density of pinholes per unit area. This expression was then used in a binomial formula for calculating the probability of having at least M good transistors on a chip with N transistors on it,

$$Y = \sum_{i=M}^{N} \binom{N}{i} e^{-iA_GD}(1 - e^{-A_GD})^{N-i}. \tag{4}$$

This was therefore, the first model for the yield of integrated circuits with fault-tolerance. Subsequently, Poisson statistics were commonly used for modeling the distribution of the number of faults per chip. According to this distribution the probability of having exactly x faults in a chip is given by,

$$Prob\ \{X = x\} = \frac{e^{-\lambda}\ \lambda^x}{x!} \tag{5}$$

where X is a random variable denoting the number of faults and λ is the average number of faults expected per chip. A fault is defined as a specific chip failure that can be caused by one or more manufacturing defects.

For chips with no redundancy the yield is therefore,

$$Y = Prob\ \{X = 0\} = e^{-\lambda}. \tag{6}$$

The average number of faults per chip is often expressed as $\lambda = AD$, the chip area A times a fault density D. This, however, is a simplification. The relationship between the average number of faults per chip and the chip area is more complicated; it depends on the circuit complexity, the density of photolithographic patterns, the number of photolithographic masks used in the process, etc. These however, are beyond the scope of this review.

It has been known since the beginning of integrated-circuit manufacture that the above yield formula is too pessimistic and leads to predicted chip yields that are too low when extrapolated from the yield of smaller chips or single circuits. It later became clear that the very low predicted yield was caused by the clustering of faults. This phenomenon has been observed in practice, but was not taken into account in equations (5) and (6).

Several modifications to the above yield formula have been proposed to account for fault clustering. The most commonly used modification is obtained by assuming the number of faults to be Poisson distributed as in (5) but to consider the parameter λ to be a random variable rather than a constant. The mere fact that λ is a random variable results in clustering of faults, no matter what type of distribution is assumed for λ.

The modified yield formula is then obtained by averaging yield formula (6) with respect to λ. Let $F(\lambda)$ be a cumulative distribution function of the average number of faults per chip. Associated with the cumulative distribution function $F(\lambda)$ is a probability density function $f(\lambda)$ given by

$$f(\lambda) = \frac{dF(\lambda)}{d\lambda} \tag{7}$$

where $f(\lambda)d\lambda$ is the probability of having an average number of faults per chip between λ and $\lambda + d\lambda$. Averaging (6) with respect to this probability density function results in a yield expression first used by Murphy[6]:

$$Y = \int_0^\infty e^{-\lambda} f(\lambda)\ d\lambda. \tag{8}$$

The function $f(\lambda)$ in this expression is known as a *compounder* or *mixing function.* A commonly used mixing function is the Gamma distribution [7–12] with two parameters α and β,

$$f(\lambda) = \frac{1}{\beta^\alpha\ \Gamma(\alpha)}\ \lambda^{\alpha-1}\ e^{-\lambda/\beta}. \tag{9}$$

Evaluating the integral in equation (8) results in the well known integrated-circuit yield formula

$$Y = (1 + \overline{\lambda}/\alpha)^{-\alpha} \tag{10}$$

where α is a clustering parameter and $\bar{\lambda} = \beta\alpha$ is the average number of faults per chip. It can be shown that $\bar{\lambda}$ is in effect the expected value of λ when the probability density function $f(\lambda)$ in (9) is used. This is therefore the grand average (average of averages) of the number of faults per chip.

The clustering parameter α also has physical significance. In the limit when $\alpha \to \infty$, the yield in expression (10) becomes equal to yield formula (6). This represents the case of random faults and complete absence of clustering. Smaller values of α indicate increased clustering. Actual values for α typically range between 0.3 and 5. Methods for determining this parameter are described in the next section.

Applying the same averaging (or compounding) procedure to the Poisson probability function for the number of faults in (5), results in the negative binomial distribution,

$$Prob\ \{X = x\} = \frac{\Gamma(\alpha + x)}{x!\ \Gamma(\alpha)} \frac{(\bar{\lambda}/\alpha)^x}{(1 + \bar{\lambda}/\alpha)^{\alpha+x}}. \tag{11}$$

The yield formula in (10) is a special case of (11) with $x = 0$.

To illustrate the effect of fault clustering consider a chip containing a varying number of identical circuits. Let us start with a single circuit that has a hypothetical yield of 0.999 and an average of 0.001 faults per circuit. If we use model (6), the yield of a chip with 600 of these circuits is equal to $e^{-600\times 0.001}$ which is approximately equal to 55%. For a chip with 40,000 logic circuits, we expect a yield of $e^{-40,000\times 0.001} = 4.248 \cdot 10^{-18}$, or for all practical purposes 0%.

If clustering is taken into account, the yield formula for a chip with M identical circuits is given by

$$Y_M = (1 + M \cdot \bar{\lambda}_1/\alpha)^{-\alpha} \tag{12}$$

where the average number of faults in a single circuit is denoted by $\bar{\lambda}_1$. Assuming again that this number is equal to 0.001, it is possible to estimate the yield for chips with any number of circuits. Calculated yields for chips with a single circuit, chips with 600 circuits, and chips with 40,000 circuits are tabulated in Table 1 for values of $\alpha = 0.5, 1, 2$ and ∞. These results show that even if $M\bar{\lambda}_1$ is high, the presence of a high degree of clustering leads to surprisingly high yields. This effect has been observed in many manufacturing lines.

In most cases a gross yield factor Y_0 must be included in the yield model. Gross yield losses usually are the result of systematic processing problems that affect whole

Table 1: Yield as a function of the number of circuits per chip and the clustering parameter α, when $\bar{\lambda}_1 = .001$.

α	Yield in % M=1	Yield in % M=600	Yield in % M=40,000
0.5	99.9	67.4	11.1
1	99.9	62.5	2.4
2	99.9	59.2	0.2
∞	99.9	55.0	0

wafers or parts of wafers. Such losses may, for example, be caused by misalignment, over- or under-etching or out-of-spec semiconductor parameters such as beta transconductance or threshold voltage. Paz and Lawson[10] have shown that fault clusters with very high fault densities can also be modeled by Y_0.

Introduction of the gross yield into the yield formula leads to

$$Y = Y_0(1 + \bar{\lambda}/\alpha)^{-\alpha} \tag{13}$$

This three parameter model has been used successfully for yield modeling since 1975. Its parameters have physical significance and can be determined by a straightforward technique described in the next section. It must be pointed out, however, that the simplicity of this model can be deceptive. Some of the hidden complexities are discussed in subsequent sections.

SIZE OF CLUSTERS AND DETERMINATION OF PARAMETERS

Defect clusters in integrated circuits can be roughly categorized into three classes. The first includes clusters much larger than the chip size. Most papers on integrated-circuit yield which take fault clustering into account have assumed, sometimes unknowingly, that clusters are larger than the chip size. This is implied by the assumption that the value of the clustering parameter α is the same when the whole chip is considered or when only part of the chip is considered. The success of these yield models can be attributed to the fact that this is not a bad assumption. According to Stapper[13] most of the clustering is expected to be caused by wafer-to-wafer variations of fault densities. In that case, the cluster area is equal to the wafer size, which is indeed larger than the area of individual chips. Another source of clustering is the radial variation in the average number of faults per chip. This effect was originally described by Yanagawa[14-15], confirmed by others[10,16] and studied more recently by Ferris-Prabhu *et al*[17], Walker[18] and Gandemer[19]. It leads to a lower chip yield along the periphery of integrated-circuit wafers. This peripheral region can therefore in effect be considered a large-area cluster.

The radial variation of chip yield has led to the use of concentric wafer zones for yield analysis [10,13,16,20]. In such analyses, it is usually assumed that the faults per chip within each zone are distributed according to Poisson's distribution. Each zone has its own average number of faults per chip λ. The yield inside a zone can therefore be estimated by using formula (6). The yield of chips in all zones from many wafers can be combined and results in a *compound* or *mixed Poisson* yield model.

Another class of clusters deals with fault clusters that are smaller than the chip area. It is sometimes believed that the faults in such small clusters should distribute themselves according to Poisson's distribution. This distribution, however, is too constrained, because it has a variance that is equal to the mean. Clusters, by their very nature, tend to increase the variability in the number of faults per chip. As a result, clustering leads to distributions with variances that are larger than the mean. Statistics applicable to this type of clustering are described by Stapper[21].

The third class of fault clusters deals with clusters that vary in dimension. This area has been investigated by Warner[22,23], Hu[24], Stapper[25], and in an approximate point defect model for wafer-scale-integration by Ketchen[26]. A simulation technique for modeling this situation has, furthermore, been described by Foard Flack[27] and Stapper[28]. These efforts, however, have not been definitive.

To understand how in practice one determines whether the large-area clustering assumption is valid (for a given size of chip) we need first to review the method used for determining the parameters of the yield model.

The values of Y_0, $\overline{\lambda}$ and α in formula (13) are usually determined with the "window method." This method was first described by Seeds[29-30] and subsequently by Okabe *et al*[31], Warner[22,23], Paz *et al*[10] and Hemmert[11]. The objective is to determine the yield as a function of chip multiples. This is done with wafer maps that show the location of functioning and failing chips at final test. The maps are analyzed using overlays with grids, or windows. These windows contain blocks of chips. Each block usually contains two, four, six, or nine chips. For each chip multiple, the number of windows containing only fault free chips can be counted. Dividing this number by the total number of windows in the sample gives us the yield for that multiple.

The results of the window analysis must next be matched to a yield formula. For the negative binomial model this has the form

$$Y_M = Y_0(1 + M\overline{\lambda}/\alpha)^{-\alpha} \tag{14}$$

where M is the chip multiple. Values for the parameters Y_0, $\overline{\lambda}$ and α are usually determined by means of a nonlinear regression analysis. Here it is implicitly assumed that the value of α is the same for all different sizes of windows, i.e., large-area clusters are assumed.

Note that high values of α obtained by this method do not necessarily mean that there is less clustering. It only implies that there is less large-area clustering. Small area clusters can still exist, but this method is impervious to them. The smaller clusters are essentially counted as single faults.

It is not difficult to use the window method. An example of a window method analysis is tabulated in Table 2. The data in that table came from 24 wafers, each one containing 89 memory chips. For each wafer a map was obtained to show the location of fault free and faulty chips. One of these maps is shown in Figure 1. Also shown on the map are locations taken up by test sites used to measure processing parameters.

The first step in evaluating the wafer map data is the determination of the chip yield. In this case 701 out of a total of 2136 chips were fault free. The yield was therefore 32.82%. Next, a transparent overlay was made with a grid containing pairs of chips. It was found that only 42 pairs could be placed on the grid of each wafer. This resulted in a sample of 1008 pairs. Only 140 of these were found to be free of faulty chips. The yield for these windows with blocks of two chips was therefore 13.89%.

The third step consisted of making an overlay grid that contained four chips in a 2 x 2 arrangement. Seventeen such windows could be fitted unambiguously on a wafer. To increase the sample size, and to include as much of the circumferential area as possible, three additional odd-shaped windows containing four chips were formed along the wafer edge. The total sample therefore contained 480 windows. For 18 of

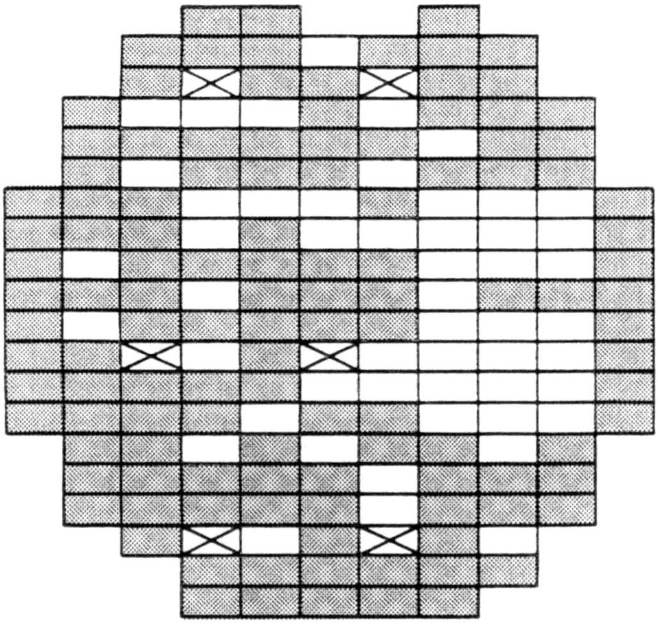

Fig. 1: Wafer map showing the locations of fault-free (light) and faulty (dark) chips. Test sites are marked with crosses.

Table 2: Illustrative use of the window method to determine model parameters. Here, $Y_0 = 1$, $\overline{\lambda} = 1.2934$, and $\alpha = 3.8274$.

Chip Multiples	Sample Size	Number Perfect	Yield in % Data	Yield in % Model
1	2136	701	32.82	32.82
2	1008	140	13.89	13.86
4	480	18	3.75	3.79

these windows it was found that all four chips were free of faults, thus resulting in a yield of 3.75%.

It is possible to obtain an additional data point by analyzing blocks of three chips. Such blocks, however, have odd-shaped windows, which makes them awkward to use. The three data points in Table 2 supply sufficient data for determining the parameters of the yield model. The values for $\overline{\lambda}$, α and Y_0 were obtained by fitting equation (14) to these data points with a computer program that minimized the sum of the squares of the differences between model and data. With three data points and three parameters in equation (14), this was equivalent to solving three nonlinear equations with three unknowns. For these data, furthermore, it was possible to set $Y_0 = 1$. This led to the values $\overline{\lambda} = 1.2934$, and $\alpha = 3.8274$ for the other two parameters. Putting these values into equation (14) led to the numbers shown in the column labeled "Model Yield" in Table 2. The experimental yields are also tabulated and are in good agreement. Because of the non-linearity, even with three data points, such agreement is not always guaranteed for this three parameter model.

The window method analysis is used regularly in the industry. A variation of such an analysis was described by Hemmert[11]. His data were obtained from wafer maps of logic chips and read only memories (ROMs). He used a least square fitting technique to determine $\overline{\lambda}$ and α in equation (14) while keeping Y_0 at 10(% yield. His results on seven manufacturing lots of wafers had an average cluster parameter of 2.2 with a standard deviation of 0.22. The values of α were therefore tightly grouped, indicating that they were stable during the fabrication of those lots.

An alternative use of equation (14) has been described in references [12,32]. The yield of different read only memory chips was analyzed as a function of the number of bits in those chips. This number was represented by M in equation (14). The values of Y_0, $\overline{\lambda}$, and α in that case were also determined with a nonlinear least square minimization technique. This analysis was performed on data from three different manufacturing lines and resulted in values for α of 1.27, 0.86, and 0.75. The lowest value, and therefore the highest degree of clustering, occurred on wafers fabricated in the manufacturing line with the highest chip yields. The highest value of α, suggesting less clustering, resulted from the wafers that were made in the line with the lowest chip yields.

The yield analysis of these read-only memory chips also showed that the gross yield Y_0 varied between 70.8% and 90.4%. Although these numbers include the yield of the support circuits on these chips, this range of gross yields is typical for most integrated circuits. The lowest value of Y_0 occurred in the low yield line and the highest value of Y_0 in the high yield line.

It must be noted here that the values of $\overline{\lambda}$ obtained by this method tend to be lower than the actual average number of faults observed on chips. This difference can be caused by the effect of clusters that are smaller than the chip. As mentioned before, such clusters are counted as single faults by this technique.

To find the chip sizes for which the large-area clustering assumption is valid particle distributions on actual wafers can be studied. This was done, for example, in reference [33], where wafer surfaces were subdivided into squares called *quadrats.* Negative binomial distribution were found in this study to be in good agreement with the frequency distribution of the number of particles in each quadrat for a wide range of quadrat sizes. The values of the cluster parameter α, however, differed for quadrats with different areas.

The data obtained with quadrat analysis can be analyzed using a maximum likelihood estimation technique described by Foard Flack[34]. This approach makes it possible to determine the variability in the estimated values of α. The results of such an analysis are shown in Figure 2. The bars around the data point indicate the range of $\pm\sigma_\alpha$, where σ_α is the standard deviation of each estimate. Note that the horizontal scale is logarithmic and represents a range of two orders of magnitude in area.

Of interest in Figure 2 are the results for the three smallest quadrat areas. The ranges of standard deviations overlap, thus suggesting that these points represent the condition for large-area clustering. The increase in values of α for the other points

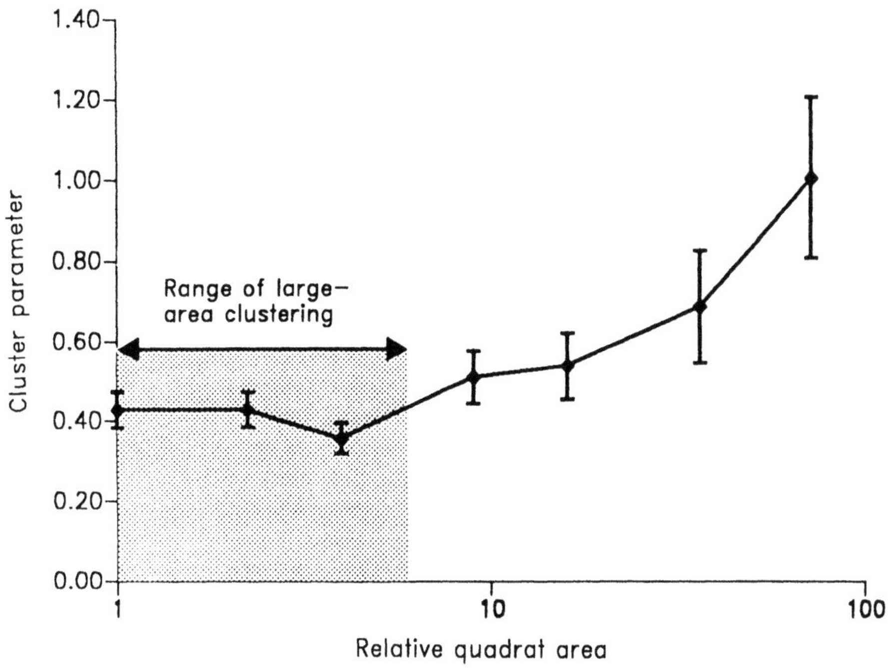

Fig. 2: Experimental dependence of the clustering parameter α on quadrat area.

on the curve indicate that the associated quadrat areas are exceeding the range for which the large-area clustering assumption is valid.

YIELD MODELS FOR CHIPS WITH REDUNDANCY

In many integrated circuit chips, identical blocks of circuits are often replicated. In memory chips, these are blocks of memory cells which are also known as *sub-arrays*. In processor arrays these basic circuit blocks are referred to as *processing elements*, or *PEs*. In other digital chips they are referred to as *macros*. The terminology depends not only on the type of circuitry that is used, but also by whom it is used. The designation *modules* is used in this paper. It is meant to be general and include all these designations.

Chips containing a number of identical modules (of one type or more) can often be used even if some of the modules do not function correctly. We obtain this way *partially good chips*. Alternatively, we can add a few redundant modules to our design and accept only those chips which have the necessary number of fault-free modules. We will first consider chips with a single type of identical modules and then extend our results to multiple module-type chips.

Let N be the number of identical circuit modules. Define the following probability

$$a_{M,N} = Prob \, \{\text{Exactly M out of the N modules are fault} - \text{free}\} \tag{15}$$

This probability can be used to calculate the yield of chips with redundancy and that of partially good chips. For example, if R out of the N modules are spares meaning that a chip with at least $(N - R)$ fault-free modules is acceptable, then the yield of the chip is given by

$$Y = \sum_{M=N-R}^{N} a_{M,N} \tag{16}$$

Two methods have been used to calculate the probability $a_{M,N}$. In the first method it is assumed that for any given subset of n modules we can compute the probability that this subset is fault-free. The latter is the yield of this subset and is denoted by

$$y_n = Prob \, \{X_n = 0\}; \qquad n = 1, 2, ..., N \tag{17}$$

where X_n is a random variable denoting the number of faults in n modules. For example, if a Poisson distribution is assumed then,

$$y_n = e^{-n\lambda} \tag{18}$$

while if the negative binomial distribution is assumed then,

$$y_n = (1 + n\bar{\lambda}/\alpha)^{-\alpha} \tag{19}$$

where λ (and similarly, $\bar{\lambda}$) is the average number of faults per module. Note that equation (19) is based on the large-area clustering assumption.

If the faults occurring in different modules are independent (as in the case where faults follow the Poisson distribution) then $y_n = y^n$ where $y = y_1$ is the yield of a single module, i.e., the probability that the module is fault-free. We can use in this case the binomial distribution to obtain the following expression for $a_{M,N}$

$$a_{M,N} = \binom{N}{M} y^M (1 - y)^{N-M} \tag{20}$$

If however, the faults in different modules are dependent (as in the case where faults follow the negative binomial distribution), then the Inclusion and Exclusion principle must be used to calculate the probability $a_{M,N}$. Defining the event - the i-th module is fault-free, then $a_{M,N}$ is the probability of exactly M such events occurring simultaneously, i.e.,

$$a_{M,N} = \binom{N}{M} \sum_{k=0}^{N-M} (-1)^k \binom{N-M}{k} y_{M+k} \tag{21}$$

Notice that for the Poisson distribution equations (20) and (21) are equivalent.

In the second method for calculating $a_{M,N}$ we compute the probability that a given number of faults occur in the complete chip (containing N modules) and then distribute these faults uniformly among the N modules. (Variants of this scheme were used in references[1-3,20,35-40].) Again, this is justified if large-area clustering is assumed. Thus, the probability that exactly $(N-M)$ modules will contain faults is[39,40]

$$a_{M,N} \;=\; \sum_{x=N-M}^{\infty} Q_{x,(N-M)}^{(N)} \;\cdot\; Prob\,\{X_N = x\} \tag{22}$$

where $Prob\,\{X_N = x\}$ is the probability that the chip has x faults and $Q_{x,j}^{(N)}$ is the probability that the x faults are distributed into exactly j out of N modules given that there are x faults. Assuming that faults are distinguishable, the latter equals[40],

$$Q_{x,j}^{(N)} \;=\; \sum_{k=0}^{j} (-1)^k \binom{N}{k,\, j-k,\, N-j} \left[\frac{j-k}{N}\right]^x \qquad \text{for } x \geq j \text{ and } 0 < j \leq N \tag{23}$$

where $\binom{N}{k,j-k,N-j} = \binom{N}{k}\binom{N-k}{N-j}$ is the multinomial coefficient.

For $x < j$ we have $Q_{x,j}^{(N)} = 0$ and for $x = j = 0$ $Q_{0,0}^{(N)} = 1$ and consequently, we can rewrite (22) as follows,

$$a_{M,N} \;=\; \sum_{x=0}^{\infty} Q_{x,(N-M)}^{(N)} \;\cdot\; Prob\,\{X_N = x\} \tag{24}$$

We show next that the two methods in (21) and (24) yield the same expression for $a_{M,N}$ when the probability of having x faults follows the Poisson distribution, i.e.,

$$Prob\,\{X_N = x\} = \frac{e^{-N\lambda}(N\lambda)^x}{x!}\;. \tag{25}$$

Substituting (23) and (25) in (24) results in,

$$\begin{aligned}
a_{M,N} &= \sum_{x=0}^{\infty} Prob\,\{X_N = x\} \;\cdot\; Q_{x,(N-M)}^{(N)} \\
&= \sum_{x=0}^{\infty} \frac{e^{-N\lambda}(N\lambda)^x}{x!} \cdot \sum_{k=0}^{N-M} (-1)^k \binom{N}{k,\, M,\, N-M-k} \left[\frac{N-M-k}{N}\right]^x \\
&= \sum_{k=0}^{N-M} (-1)^k \binom{N-M}{k} \binom{N}{M} e^{-N\lambda} \sum_{x=0}^{\infty} \frac{[\lambda(N-M-k)]^x}{x!} \\
&= \sum_{k=0}^{N-M} (-1)^k \binom{N-M}{k} \binom{N}{M} e^{-N\lambda} e^{(N-M-k)\lambda} \\
&= \binom{N}{M} e^{-N\lambda} e^{(N-M)\lambda} \sum_{k=0}^{N-M} (-1)^k \binom{N-M}{k} e^{-k\lambda} \\
&= \binom{N}{M} e^{-M\lambda}(1-e^{-\lambda})^{N-M} = \binom{N}{M} y^M (1-y)^{N-M}
\end{aligned}$$

The equivalence of the above two methods is not restricted to the simple case of the Poisson distribution for faults but holds for other distributions as well, in particular for the negative binomial distribution as shown in what follows. Substituting (23) in (24) we obtain,

$$a_{M,N} = \sum_{x=0}^{\infty} Prob\,\{X_N = x\} \;\cdot\; Q_{x,(N-M)}^{(N)}$$

$$= \sum_{x=0}^{\infty} Prob\ \{X_N = x\} \sum_{k=0}^{N-M} (-1)^k \binom{N-M}{k} \binom{N}{M} \left[\frac{N-M-k}{N}\right]^x$$

$$= \sum_{k=0}^{N-M} (-1)^k \binom{N-M}{k} \binom{N}{M} \sum_{x=0}^{\infty} Prob\ \{X_N = x\} \left[\frac{N-M-k}{N}\right]^x$$

$$= \sum_{k=0}^{N-M} (-1)^k \binom{N-M}{k} \binom{N}{M} G\left(\frac{N-M-k}{N}\right)$$

where $G(s)$ is the generating function of the probability distribution $Prob\ \{X_N = x\}$. For the negative binomial distribution we have

$$G(s) = \left[1 + \frac{(1-s)N\overline{\lambda}}{\alpha}\right]^{-\alpha} \tag{26}$$

Substituting (26) into the above equation yields,

$$a_{M,N} = \sum_{k=0}^{N-M} (-1)^k \binom{N-M}{k} \binom{N}{M} \left[1 + \frac{\left(1 - \frac{N-M-k}{N}\right) N\overline{\lambda}}{\alpha}\right]^{-\alpha}$$

$$= \sum_{k=0}^{N-M} (-1)^k \binom{N-M}{k} \binom{N}{M} \left[1 + \frac{(M+k)\overline{\lambda}}{\alpha}\right]^{-\alpha} \tag{27}$$

which is identical to (21) after substituting y_{M+k} by its proper expression from (19). Note that the equivalence of the two schemes for the Poisson distribution can be proved similarly using the generating function for the Poisson distribution which is given by

$$G(s) = e^{N\lambda(s-1)} \tag{28}$$

The negative binomial distribution is obtained from the Poisson distribution by averaging over all values of λ, using the Gamma distribution function. This compounding procedure can be applied to any statistical measure. We can derive an expression for the desired measure assuming the very convenient Poisson distribution (whose most useful property is the statistical independence between faults in different modules), and then apply the compounding procedure to obtain the required expression for the negative binomial model.

To illustrate this procedure we show next that $a_{M,N}$ in (27) can be obtained by compounding (20) when $y = e^{-\lambda}$. Equation (20) can be rewritten in the form of (21) by expanding $(1-y)^{N-M}$ into the following binomial series

$$(1-y)^{N-M} = \sum_{k=0}^{N-M} (-1)^k \binom{N-M}{k} y^k \tag{29}$$

Substituting this series expansion into (20) results in,

$$a_{M,N} = \binom{N}{M} \sum_{k=0}^{N-M} (-1)^k \binom{N-M}{k} y^{M+k} \tag{30}$$

By compounding (30) with the Gamma distribution in (9) we obtain,

$$a_{M,N} = \binom{N}{M} \sum_{k=0}^{N-M} (-1)^k \binom{N-M}{k} \int_0^{\infty} e^{-(M+k)\lambda} \cdot f(\lambda) \cdot d\lambda$$

$$= \binom{N}{M} \sum_{k=0}^{N-M} (-1)^k \binom{N-M}{k} \left[1 + \frac{(M+k)\overline{\lambda}}{\alpha}\right]^{-\alpha} \quad (31)$$

This very powerful compounding procedure was employed to derive yield expressions for interconnection buses in VLSI chips[41], for wafer scale cube-connected-cycles[42], and for partially good memory chips[28,37].

The expression for $a_{M,N}$ in (31) can be used to calculate the yield of chips with redundancy, using equation (16), and partially good chips as will be shown subsequently. To calculate the yield of a chip with a single type of redundant modules we can derive a somewhat simpler expression involving one summation instead of two. Employing the previous notation, the required yield is

$$Y = Prob\ \{\text{There are at least (N} - \text{R) fault} - \text{free modules}\} \quad (32)$$

Define the event - the i-th module is fault-free, then Y is the probability that at least $(N - R)$ such events occur simultaneously. According to the Inclusion and Exclusion principle, Y can be written as,

$$Y = \sum_{n=N-R}^{N} (-1)^{n-N+R} \binom{n-1}{N-R-1} \binom{N}{n} y_n \quad (33)$$

It should be noted that we may use either equation (18) or (19) for y_n in (33) and obtain a yield formula for evenly distributed faults or clustered faults, respectively. Replacing the index in (33) by $i = N - n$ yields,

$$Y = \sum_{i=0}^{R} (-1)^{R-i} \binom{N}{N-i} \binom{N-1-i}{N-R-1} y_{N-i} \quad (34)$$

This is the yield formula (with a different notation, i.e., $M = N - R$) presented by Harden and Strader[43] and deduced from several special cases of N and R. The yield expressions in (34) and (16) are equivalent.

We conclude this section with an example illustrating the effect of clustering on the yield of chips with redundancy. Consider a chip on which ten identical circuit modules must be functioning correctly if the chips are to be usable. Let the yield of the ten circuit modules be equal to 10%. We can then investigate how the chip yield is affected when we have one to five redundant circuits. This is done in Table 3, where yields (in %) correspond to different values of the clustering parameter α.

The pure random fault model corresponds to $\alpha = \infty$. In this case, according to Table 3, the use of five redundant circuits increases the yield from 10% to 93.1%. If, however, the clustering parameter is $\alpha = 0.5$, the yield is expected to improve from 10% to 24.1%. This indicates that the yield prediction for purely random faults is four times higher than the prediction for clustered faults. Miscalculations by a factor of four in the productivity of semiconductor manufacturing plants can be very costly.

Table 3: Yield (in %) with different amounts of redundancy for varying degrees of large-area fault clustering.

α	R=0	R=1	R=2	R=3	R=4	R=5
0.5	10	14.6	17.8	20.3	22.4	24.1
1	10	18.3	25.2	31.1	36.1	40.6
2	10	22.4	34.3	44.8	53.7	61.1
∞	10	30.6	53.8	73.0	85.8	93.1

Inclusion of clustering in redundancy yield calculation is therefore of considerable importance.

Partially Good Chips

Partially good chips are chips which are usable even if only some of their identical modules are fault-free. Consider for example, chips consisting of four identical modules. These chips are known as *perfect* if all four modules are fault-free. The fraction of chips falling in this category represents the perfect chip yield. The chips with three operating modules and one defective module are referred to as being *three-quarter-good.* The yield of these chips is known as the three-quarter-good yield and is equal to $a_{3,4}$. Similar designations apply to chips that are *half-good* and *quarter-good* and their yields are $a_{2,4}$ and $a_{1,4}$, respectively. In general, the yield of a partially-good chip with exactly M fault-free modules out of N, is given by $a_{M,N}$. The difference between a partially-good chip and a chip with redundancy is that *all* fault-free modules in a partially-good chip are considered to be usable while only $(N-R)$ fault-free modules in a chip with redundancy are expected to be used. Thus, a three-quarter-good chip is more valuable than a half-good chip. Consequently, instead of summing up all $a_{M,N}$'s to obtain the yield as is done for chips with redundancy, we multiply each $a_{M,N}$ by a weight equal to M/N to obtain what is called the *equivalent yield,*

$$Y_{EQ} = \sum_{M=J}^{N} \frac{M}{N} a_{M,N} \tag{35}$$

where J is the minimum number of modules which have to be fault-free.

We show in what follows that if $J = 1$ then $Y_{EQ} = y$, i.e., the equivalent yield in this case is equal to the yield of the individual circuit module. This implies that the use of partially good chips results in utilization of all the fault-free modules; none have been wasted. We first prove it for the Poisson distribution using equation (20),

$$Y_{EQ} = \sum_{M=1}^{N} \frac{M}{N} a_{M,N} = \sum_{M=1}^{N} \frac{M}{N} \binom{N}{M} y^M (1-y)^{N-M} = y \sum_{M=1}^{N} \binom{N-1}{M-1} y^{M-1} (1-y)^{N-M}$$

Substituting $m = M - 1$ and $n = N - 1$ yields,

$$Y_{EQ} = y \sum_{m=0}^{n} \binom{n}{m} y^m (1-y)^{n-m} = y \tag{36}$$

Applying the compounding procedure to the above equation proves our claim for the negative binomial distribution as well.

Some practical modifications

The simple architecture analyzed in the preceding section is an idealization because actual chips rarely consist entirely of identical circuit modules. In all chips there are support circuits in addition to such modules. These support circuits are shared by the replicated modules. The chips, however, become unusable if such support circuits are damaged beyond use. In principle, this effect can be included in the expression for $a_{M,N}$ by multiplication with the yield of the support circuits. Doing so, however, would assume that the clustering of the support circuit faults is completely independent of the clustering of the module faults. In most practical cases there is a dependence between the average number of faults in different circuits. This effect can be taken into account by including in formula (19) the average number of faults that cause these support circuits to be defective. This results in

$$y_n = [1 + (\overline{\lambda}_{CK} + n\overline{\lambda})/\alpha]^{-\alpha} \tag{37}$$

where $\overline{\lambda}_{CK}$ is the average number of fatal faults or *chip-kill* faults in the support circuits. Chips with these faults cannot be used. Substituting expression (37) into formula (31) makes it possible to take these types of faults into account when calculating the yields of partially-good chips (or chips with redundancy) with support circuitry.

Another effect that must be included in yield estimates is the gross yield. Unless the chips are very large, this yield is independent of chip area. It is used as a yield multiplier, which has been denoted by Y_0 in the preceding sections. Introducing it into yield formula (37) results in

$$y_n = Y_0[1 + (\overline{\lambda}_{CK} + n\overline{\lambda})/\alpha]^{-\alpha} \tag{38}$$

Introduction of this expression into equation (31) results in a formula that can be used to estimate yields of partially-good chips and chips with redundancy, with support circuits and gross yield losses.

Multiple Module-type Chips

The discussion above was restricted to the case where redundancy is provided to tolerate faults in a single type of circuit modules. In this section we extend the previous results to fault tolerant chips with multiple types of modules. We derive yield expressions for chips with two different types of modules, say, Type 1 and Type 2. The extension to a larger number of module types is straightforward and is therefore, not presented here.

Suppose that there are redundant modules of both types and that the modules of each type can be reconfigured separately when necessary. Then, we can calculate the yield of each module type separately (assuming that faults follow the Poisson distribution), and multiply the two results to obtain the overall yield,

$$Y = Y_1 \cdot Y_2 \tag{39}$$

where Y_i $(i = 1, 2)$ is the yield of the set of N_i modules of type i. This yield can be calculated using

$$Y_i = \sum_{M_i=N_i-R_i}^{N_i} a_{M_i,N_i} \tag{40}$$

The expression for Y will consist of $R_1 \cdot R_2$ terms. However, in many practical chip architectures there is no such ***architectural independence*** and a fault in a module of type 1 may affect the usefulness of type 2 modules, i.e., a type 2 module may become useless when a type 1 module is defective. In such a case not all $R_1 \cdot R_2$ possible terms should be included in the expression for Y. In well-structured architectures we can easily identify those terms that should be included and we can therefore define a "coverage factor" as follows, $c_{M_1,M_2} = 1$ if the chip is acceptable with M_1 and M_2 fault-free modules of type 1 and 2, respectively. Otherwise, $c_{M_1,M_2} = 0$. Consequently,

$$Y = \sum_{M_1=N_1-R_1}^{N_1} \sum_{M_2=N_2-R_2}^{N_2} a_{M_1,N_1} \cdot a_{M_2,N_2} \cdot c_{M_1,M_2} \tag{41}$$

c_{M_1,M_2} serves to select all the *fixable* combinations out of all combinations of fault-free modules of type 1 and type 2.

In less structured architectures, the number of fault-free modules of either type may be insufficient to determine whether the chip is fixable or not; we may also need to know the exact position of the fault-free modules. In such a case, c_{M_1,M_2} will not be a factor assuming only the values 0 and 1, but the fraction of fixable patterns out of all patterns consisting of M_1 and M_2 fault-free modules of type 1 and 2, respectively.

The final expression for the yield of a chip with two types of identical modules and support circuitry when the Poisson distribution is assumed, is as follows,

$$Y = Y_0 \sum_{\substack{M_1=\\N_1-R_1}}^{N_1} \sum_{\substack{M_2=\\N_2-R_2}}^{N_2} \sum_{k_1=0}^{N_1-M_1} \sum_{k_2=0}^{N_2-M_2} (-1)^{k_1}(-1)^{k_2} \binom{N_1}{M_1}\binom{N_1-M_1}{k_1}\binom{N_2}{M_2}\binom{N_2-M_2}{k_2}$$

$$\cdot\, e^{-(M_1+k_1)\lambda_1} e^{-(M_2+k_2)\lambda_2} e^{-\lambda_{CK}} \cdot c_{M_1,M_2} \tag{42}$$

Next we have to apply the compounding procedure to calculate the yield when clustering of faults is allowed. We should not however, perform three separate compounding steps (for the two types of modules and the support circuits) since the clustering of faults in one type of circuits is not independent of the clustering in the other two. We must therefore, perform a single compounding step using the average number of faults in the complete chip, i.e.,

$$\lambda = \lambda_1 N_1 + \lambda_2 N_2 + \lambda_{CK} \tag{43}$$

To simplify the integration of the various summands in (42) which contain different multiples of λ_1 and λ_2 we define,

$$\delta_1 = \frac{\lambda_1 N_1}{\lambda}, \quad \delta_2 = \frac{\lambda_2 N_2}{\lambda} \quad \text{and} \quad \delta_3 = \frac{\lambda_{CK}}{\lambda}$$

Note that δ_1 (and similarly, δ_2) is a constant which depends mainly on the area ratio of a type 1 module to the whole chip. The exponential terms in (42) now take the form,

$$e^{-(M_1+k_1)\lambda_1-(M_2+k_2)\lambda_2-\lambda_{CK}} = e^{-[(M_1+k_1)\delta_1/N_1+(M_2+k_2)\delta_2/N_2+\delta_3]\lambda}$$

Substituting the above in (42) and compounding with respect to λ results in,

$$Y = Y_0 \sum_{\substack{M_1= \\ N_1-R_1}}^{N_1} \sum_{\substack{M_2= \\ N_2-R_2}}^{N_2} \sum_{k_1=0}^{N_1-M_1} \sum_{k_2=0}^{N_2-M_2} (-1)^{k_1}(-1)^{k_2} \binom{N_1}{M_1} \binom{N_1-M_1}{k_1} \binom{N_2}{M_2} \binom{N_2-M_2}{k_2}$$

$$\cdot \left[1 + \frac{((M_1+k_1)\delta_1/N_1 + (M_2+k_2)\delta_2/N_2 + \delta_3)\,\overline{\lambda}}{\alpha}\right]^{-\alpha} \cdot c_{M_1,M_2} \tag{44}$$

Finally, we define $\overline{\lambda}_1 = \overline{\lambda}\delta_1/N_1$, $\overline{\lambda}_2 = \overline{\lambda}\delta_2/N_2$, $\overline{\lambda}_{CK} = \overline{\lambda}\delta_3$ and obtain,

$$Y = Y_0 \sum_{\substack{M_1= \\ N_1-R_1}}^{N_1} \sum_{\substack{M_2= \\ N_2-R_2}}^{N_2} \sum_{k_1=0}^{N_1-M_1} \sum_{k_2=0}^{N_2-M_2} (-1)^{k_1}(-1)^{k_2} \binom{N_1}{M_1} \binom{N_1-M_1}{k_1} \binom{N_2}{M_2} \binom{N_2-M_2}{k_2}$$

$$\cdot \left[1 + \frac{\left((M_1+k_1)\overline{\lambda}_1 + (M_2+k_2)\overline{\lambda}_2 + \overline{\lambda}_{CK}\right)}{\alpha}\right]^{-\alpha} \cdot c_{M_1,M_2} \tag{45}$$

A simpler expression for the yield of a multiple module-type chip can be derived if it is possible to determine for any single fault whether it can be tolerated or not. In this case, instead of considering modules which may have any number of faults, we examine individual faults. An example is a memory chip which can have the following types of faults: single cell faults, adjacent cell faults, single word line faults, adjacent word line faults, single bit line faults and adjacent bit line faults[37]. A fault-tolerant memory chip has three types of identical circuits for which some form of redundancy is provided, namely, memory cells, word lines and bit lines. However, only two types of redundant circuits are added to the memory chip: spare word lines and bit lines. These two types of redundant circuits are used to replace defective word lines, bit lines and memory cells.

We first derive a yield expression for a chip with two types of faults, type a and type b, and then extend our result to memory chips with a larger number of fault types. Let λ_a and λ_b denote the average number of faults of type a and b, respectively. Then,

$$Y = Y_0 \sum_{k_a,k_b} Prob\,\{\text{There are } k_a \text{ faults of type } a \text{ and } k_b \text{ faults of type } b\} \cdot h_{k_a,k_b}$$

where h_{k_a,k_b} is the probability that the combination of k_a and k_b faults can be tolerated. The above yield expression can be derived by first assuming independency between the two types of faults (i.e., faults follow the Poisson distribution) and then applying the compounding procedure to allow for fault clustering. We also take into account the support circuitry and denote by λ_{CK} the average number of faults in this part of the chip. Consequently,

$$Y = Y_0 \sum_{k_a,k_b} \frac{\lambda_a^{k_a}\, e^{-\lambda_a}}{k_a!} \cdot \frac{\lambda_b^{k_b}\, e^{-\lambda_b}}{k_b!} \cdot e^{-\lambda_{CK}} \cdot h_{k_a,k_b}$$

$$= Y_0 \sum_{k_a,k_b} \frac{\lambda_a^{k_a}\lambda_b^{k_b}}{k_a!k_b!}\, e^{-(\lambda_a+\lambda_b+\lambda_{CK})} \cdot h_{k_a,k_b} \tag{46}$$

The compounding procedure has to be applied only once due to the dependence between the clustering of the three types of faults. Therefore, we define as before

$$\lambda = \lambda_a + \lambda_b + \lambda_{CK}, \quad \delta_1 = \frac{\lambda_a}{\lambda}, \quad \delta_2 = \frac{\lambda_b}{\lambda}, \quad \text{and} \quad \delta_3 = \frac{\lambda_{CK}}{\lambda}$$

Substituting these in (46) yields

$$Y = Y_0 \sum_{k_a, k_b} \frac{\lambda^{k_a+k_b} \delta_1^{k_a} \delta_2^{k_b}}{k_a! k_b!} e^{-\lambda} \cdot h_{k_a, k_b} \tag{47}$$

Compounding now with respect to λ results in,

$$Y = Y_0 \sum_{k_a, k_b} \frac{\Gamma(k_a + k_b + \alpha)}{k_a! k_b! \, \Gamma(\alpha)} \cdot \frac{(\frac{1}{\alpha})^{k_a+k_b} \delta_1^{k_a} \delta_2^{k_b} \bar{\lambda}^{k_a+k_b}}{(1 + \frac{\bar{\lambda}}{\alpha})^{k_a+k_b+\alpha}} \cdot h_{k_a, k_b} \tag{48}$$

Defining $\bar{\lambda}_a = \bar{\lambda}\delta_1$ and $\bar{\lambda}_b = \bar{\lambda}\delta_2$ and substituting in (48) we obtain

$$Y = Y_0 \sum_{k_a, k_b} \frac{\Gamma(k_a + k_b + \alpha)}{k_a! k_b! \, \Gamma(\alpha)} \cdot \frac{(\frac{1}{\alpha})^{k_a+k_b} \bar{\lambda}_a^{k_a} \bar{\lambda}_b^{k_b}}{(1 + \frac{\bar{\lambda}}{\alpha})^{k_a+k_b+\alpha}} \cdot h_{k_a, k_b} \tag{49}$$

We now extend the above result to memory chips with seven types of faults: single cell faults, adjacent cell faults along word lines, adjacent cell faults along bit lines, single word line faults, adjacent word line faults, single bit line faults and adjacent bit line faults. The corresponding fault averages are denoted by λ_{sc}, λ_{acwl}, λ_{acbl}, λ_{swl}, λ_{awl}, λ_{sbl} and λ_{abl}. Following the same procedure as outlined in equations (46) through (49) we arrive at the following yield expression,

$$Y = Y_0 \sum_{i,j,k,l,m,n,q} \frac{\Gamma(i + j + k + l + m + n + q + \alpha)}{i! \, j! \, k! \, l! \, m! \, n! \, q! \, \Gamma(\alpha)}$$

$$\cdot \frac{(\frac{1}{\alpha})^{i+j+k+l+m+n+q} \bar{\lambda}_{sc}^{i} \bar{\lambda}_{acwl}^{j} \bar{\lambda}_{acbl}^{k} \bar{\lambda}_{swl}^{l} \bar{\lambda}_{awl}^{m} \bar{\lambda}_{sbl}^{n} \bar{\lambda}_{abl}^{q}}{(1 + \frac{\bar{\lambda}}{\alpha})^{i+j+k+l+m+n+q+\alpha}} \cdot h_{i,j,k,l,m,n,q} \tag{50}$$

where the definitions of $\bar{\lambda}_{sc}$, $\bar{\lambda}_{acwl}$, etc are similar to those above.

Finally, note that we must assume that the probability of multiple faults (of any of the seven types) occurring in a single word line (or bit line) is negligible. Otherwise, the summation in (50) will be infinite. In VLSI memory chips with thousands of word lines and bit lines and a small number of expected faults, this assumption is well justified.

Also, it must be noted here that the above approach is valid only if there is a perfect correlation between the distribution of faults of different types. Such a perfect correlation was assumed in the derivation of (45), (49) and (50). If there is no correlation, independent negative binomial distributions, each with their own clustering parameter α, have to be used for each fault type. Separate compounding steps for each type of faults have to be performed, as was done in reference[41]. Small correlation between failure types requires the use of a multivariate model like the one described by Stapper *et al*[20].

CONCLUSIONS

The statistical models used to estimate the manufacturing yield of defect tolerant VLSI chips have been reviewed in this paper. We have shown how to take fault clustering into account when deriving yield expressions for VLSI chips with redundancy or partially-good chips, with one or more types of circuit modules. Also, the method used to determine the parameters of the yield model was reviewed and the assumed size of fault clusters was discussed.

ACKNOWLEDGMENTS

The work of the first co-author was supported in part by NSF under contract MIP-8805586.

The figures and tables in this manuscript also appeared in the IBM Journal of Research and Development, Vol. 33, No. 2., March 1989 in a paper by the second co-author entitled "Large Area Fault Clusters and Fault Tolerance in VLSI Circuits: A Review." These figures and tables are copyrighted in 1989 by the International Business Machines Corporation and are reprinted here with permission.

REFERENCES

1. T.E. Mangir, "Use of On-Chip Redundancy for Fault-Tolerant Very Large Scale Integrated Circuit Design," Ph.D. Dissertation, Univ. of Calif., Los Angeles, 1981.
2. K.S. Hedlund, "Wafer Scale Integration of Configurable, Highly Parallel Processors," Ph.D. Dissertation, Purdue Univ., Indiana, 1982.
3. J.C. Harden, "A Wafer Scale Cellular Tree Architecture," Ph.D. Dissertation, Texas A & M University, College Station, TX, 1985.
4. T. J. Wallmark, "Design Considerations for Integrated Electron Devices," *Proc. IRE,* Vol. 48, pp. 293-300, March 1960.
5. S. R. Hofstein and F. P. Heiman, "The Silicon Insulated-Gate Field Effect Transistor," *Proc. IEEE,* Vol. 51, pp. 1190-1202, Sept. 1963.
6. B.T. Murphy, "Cost-size Optima of Monolithic Integrated Circuits," *Proc. IEEE,* Vol. 52, pp. 1537-1545, Dec. 1964.
7. C.H. Stapper, "Defect Density Distribution for LSI Yield Calculations," *IEEE Trans. Electron Devices*, Vol. ED-20, pp. 655-657, July 1973.
8. A.P. Turley and D.S. Herman, "LSI Yield Projections Based Upon Test Patterns Results: An Application to Multilevel Metal Structures," *IEEE Trans. Parts, Hybrids, Packag.*, Vol. PHP-10, pp. 230-234, Dec. 1974.
9. C.H. Stapper, "On a Composite Model of the IC Yield Problem," *IEEE J. Solid-State Circuits*, Vol. SC-10, pp. 537-539, Dec. 1975.
10. O. Paz and T.R. Lawson, Jr., "Modification of Poisson Statistics: Modeling Defects Induced by Diffusion," *IEEE J. Solid-State Circuits*, Vol. SC-12, pp. 540-546, Oct. 1977.
11. R.S. Hemmert, "Poisson Process and Integrated Circuit Yield Prediction," *Solid-State Electronics*, Vol. 24, pp. 511-515, June 1981.

12. C.H. Stapper and R.J. Rosner, "A Simple Method for Modeling VLSI Yields," *Solid-State Electronics*, Vol. 25, pp. 487-489, June 1982.

13. C.H. Stapper, "The Effects of Wafer to Wafer Density Variations on Integrated Circuit Defect and Fault Distributions," *IBM J. Res. Develop.*, Vol. 29, pp. 87-97, January 1985.

14. T. Yanagawa, "Influence of Epitaxial Mounds on the Yield of Integrated Circuits," *Proc. IEEE*, Vol. 57, pp. 1621-1696, Sept. 1969.

15. T. Yanagawa, "Yield Degradation of Integrated Circuits Due to Spot Defects," *IEEE Trans. Electron Devices*, Vol. ED-19, pp. 190-197, Feb. 1972.

16. C.H. Stapper, "LSI Yield Modeling and Process Monitoring," *IBM J. Res. Develop.*, Vol. 20, pp. 228-234, May 1976.

17. A.V. Ferris-Prabhu, L.D. Smith, H.A. Bonges and J.K. Paulsen, "Radial Yield Variations in Semiconductor Wafers," *IEEE Circuits and Devices Magazine*, Vol. 3, pp. 42-47, March 1987.

18. D.M.H. Walker, *Yield Simulation for Integrated Circuits*, Boston: Kluwer Academic Publishers, 1987, Ch. 4, pp. 45-49, Ch. 8, pp. 158-160.

19. S. Gandemer, "Modelisation de l'Impact des Defauts de Fabrication sur le Rendement des Microcircuits Integres Fabriques en Technologie Silicium," Doctoral Dissertation, Ecole Nationale Superieure des Telecommunications, Sept. 1987.

20. C.H. Stapper, F.M. Armstrong and K. Saji, "Integrated Circuit Yield Statistics," *Proc. IEEE*, Vol. 71, pp. 453-470, April 1983.

21. C.H. Stapper, "Small-Area Fault Clusters and Fault-Tolerance in VLSI Circuits," *IBM J. Res. Develop.*, Vol. 33, March 1989.

22. R.M. Warner, "Applying a Composite Model to the IC Yield Problem," *Solid-State Circuits,* Vol. SC-9, pp. 86-95, June 1974.

23. R.M. Warner, "A Note on IC Yield Statistics," *Solid-State Electronics*, Vol. 24, pp. 1045-1047, Dec. 1981.

24. S.M. Hu, "Some Considerations in the Formulation of IC Yield Statistics," *Solid-State Electronics*, Vol. 22, pp. 205-211, Feb. 1979.

25. C.H. Stapper, "Comments on 'Some Considerations in the Formulation of IC Yield Statistics'," *Solid-State Electronics*, Vol. 24, pp. 127-132, Feb. 1981.

26. M.B. Ketchen, "Point Defect Yield Model for Wafer Scale Integration," *IEEE Circuits and Devices Magazine*, Vol. 1, No. 4, pp. 24-34, July 1985.

27. V. Foard Flack, "Introducing Dependency into IC Yield Models," *Solid-State Electronics*, Vol. 28, No. 6, pp. 555-559, June 1985.

28. C.H. Stapper, "Block Alignment: A Method for Increasing the Yield of Memory Chips that are Partially Good," *Defect and Fault Tolerance in VLSI Systems,* I. Koren (ed.), pp. 243-255, New York: Plenum, 1989.

29. R.B. Seeds, "Yield, Economic, and Logistic Models for Complex Digital Arrays," in *1967 IEEE Int. Conv. Rec.*, 1967, pt. 6, pp. 61-66.

30. R.B. Seeds, "Yield and Cost Analysis of Bipolar LSI," presented at the 1967 Int. Electron Device Meeting Keynote Session, p. 12, Oct. 1967.

31. T. Okabe, M. Nagata, and S. Shimada, "Analysis on Yield of Integrated Circuits and a New Expression for the Yield," *Elec. Eng. Japan*, Vol. 92, pp. 135-141, Dec. 1972.

32. C.H. Stapper, "The Defect-Sensitivity Effect of Memory Chips," *IEEE J. Solid-State Circuits*, Vol. SC-21, pp. 193-198, Feb. 1986.

33. C.H. Stapper, "On Yield, Fault Distributions and Clustering of Particles," *IBM J. Res. Develop.*, Vol. 30, pp. 326-338, May 1986.
34. V. Foard Flack, "Estimating Variations in IC Yield Estimates," *IEEE J. of Solid-State Circuits*, Vol. SC-21, pp. 362-365, April 1986.
35. T.E. Mangir, "Sources of Failures and Yield Improvement for VLSI and Restructurable Interconnects for RVLSI and WSI: Part I - Sources of Failures and Yield Improvement for VLSI," *Proc. IEEE*, Vol. 72, pp. 690-708, June 1984.
36. J.C. Harden, "Comments on 'Sources of Failures and Yield Improvement for VLSI and Restructurable Interconnects for RVLSI and WSI'," *Proc. IEEE*, Vol. 74, pp. 515-516, March 1986.
37. C.H. Stapper, A.N. McLaren, and M. Dreckmann, "Yield Model for Productivity Optimization of VLSI Memory Chips with Redundancy and Partially Good Product," *IBM J. Res. Develop.*, Vol. 20, pp. 398-409, 1980.
38. I. Koren and D.K. Pradhan, "Introducing Redundancy into VLSI Designs for Yield and Performance Enhancement," *Proc. 15th Annual Int. Symp. on Fault-Tolerant Computing*, pp. 330-335, 1985.
39. I. Koren and D.K. Pradhan, "Yield and Performance Enhancement through Redundancy in VLSI and WSI Multiprocessor Systems," *Proc. IEEE*, Vol. 74, pp. 699-711, May 1986.
40. I. Koren and D.K. Pradhan, "Modeling the Effect of Redundancy on Yield and Performance of VLSI Systems," *IEEE Trans. on Computers*, Vol. C-36, pp. 344-355, March 1987.
41. I. Koren, Z. Koren and D.K. Pradhan, "Designing Interconnection Buses in VLSI and WSI for Maximum Yield and Minimum Delay," *IEEE J. of Solid-state Circuits*, pp. 859-866, June 1988.
42. J-J. Shen and I. Koren, "Yield Enhancement Designs for WSI Cube Connected Cycles," *Proc. of Int. Conf. on WSI*, pp. 289-298, Jan. 1989.
43. J.C. Harden and N.R. Strader, "Architectural Yield Optimization for WSI," *IEEE Trans. on Computers*, Vol. C-37, pp. 88-110, Jan. 1988.

2 WAFER SCALE REVISITED

WAFER SCALE REVISITED

Douglas L. Peltzer

Tactical Fabs, Inc.
San Jose, CA 95112

In 1984, Jack Kilby and N.R. Strader presented a paper at the SRC workshop on wafer scale processing entitled, "Wafer Scale Integration - Historical Perspective."[1] During the discussion Dr. Kilby observed that the "semiconductor industry has never been compensated for the cost of interconnect". He illustrated this concept with the example of RAM chips. RAM chips of greater integration become attractive only after the price per bit is lower than the price per bit of RAMs of lesser integration, irrespective of the additional difficulty of producing the larger RAMs and the advantages accruing to the user of smaller space, lower assembly costs, and higher reliability.

The corollary to Kilby's statement is that the cost per function in wafer scale must be lower than the cost per function in lower integration to be economic. This is not easily achieved. Considerations include yield, manufacturing cost, packaging cost, ease of use, and speed of operation. Numerous approaches have been tried and generally have been superceded by improvements in manufacturing enabling larger integration on larger wafers. However, like gate arrays, one of the early wafer scale techniques may yet be proven viable.

In the late 1960's, TI implemented discretionary wiring to achieve full wafer circuitry in the fabrication of a full-slice computer. Other approaches included Pad Relocation at Hughes Aircraft, Adapative Wafer Scale Integration at McDonnell Douglas, Spiral Arrays at Middlesex Polytechnic, Restructurable VLSI at MIT Lincoln Labs, and Triple Redundancy at Trilogy.[1]

As a recent example of wafer scale fabrication, a little more might be said about the Trilogy approach. A drawing of the Trilogy chip is shown in Figure 1. The chip was 6.2 cm on a side with features as shown in Table 1. The technology was double polysilicon bipolar with 4 1/2 layers of metal. The chip had 2244 connections contacted by gold bumps. Power dissipation approximated 1000 watts and required special testers to handle and test the unpackaged parts.

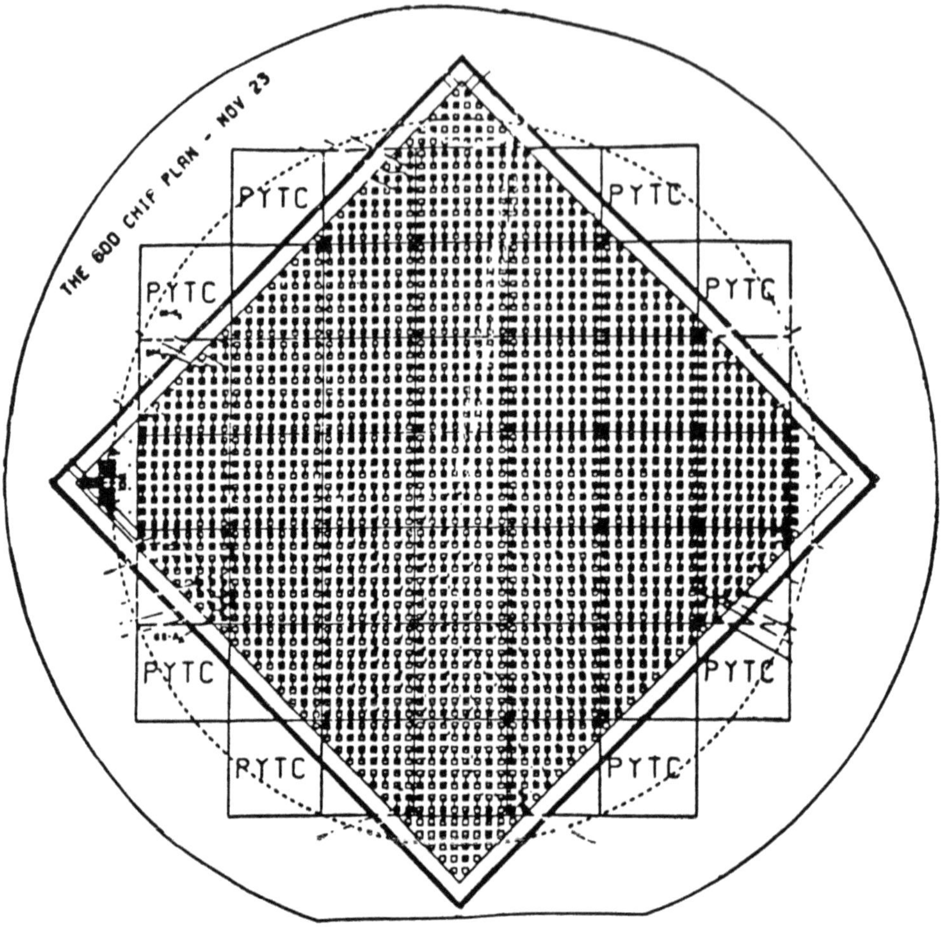

Figure 1: Trilogy Chip

Table 1: Wafer Statistics

Power/Chip	400W-1200W
Amperes/Chip	80A-240A
Total Connections	2244
Signal Connections	1632
Voltage	272
Ground	289
Spare	51
Total Array Sites	12300
Array Sites/cm^2	320
Internal sites/cm^2	~280
I/O Sites/cm^2	~48
Transistors/Array Site (Includes Bias Drivers)	75
Total Transistors	922500

A special test pattern was fabricated to check the defect distribution of the Trilogy process. The pattern consisted of 5494 ring oscillators tested for operation within specified frequency limits. The conclusions were that low defect wafers exhibitied random defect distributions and that high defect wafers had clustered defects. A sample of the test output from a wafer is shown in Figure 2. The small dots are good ring oscillators and the large dots are failures.

Figure 2 points out some of the fabrication difficulties of wafer scale which include scratches and other extensive defects as fatal flaws. In contrast, wafers containing discrete die would be able to obtain some yield even when extensively scratched. Figure 3 illustrates a feature of this problem. The fab line can remove wafers with

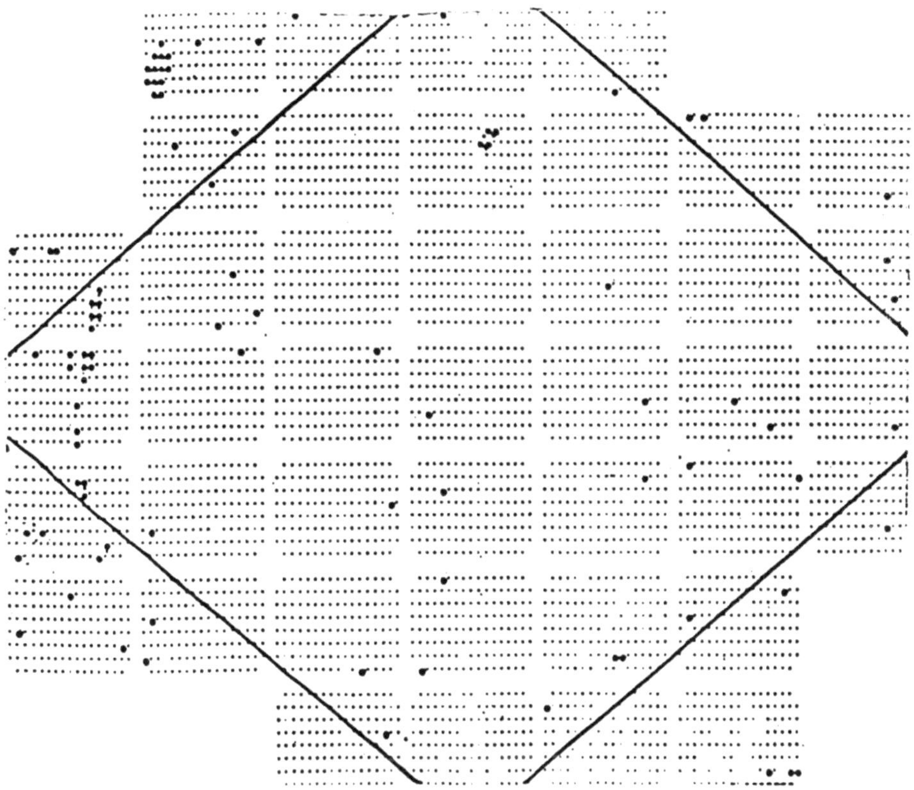

Figure 2: Ring Oscillator Test Output

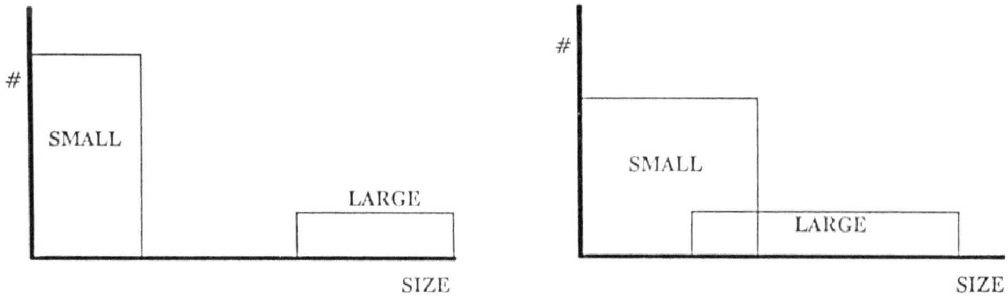

Figure 3: Types of Defects

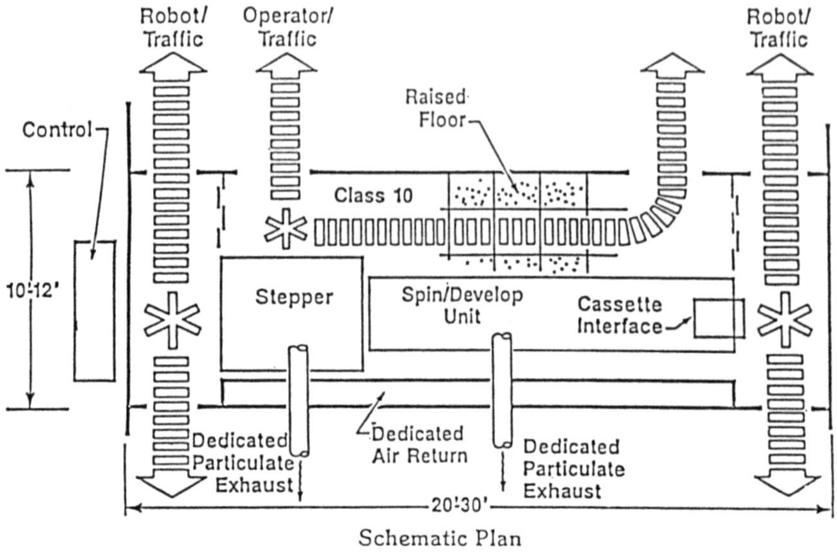

Figure 4: Ganged Photolithography System

large defects at early stages by inspection. Manufacturing improvements, such as automated handling, could then be introduced to eliminate the cause for yield loss. At test, small defects are rendered innocuous by fault-tolerant design. It is desirable for the defects capable of being handled by the fault tolerant design to be larger than the minimum sized defects discoverable by inspection. In this way, there are fewer classes of defects that escape discovery to result in wafer sort yield loss.

Recent advancement in manufacturing technology have vastly improved the manufacturability of semiconductors. CMOS designs are now widely employed and speed is achieved by parallel processing techniques. Defect levels of .5 defects per square centimeter, and better, have been rumored on large and complex die. One of the new manufacturing technologies used is to gang equipment into equipment sets as shown in Figure 4. Here, a stepper is joined to a scrub/spin/bake unit and a develop/bake unit for the processing of photoresist patterns. Close coupling of equipment decreases throughput time, recudes inventory and defect levels, reduces start-up time, improves machine utilization, and improves process control.[2]

This combination of process and manufacturing technology has produced exciting new products only now reaching the consumer. However, the demand for ever larger integration still exists. Large computers are running out of bandwidth when communicating with disk drives. The introduction of the low cost optical disk will strain communications between existing disk drives and present memories. A semiconductor replacement for rotating disk as a new level of auxiliary storage is very attractive at the right price. A large market would be available and the products would readily accept the new form factors associated with full slice semiconductor products. Any price reduction would give a significant advantage.

A full slice memory chip could be manufactured economically. As shown in Table 2, most DRAM dice are approximately 50% occupied by bit area.[3–6] The rest is composed of logic and pads. If a more efficient design were constructed, utilizing a

Table 2: DRAM Density

256K DRAM	Bit Area	Total Bit Area	Die	%
Intel (84)	$70.42\mu^2$	$18.5mm^2$	$39.6mm^2$	46.7
Siemens (84)	$90.0\mu^2$	$23.6mm^2$	$45mm^2$	52.4

1M DRAM	Bit Area	Total Bit Area	Die	%
T1 (86)	$21.32\mu^2$	$22.3mm^2$	$50.31mm^2$	44.3
Toshiba (86)	$29.22\mu^2$	$30.6mm^2$	$54.2mm^2$	56.3

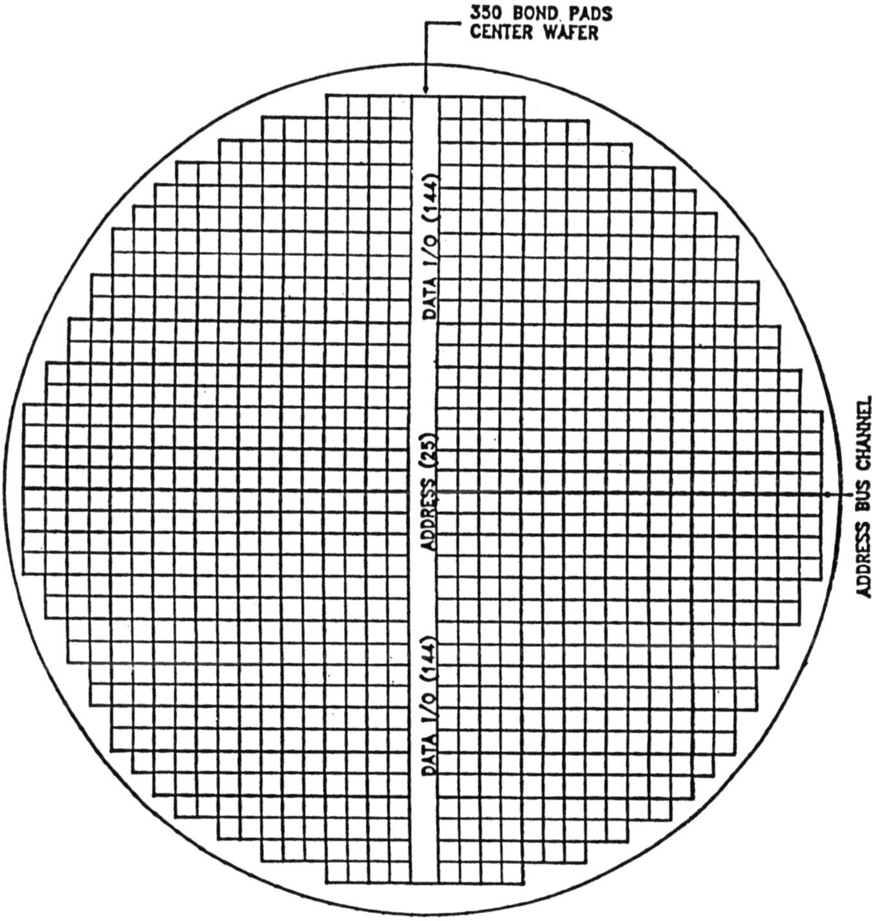

Figure 5: 112 Megabyte RAM

Table 3: 112 Megabyte RAM

ORGANIZATION
3.5MWORD (512 X 9) X 288 BITES (32 BYTES)
BUS ORGANIZED WITH FLEXIBLE MEMORY ALLOCATION
TECHNOLOGY
TRENCH CELL CMOS (12 sq. micron cell)
1 POLYCIDE, 4 METAL INTERCONNECT
LAYER 1, 3μ PITCH
LAYER 2, 3.5μ PITCH
LAYER 3, 12μ PITCH
LAYER 4, 12μ PITCH
MODE
FULL PARALLEL ADDRESSING
PARALLEL 288 BIT READ/WRITE

Table 4: 112 Megabyte RAM

POWER
5 VOLT EXTERNAL (10 AMP)
POWER DISSIPTATION: ACTIVE <50 W
STANDBY < 10
PERFORMANCE
CYCLE TIME < 300 nsec
ADDRESS TIME < 120 nsec
PHYSICAL SIZE
CHIP = 150 mm ROUND
PACKAGE 350 PINS; 8"x 7"x 0.5"

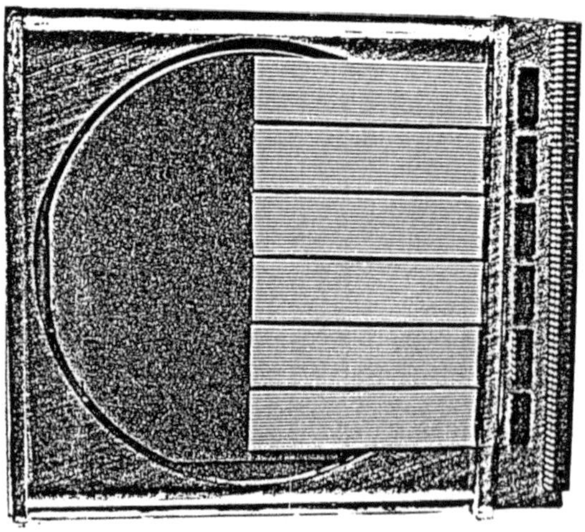

Figure 6

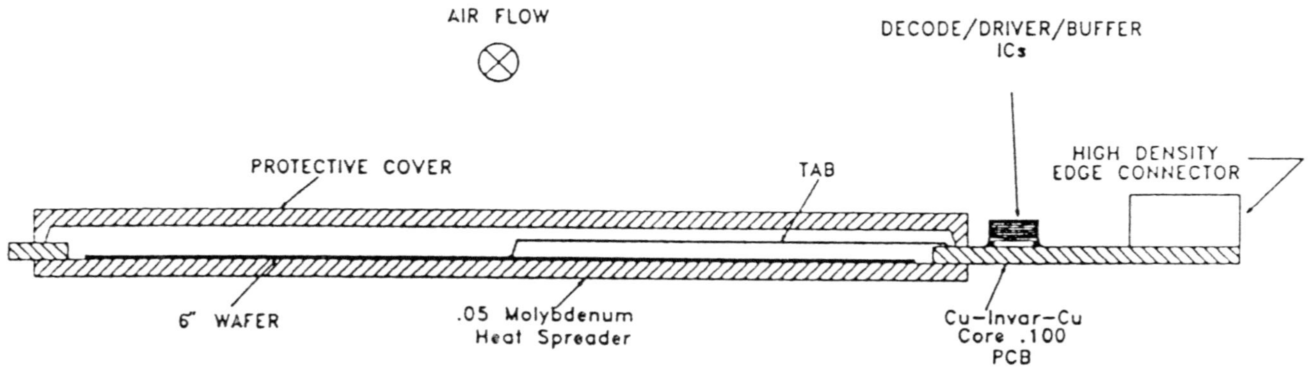

PAK DIMENSION: 8" X 7" X 0.7"

CARD CAGE PITCH: 0.75"

Figure 7

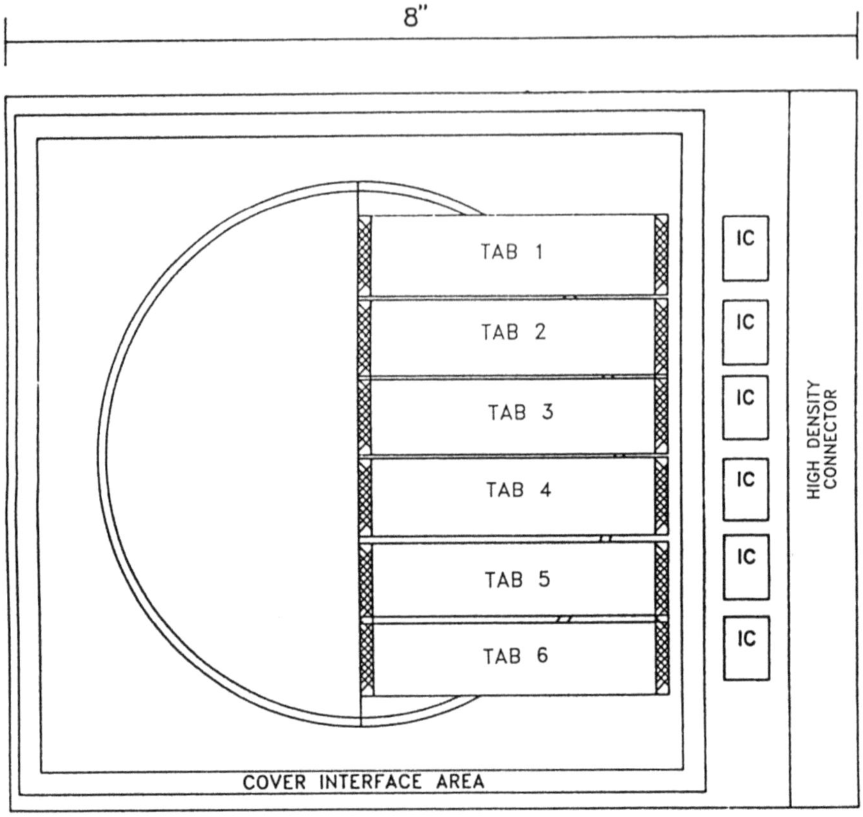

Figure 8

larger percentage of available silicon area without substantially increasing processing costs or decreasing yields, a reduction in cost per bit could be obtained. In this commodity market, any price reduction would give a significant advantage.

A conceptual study of a full slice memory device has been completed. Figure 5 shows a drawing of a 112 megabyte RAM on a six inch wafer. Pads and I/O locations are arranged down the center of the wafer to allow for ease of packaging using strip line, or flat lead. Table 3 shows the organization of this large RAM.

The four layer metal structure utilizes a twelve square micron cell such as used in megabit and four megabit DRAMs. If smaller cells such as the four micron cells are used, larger DRAM capacity could be obtained.

Table 4 shows the power, performance, and physical size of the wafer. The low power allows for an air cooled package. Two wafers can be included in the package back to back, doubling the packaged capacity and heat dissipation.

A model of the package design is shown in Figure 6. Here the wafer is attached to a molybdenum heat sink on the backside. Stripline is brought to the center connection to contact the pads. Address logic and auxiliary logic are attached to the board and the entire structure is brought to an edge connector for insertion into a frame. The top and side views of the package are shown in Figures 7 and 8 respectively. The package is designed to fit into a half rack mount.

The surface area occupied by memory bits in the full slice DRAM exceeds 80% of the wafer. This allows the cost per bit to drop by nearly a factor of two in high volume. Packaging costs would be expected to be competitive with current packaging technology and hybrid technology. In summary, the drive to achieve full slice integration still exists. Economics dictate that the costs per function of these large devices be less than the cost per function of lesser integrated devices. This is achievable in certain classes of products, such as RAMs. Many problems, such as power dissipation and manufacturing defect levels, have been substantially reduced by the adoption of CMOS technology and advanced manufacturing methods. Markets for these large RAMs can be developed and these markets will probably be enhanced by the advent of cheap optical disk technology. Future fault tolerant designs may achieve economic viability in the area of large RAMs.

REFERENCES

1. N.R. Strader and J.S. Kilby, "Wafer Scale Integration - Historical Perspective", SRC Wafer Scale Integration Workshop, pp. 2-34, Sept. 1984.
2. A.F. Moore and Z. Lam, "Process Control Improvements through Robotic Interface of Lithography Equipment", Technical Proceedings Semicon West, pp. 2-8, May 1986.
3. Roger Kung, et al., "A Sub 100 ns 256K DRAM in CMOS III Technology", ISSCC Digest of Technical Papers, p. 278, 1984.
4. Dieter Kantz, et al., "A 256K DRAM with Descrambled Redundancy Test Capability", ISSCC Digest of Technical Papers, p. 272, 1984.
5. Joseph Neal, et al., "A 1Mb CMOS DRAM with Design-for-Test Functions", ISSCC Digest of Technical Papers, p. 264, 1986.
6. Syuso Fujii, et al., "A 50 Microamps Standby 1MW x 1b/256kw x 4b CMOS DRAM", ISSCC Digest of Technical Papers, p. 266, 1986.

3 MODELS FOR DEFECTS AND YIELD

DEFECTS, FAULTS AND SEMICONDUCTOR DEVICE YIELD

A.V. Ferris-Prabhu

IBM General Technology Division
Essex Junction, VT 05452

ABSTRACT

Semiconductor device yield, defined as the average fraction of devices on a wafer that passes final test, is directly measured. Yield is limited by the occurrence of faults. The average number of faults per chip is in principle obtainable from direct measurement, but in practice is inferred from the observed yield. Faults result from the interaction between the device structure and defects of all sizes. Defects, or defect density, is ill-defined, and in turn needs to be inferred from the average number of faults. This paper discusses the relation between defects, faults and semiconductor yield, and attempts to illuminate areas that invite closer attention.

INTRODUCTION

Semiconductor device yield at the wafer level can be defined as the fraction of devices on a wafer that passes final test, averaged over a number of wafers. Averaging over too few wafers can result in excessive scatter. Averaging over too long a period can result in loss of information. In any event, yield (or yield loss) is directly measurable: the number of chips that pass (or fail) test is counted directly.

Yield loss, and its causes, may be described under three categories. Parametric yield loss results from variation in process parameters to the extent that some device electrical parameters do not lie within the operational window. Gross yield loss is caused by relatively large scale errors such as scratches, or chip edge damage due to handling or incorrect tool placement. Random yield loss arises from faults that are tool, process, technology, or "dirt" related. This paper will discuss only random yield, which will be referred to hereafter merely as yield.

Faults are fatal defects. Knowledge of their spatial distribution enables one to assess the probability of finding a chip with no fault. Thus embedded in the relation between yield and faults, are the assumptions made about their spatial distribution. In principle, faults should be directly observable. In practice, their existence is inferred from the electrical signature of tests that have failed and their average number per chip is inferred from the observed yield.

The interaction between defects and device can give rise to faults. In a sense, a defect is an ill-defined quantity. It is not always known for certain that a defect *can* be fatal until it *is* fatal, and it often cannot be known to have been fatal until determined by electrical test. Physical failure analysis of special test structures or of actual failed devices is needed to identify the type of defect associated with a fault that is defined through its electrical signature. Defects may be classified as point defects or defects of finite size. A point defect *will* be fatal if it occurs where it *can* be fatal, e.g., pinholes in insulating layers or crystallographic imperfections that provide conductive paths where there should not be any. Within the region where they can be fatal, the spatial distribution of point defects is the same as that of the faults resulting from them. For other types of defects to be faults, not only must they occur where they can be fatal, but they must also be large enough to be fatal. For example, a defect in a metal layer will become fatal only if large enough to break a pattern designed to be continuous, or to connect two patterns designed to be disjoint. The spatial distribution of faults resulting from such defects is, in a sense, a convolution of the spatial positioning of the device patterns with the spatial and size distribution of the defects that interact with them. The interaction between device and defect is described by a kernel that is unique to the device structure. For any given device, the pattern geometry is known and the kernel can be found, but extracting information about the defect size distribution from the inferred number of faults involves a deconvolution problem that does not appear to have been recognised. It is therefore useful to realise that although "defect density" is a widely used term, in practice it is an inferred and not a measured quantity. For this reason it is important for yield models to be examined not merely for statistical sophistication but also for physical content.

YIELD MODELS AND YIELD

Yield models have at least two objectives: from yield data on past or current product to estimate the future yield of current or new product, and to estimate the contribution to yield loss from each of the process steps so that measures can be taken to reduce this loss. To meet the first objective it is necessary to assess the average probability of finding on a wafer a chip with no fault. The value of this probability depends upon the manner in which faults, on the average, are distributed across the wafer, that is to say upon their spatial distribution. To meet the second objective, it is necessary to know not only the type of defects generated at each process step, but also their incidence, distribution and the nature of the interaction with the device structure that makes them fatal. Two models that make considerably different assumptions regarding the spatial distribution of faults are the Poisson, and the negative binomial which has had a long history in a variety of applications[1-8].

The Poisson Model

The Poisson model assumes that the spatial distribution of faults is random in the sense that it is as likely for a fault to occur at one location as at another, and that the occurrence of a fault at any location is independent of the occurrence of any other fault. Of the non-defective devices that enter the i-th step, the fraction leaving it that are still non-defective can be written as

$$Y_i = e^{-\lambda_i}, \tag{1a}$$

which can be estimated by the relation

$$\hat{Y}_i = 1 - n_i/N_i, \tag{1b}$$

where N_i is the number of good devices that enter the i-th set of operations, or i-th mask level, n_i is the number of devices that are defective at the end of that step, and λ_i is the average number of faults per chip generated during that step.

The fraction of all devices on a wafer that are not defective at the end of all processing steps can similarly be written as

$$Y = e^{-\lambda}, \tag{2a}$$

which can be estimated by the relation

$$\hat{Y} = \prod_{i=1}^{n} (1 - n_i/N_i) \tag{2b}$$

$$= 1 - \sum n_i/N_1, \tag{2c}$$

where λ is the average number of faults per chip generated over all processing steps.

The smallest number of devices that can be processed at a time are those contained on a single wafer, so it is not physically possible to exclude from subsequent processing operations those chips that become defective at some particular mask level. Nor is it always possible to identify defective chips at the end of each mask step. If later failure analysis of failed devices attributes a loss of n_i chips to the i-th mask step, it is not uncommon to compute the ratio $(1 - n_i/N_1)$ and to call this the yield of that mask level. This is often convenient, but it is not correct[9].

From the observed yield and eqs. (1) and (2), the average number of faults per chip introduced at the i-th mask level and the average number of faults per chip introduced over all levels, is estimated to be

$$\hat{\lambda}_i = -\ln Y_i = -\ln(1 - n_i/N_i), \tag{3}$$

and

$$\hat{\lambda} = -\ln Y = -\ln(1 - \sum n_i/N_1). \tag{4}$$

The definition of net yield as the product of the intermediate yields leads to the relation

$$e^{-\lambda} = e^{-\Sigma \lambda_i}, \tag{5}$$

or equivalently

$$\lambda = \Sigma\lambda_i, \tag{6}$$

suggesting that the average number of all faults per chip is equal to the sum of the average number of faults per chip generated at each mask level.

Now it has been known for some time that when the average number of all faults per chip, extracted from the observed yield of devices of a given area, is applied to larger area devices, the Poisson model gives estimates that are lower than the observed yield. The cause of this deviation is attributed to the spatial clustering of faults[10,11]. However, the published literature does not make clear whether clustering refers to the occurrence of more than one fault on the same chip or to the adjacency of failed chips. Nor is it clear whether the faults have resulted from the same or different process steps, or from different mechanisms at the same process step. Faults on the same chip that are generated at different process steps should not be multiply counted because a chip that is defective at the end of a particular process step should be excluded from the sample size of the next. And when adjacent chips have failed, the average number of faults per chip used for the estimation of the yield of chips of larger area needs to be adjusted to account for the fact that were a larger chip to be fabricated with the same spatial positioning of faults, two or more or them may well lie on the same chip. A simple and physically meaningful model that makes just such a reduction has been used at IBM Burlington, Vermont since the mid 1970s. According to this method, variants of which have been used at other IBM factories, the yield of a chip of area A is estimated from the observed yield of a chip of area A_0, $(A > A_0)$, by using the Poisson equation with

$$\lambda(A) = \lambda(A_0) \times (A/A_0)^{1-b}, \qquad 0 < b < 1 \tag{7}$$

and

$$\lambda_i(A) = \lambda_i(A_0) \times (A/A_0)^{1-b_i}, \qquad b_i \simeq b. \tag{8}$$

When $b = 0$ the unmodified Poisson yield results, but when $b > 0$, the estimated yield is higher than predicted by the unmodified Poisson equation. As the effective average number of faults per chip is reduced by the exponent b, its magnitude can be interpreted as a measure of clustering or deviation from Poisson statistics. The value of $\lambda(A_0)$ is extracted from the observed yield $Y(A_0)$ by using eq. (4), and the value of b is determined by the so-called window method[12]. In this method, chips with areas in multiples of the area of the data-base chip are constructed on wafer maps on which the location of defective chips is shown. The number of good chips is counted and the ratio $\lambda(A)/\lambda(A_0)$ plotted against the ratio A/A_0. By means of a least-square-error or similar criterion, the value of b is found that best fits eq. (7) to the data. Such wafer maps do not in general indicate the number of faults on a defective chip, nor at which process step the fault(s) occurred. And though there is no theoretical justification for it, due to the scarcity of wafer map data at each intermediate process step, b_i is often set equal to b.

Inserting eqs. (7) and (8) into eqs. (2) and (1) shows that the definition of net yield as the product of the intermediate yields is met, and that the sum of the average number of faults per chip at each of the intermediate steps is equal to the average number of all faults per chip. Scaling the number of faults per chip in this manner allows the Poisson equation to be retained. It preserves the physical meaning of λ_i and of λ as the average number of faults per chip at step i and of all faults per chip respectively. And it has been used to predict the yield of each individual mask level, as well as the net yield, with results that agree well with actual yield data[13]. Nevertheless, the yield model is still based on the Poisson equation with its inherent assumption that the spatial distribution of defects is random.

The Negative Binomial Model

The negative binomial as derived by Rogers[8] explicitly assumes that the likelihood of an event occurring at a given location increases linearly with the number of events that have already occurred at that location. Although no quantitative physical model governing such a behaviour has been presented, this assumption has been translated to yield theory[14] to describe the spatial clustering of fatal defects. The fraction of devices on a wafer that can be expected not to be defective after all process steps can then be written as

$$Y = (1 + \beta\lambda)^{-1/\beta}, \qquad 0 < \beta < 1. \tag{9}$$

From the observed yield the average number of faults of all types can be inferred from the expression

$$\lambda = (Y^{-\beta} - 1)/\beta. \tag{10}$$

The term β, which is defined in statistics as the square of the coefficient of variation, has been interpreted[15] as an index of spatial clustering or extent of deviation from the Poisson statistics and can be determined empirically by the window method mentioned earlier. In the limit as $\beta \rightarrow 0$, eq. (9) passes over into the Poisson model.

From eq. (9), the yield of a chip of area A can be estimated from the observed yield of a chip of area A_0, ($A > A_0$), with

$$\lambda(A) = \lambda(A_0) \times (A/A_0), \tag{11}$$

using eq. (10) to infer the value of $\lambda(A_0)$ from the observed yield $Y(A_0)$.

It should be borne in mind that no matter what model is used, the yield of a new product, usually of larger area, is necessarily estimated from the known yield of existing product by area scaling the inferred average number of faults per chip of the known product. As eq. (7) and eq. (11) both show, $\lambda(A)$ is scaled to the inferred value of $\lambda(A_0)$ by the area ratio (A/A_0). In the modification of the Poisson equation used in IBM Burlington, the area ratio is raised to an exponent less than unity, accounting for clustering by lowering the effective average number of faults per chip. In the negative binomial equation, the area ratio is multiplied by the factor β, also less than unity, which is related to the extent of clustering. However, in the latter case, the term β occurs as well in the exponent of the entire argument. It is this feature that gives the negative binomial its flexibility. By a judicious determination of the cluster factors b and β, either the modified Poisson or the negative binomial can be shown to exhibit similar behaviour, in particular the upward curvature predicting for larger chips a higher yield than that predicted by the simple Poisson model.

All previous uses of the negative binomial known to this author apply it to the net result of a sequence of operations, the separate identity of each of which is neglected. In such treatments, interpretation of β as the square of the coefficient of variation, or a measure of the over-dispersion, is clear. However the manufacture of semiconductor devices involves many sequential and clearly separate operations, each with a distinct identity. The assumption that the faults generated at each independent operation are also spatially clustered in the sense of Rogers[8], suggests that the negative binomial is applicable to each of the sequential processes. In that case, of the good devices entering a process step, the fraction that is not defective at the end of it can be written as

$$Y_i = (1 + \beta_i \lambda_i)^{-1/\beta_i}, \tag{12}$$

with the concomitant inference

$$\lambda_i = (Y_i^{-\beta_i} - 1)/\beta_i. \tag{13}$$

Now the definition of net yield as the product of the independent intermediate yields requires that the expression

$$(1 + \beta\lambda)^{-1/\beta}$$

equal the product

$$\Pi(1 + \beta_i \lambda_i)^{-1/\beta_i}.$$

So if the cluster parameters β_i are set equal to β as in practice they often are, this is equivalent to requiring that

$$\lambda = \Sigma \lambda_i + \beta \Sigma \Sigma' \lambda_i \lambda_j + \beta^2 \Sigma \Sigma' \Sigma'' \lambda_i \lambda_j \lambda_k + \cdots, \tag{14}$$

where the prime(s) on the summation signs means that the sum should not be taken over the same indices.

Equation (14), which results from equating a single application of the negative binomial to a set of sequential applications, suggests that β behaves as a coupling parameter, coupling faults generated at different and presumably independent process operations. In fact, if the equality holds, it would imply that the product of negative-binomial distributions is itself negative binomial, but the conditions under which this would hold are not clear. This is not to say, however, that the sum of random variables that are negative-binomially distributed is not itself negative-binomially distributed.

If the coupling parameter β is zero, that is, if the occurrence of a fatal defect at a given location is independent of the occurrence of another at the same or adjacent locations, then the Poisson model is recovered and the average number of all faults per chip is the sum of the average number of faults per chip generated at each step. But if there are underlying causes that promote the clustering of defects generated during presumably independent manufacturing operations, then $\beta \neq 0$ and $\lambda > \Sigma\lambda_i$ suggesting a synergistic effect in which the whole is greater than the sum of its parts. However, the physical basis of such cooperative interactions between presumably independent manufacturing operations is not clear. Mathematically there is no necessity for the coupling parameter to be zero, and a statistical model has been described[16] that introduces essentially a nearest-neighbour coupling to effect a localised dependency between the number of defects on adjacent chips. Although simulations show a clustering of defective chips, it is not clear whether these defects arise from a sequence of independent operations, in which case the question of clustering at each such step should be examined, or whether the clustering arises from a single operation. In any event, there has not yet been proposed a physical description of the coupling parameter in quantitative terms from which an estimate of its magnitude can be made.

If λ_i is interpreted as the average number of faults per chip generated at the i-th mask level, eqs. (6) and (14) show that for a given yield, the negative-binomial model has a larger value for λ than does the Poisson model, as has also been noted previously[17]. But then the interpretation of λ as the average number of faults of all of types per chip is no longer as clear and the applicability of the negative-binomial model to estimating the yield of sequential semiconductor processes is open to discussion.

As a major objective of yield modelling is to relate the incidence of defects generated at each manufacturing operation to the loss in yield that they cause, so that specific actions can be taken to lower the number of defects and increase yield, it should be possible to invest the argument of the yield equation with an unambiguous physical meaning. If λ_i indeed represents the average number of faults per chip at the i-th mask level, then the observation that $\lambda > \Sigma\lambda_i$ may suggest a multiple counting of faults, perhaps due to the inability to exclude from the sample size those chips on a wafer that have failed at earlier process steps. It is not clear from the literature whether the reported agreement between the estimates of the negative binomial model with the actual yield is due to this artifact of the calculation. This is an area that may benefit from closer examination. It would be informative, for each unique type of fault, to compare the predicted yield and spatial distribution of faults at each process step with that shown by actual, as opposed to simulated or hypothetical data. There is also a need to develop techniques that can be used to determine yield loss after each process step as well as for test structures that correctly represent each process step and are selective in the type of defect to which they are susceptible. In addition to these unresolved matters, there is another area that invites attention, and that is the relation of faults to defects, the size distribution of the latter, and how (or even whether) such information can be extracted solely from yield data.

FAULTS AND DEFECTS

As indicated earlier, yield models reflect assumptions about the spatial distribution of faults. Although faults are fatal defects, not all defects are fatal. Faults result from the interaction with the device pattern geometry[18] of all sizes of defects related to the environment, technology, processes, chemicals and tools involved in semiconductor device fabrication.

The spatial distribution of point defects is the same as that of the faults resulting from them. The spatial distribution of potentially fatal defects of finite size is not as straightforward. In fact it is first necessary to determine the size distribution of these defects as it is their interaction with the device pattern geometry that leads to faults, the spatial distribution of which is reflected in the yield model. A model has been described[19] that includes the effect of the size distribution of mask defects as well as line-width fluctuations on yield. Other treatments that discuss the interaction of defect with pattern geometry and the role of defect size distribution in yield modeling have also been described[20-25,31].

In brief, to break a pattern designed to be continuous, a defect must not merely be larger than the pattern, but its center (assuming it to be circular for convenience) must lie within a certain interval of the center-line of the pattern. To connect two patterns designed to be separated, the defect must be larger than the separation distance and it must occur where it can make contact with both patterns. To focus the discussion, consider first a single pattern of width w and length L, $w/L << 1$, lying in a region of length L and width X. The probability K that a defect of a size less than the pattern width will be fatal, is zero. There is also a certain critical defect size $x_c = X + w$ above which the probability of it being fatal is unity. For this example of a single pattern, the behaviour in between is linear, although for actual devices that will not necessarily be the case. The probability, then, that a defect of size x will be fatal to this pattern can be described by the linear skew-symmetric kernel

$$K(x,w) = 0 \qquad x < w, \tag{15a}$$

$$K(x,w) = (x - w)/X \qquad w \le x \le x_c, \quad x_c = X + w, \tag{15b}$$

$$K(x,w) = 1 \qquad x_c < x. \tag{15c}$$

The kernel can be interpreted as a representation of the interaction between device structure and defect that makes the defect fatal. As the kernel is a direct measure of the probability that a defect of given size will be fatal to a device of given structure[15,24,25], it is sometimes referred to as a defect sensitivity function[31]. The product of the kernel K and the total area XL in which the defect can occur, gives a so-called critical area[18,20,23]. It is that physical area of width XK(x,w), and length L parallel to the long axis of the pattern, within which the occurrence of a defect of size x is fatal to this pattern. There is, however, a distribution of defect sizes governed by some size probability density function S(x) normalised to unity, i.e., $\int dx S(x) = 1$. The average probability that the interaction of all defects with this pattern will result in a fault is therefore

$$\Phi(w) = \int dx S(x) K(x,w). \tag{16}$$

Expressed in units of faults/defect, the fault probability $\Phi(w)$ can be interpreted as the weighted average effectiveness of this pattern in converting a defect into a fault in analogy to the photon collecting efficiency of photovoltaic devices[26].

For actual product, within the active area, i.e., the area enclosed by the perimeter within which active devices are fabricated, there is a distribution of pattern widths, the layout of which depends upon the type of device and design system used. For memory devices the patterns have a high degree of symmetry, whereas for logic or Application

Specific Integrated Circuit devices the symmetry is lower. The kernel can still be described by an equation of the form of eq. (15) with a different value for the critical defect size x_c, and the fault probability can still be described by eq. (16).

However, again to focus the discussion, consider a test chip on which there are multiple patterns each of length L, width w and period s+w where s is the separation between patterns, for which the only faults that will be considered are those that result in a broken pattern. With appropriate choice[15] of critical defect size x_c, eqs. (15) and (16) can again be used to define the kernel and the fault probability. The average number of faults per chip, resulting from the interaction by defects of all sizes with all patterns on this chip, can then be written in one of three forms:

$$\lambda = \Phi N_D, \tag{17a}$$

$$\lambda = \Phi A D, \tag{17b}$$

$$\lambda = A_c D, \tag{17c}$$

where N_D is the average number and D is the average density of defects per chip of active area A. Although the critical area A_c, defined as the product of the fault probability Φ and the active area A, still has the meaning of an area within which the occurrence of a defect is fatal, its physical visualisation is fuzzy. It also has the hybrid units of faults-area/defect-chip, so it is preferable to use the term fault probability and to use eq. (17a) or eq. (17b) when relating defects to faults.

In the spirit of this treatment suppose that on the average n out of the N such test chips or monitors on a wafer have at least one broken pattern. The yield can then be written as $Y = f(\lambda) = (1 - n/N)$. The relation between yield, faults and defects can now be written as

$$f(N_D \int dx S(x) K(x,w)) = [(1 - n/N)]_w, \tag{18}$$

where the subscript on the observed yield is to show that similar equations can be written with the observed yields of similarly fabricated monitors of different pattern widths.

Equation (18) is an interesting one. The expression on the right side is the actual fraction of chips on a wafer that do not have a fault. The expression on the left side is the probability of finding a chip with no fault. Assumptions about the spatial distribution of faults are embedded in the form of the function, the argument of which contains additional assumptions about the average number of defects on a chip and their size distribution. So it is entirely possible that the expectation of being able to extract the size density function from this equation may not be met.

Eq. (18) can also be written as

$$N_D \int dx S(x) K(x,w) = f^{-1}[(1 - n/N)]_w. \tag{19}$$

In this form it is recognised as a singular Fredholm convolution integral equation of the first kind, general solutions of which are not always known to exist. If the upper limit on the integral is replaced by x_M, the size of the largest defect that can be expected to occur, the equation is no longer singular, but that does not make its solution particularly more tractable. The subtleties of deconvolution involved in such equations are treated at length in standard texts[27,28]. Here we will discuss only those features that can readily be seen to affect attempted solutions.

The first feature is that embedded in eq. (19) are the assumptions about the spatial distribution of the faults that result from the interaction with the pattern geometry of defects of all sizes. If the fault distribution is assumed to be random in the sense of the Poisson model, eq. (19) can be reformulated as

$$N_D \int dx S(x) K(x,w) = [-\ln(1 - n/N)]_w, \tag{20}$$

whereas if the faults are assumed to be spatially clustered in the sense of Rogers, eq. (19) is reformulated as

$$N_D \int dx S(x) K(x,w) = [\{(1 - n/N)^{-\beta} - 1\}/\beta]_w. \tag{21}$$

If the faults are spatially clustered, eq. (21) is appropriate. If they are not, eq. (20) is indicated. But it is inconsistent, if not incorrect, to assume a random distribution of faults to estimate the defect-size distribution, and then to assume a spatially clustered distribution of faults to estimate the yield. Furthermore, actual product undergoes several mask operations associated with each of which there is a yield loss. If the monitor data are to be translated accurately to product, one must necessarily assume that the defect spatial and size distributions are the same for the processes used on monitor wafers as on product wafers, and that the net yield is the product of the individual mask level-limited yields. In that case, the same yield model, either eq. (1) or eq. (12) should be used for estimating defect-size density functions from monitor yield data, as well as for estimating product yield with defect-size densities extracted from monitor yield data.

The second feature is that the value of the term N_D representing the average number of defects per chip is not known. In fact the spatial distribution of defects may not be determinable because faults result from the interaction of defects and the device pattern geometry, which as has been pointed out[18], is itself subject to a distribution of line registration errors. It is not evident that N_D, or its effective equivalent the defect density D, is the same for all monitors, as a result of which it may not be possible to eliminate it from eq. (18) by normalising each of a set of similar equations to one particular pattern width. In any event such a normalisation cannot be made without assuming some representation for S(x).

A third feature is that if a spatially clustered fault distribution is assumed to hold, then the value of the cluster factor β is not known. Though estimates of its value can be made by the window method applied to test structures, it has not been shown that the same value will hold for larger area structures. This again suggests the necessity for a quantitative physical model for this coupling parameter.

A fourth feature is that the right side of eq. (19) is known only as a set of numbers, one for each value of pattern width w. Some sort of curve fit needs to be applied to obtain a functional representation in terms of pattern width. The choice of this representation can affect the form of S(x). Moreover, numerical solutions of such equations are known to be unstable with respect to small perturbations in the values of the coefficients.

A method that has been described[23], using hypothetical values in an attempt to solve eq. (19) for S(x), assumes a Poisson distribution of faults, then in effect assumes different representations for S(x), evaluates both sides of eq. (19) separately, and chooses that form of S(x) that gives the best correlation between the left and right sides. For the case reported, it appears that the size density function of defects that result in broken patterns is inverse cube. However, inconsistencies have been noted[18] with respect to the numerical values reported, which invite further examination of the work and suggest the application of other methods of obtaining the size density function that do not rely on deconvolving yield data but can count defects directly. One such method[29] employs a double-bridge test structure that makes use of the difference in the electrical resistance of monitors to detect the occurrence of faults. The initial data reported suggest that the number of observed defects varies inversely as the square of their size. Allowing for experimental or other error, the difference between inverse square and inverse cube may not be particularly significant if it were only for one defect type. But each defect mechanism has its own density[14], and presumably size distribution, and in actual product there are many different processing operations, which generate more than one type of yield limiting defect. Moreover, the net yield is the product of the yield at each independent step. Consequently, errors in the determination of the size density function of the different defect types can have a pronounced effect on yield estimates[15,18,23,30,31]. The double-bridge method does not have the deconvolution and other constraints inherent in eq. (19), but it too relies on irreversible change in some function, in this case, electrical resistance, in order to detect a defect and to determine its size. It does not detect defects that do not affect the resistance and like all methods has a threshold of resolution. Thus it only detects and measures the size of what are, in essence, fatal defects. It provides no information about defects that are not fatal. But as design dimensions become smaller, and tools, processes and environments cleaner, the definition of what constitutes a defect becomes elusive, and a determination of their size distribution becomes more so. Consequently direct methods that determine the size of fatal defects appear to be needed. It does not appear to have been recognised that to use yield data alone may lead to results of dubious validity.

SUMMARY AND CONCLUSION

This paper has attempted to show the relationship between defects, faults and semiconductor yield, and that assumptions regarding the spatial distribution of faults can affect the form of the size distribution of defects extracted from yield data. Two yield models have been discussed: the Poisson, which assumes that fatal defects are randomly distributed; and the negative binomial, which assumes that they are spatially clustered in the sense of Rogers. It was shown that by scaling the average number of fatal defects inferred from the observed yield of a given product, either model can be used to estimate the yield of a larger area product. For the Poisson, the scale factor is merely the ratio of areas raised to a positive power that is lowered from unity by a positive exponent b, the magnitude of which is a measure of spatial clustering. For the negative binomial, the index of spatial clustering β, which also lies between zero and unity, plays two roles. It enters into the scale factor as a multiplier of the area ratio, and also serves as an exponent for the entire yield equation. As a result the negative binomial model is flexible and has interesting features. However, a quantitative model with physical insight is needed to determine the value of β, which appears to play the role of a parameter coupling presumably independent manufacturing operations. When the negative binomial is applied to sequential semiconductor manufacturing operations, the definition of net yield as the product of the yield of each one of a sequence of presumably independent operations has been shown to support the physical interpretation of β as a coupling parameter. As a consequence, the sum of the average number of faults per chip λ_i generated at each of the independent manufacturing operations is less than λ, which is then difficult to interpret as the average number of faults per chip generated over all operations. This poses questions concerning the independence of the operations and the physical meaning of the coupling term, and even the applicability of the negative binomial to sequential processes. The cluster modification of the Poisson model does not have these difficulties.

Also of concern is the determination of the size distribution of defects. Assumptions concerning the spatial distribution of fatal defects are contained in the functional form of the yield equation, whereas assumptions concerning the size distribution of defects are contained in the argument of the yield equation, with the interaction between defect and device that makes a defect fatal being described by a kernel that is unique to the device structure. Different assumptions for the form of the yield function can lead to different representations of the size density function. These unresolved difficulties invite the question as to whether it is possible to extract information about the size distribution of (potentially fatal) defects solely from yield data.

Although yield models of increasing statistical sophistication are being developed, there is need for a better understanding of the physical bases underlying the statistical parameters contained in these models. There also does not appear to have been as much attention to the determination of defect size distributions and their relation to tools, technologies and environment. In particular there is need for actual data, more of it, and increased emphasis on direct methods of defect size determination.

ACKNOWLEDGEMENT

I wish to thank L. Lareau for explaining some of the finer points of statistical analysis. Any errors of interpretation are my own. It is also a pleasure to thank K. Ohlson and S. Martin for invaluable help in the formatting of this paper.

REFERENCES

1. Student, "On the error of counting with a haemocytometer", *Biometrika*, vol. 5, pp. 351-360 (1907). This appears to be the first application of the negative binomial to data fitting.
2. R. A. Fisher, "The negative binomial distribution", *Annals of Eugenics*, vol. 11, pp. 182-187 (1941).
3. W. Feller, *"Introduction to Probability Theory and its Applications"*, Vol. I, Second Edition, Wiley (1950).
4. H. S. Sichel, "The estimation of the parameters of a negative binomial distribution with special reference to psychological data" data, *Psychometrika*, vol. 16, pp. 102-127 (1951).
5. J. Gurland, "Some applications of the negative binomial and other contagious distributions", *American Journal of Public Health*, vol. 49, pp. 1388-1399 (1959).

6. G. P. Patil, "On the evaluation of the negative binomial distribution with examples", *Technometrics*, vol. 2, pp. 501-505 (1960).

7. F. M. O'Carroll, "Fitting a negative binomial distribution to coarsely grouped data by maximum likelihood", *Applied Statistics*, vol. 11, pp. 196-201 (1962).

8. A. Rogers, "*Statistical Analysis of Spatial Dispersion*", Pion Ltd., London (1974).

9. A. V. Ferris-Prabhu, "VLSI Yield Management and Prediction", *Quality Reliability Engineering International*, vol. 1 (2), pp. 219-225 (1985).

10. E. Muehldorf, "Fault Clustering: Modeling and Observation on Experimental Chips", *IEEE J. Solid-State Circuits*, vol. SC-10 (4), pp. 237-244 (1975).

11. O. Paz and T. Lawson, "Modification of Poisson Statistics: Modeling Defects Induced by Diffusion", *IEEE J. Solid-State Circuits*, vol. SC-12 (5), pp. 540-546 (1977).

12. R.S.Hemmert, "Poisson process and integrated circuit yield prediction", *Solid-State Electronics*, vol. 24 (6), pp. 511-515 (1981).

13. A. V. Ferris-Prabhu, "Forecasting Semiconductor Yield", *IEEE International Conference on Computers, Systems and Signal Processing* (Bangalore, India), Paper No. R46.13, pp. 1491-1497 (1984).

14. C. H. Stapper, F. M. Armstrong and K. Saji, "Integrated Circuit Yield Statistics", *Proc. IEEE*, vol. 71 (4), pp. 453-470 (1983).

15. A. V. Ferris-Prabhu, "Role of Defect Size Distribution in Yield Modeling", *IEEE Trans. Electron Devices*, vol. ED-32 (9), pp. 1727-1736 (1985).

16. V. F. Flack, "Introducing Dependency into IC Yield Models", *Solid State Electronics*, vol. 28 (6), pp. 555-559 (1985).

17. C. H. Stapper, "The effect of wafer to wafer defect density variations on integrated circuit defect and fault distributions", *IBM J. Res. Develop.*, vol. 29 (1), pp. 87-97 (1985).

18. W. Maly, "Modeling of Lithography Related Yield Losses for CAD of VLSI Circuits", *Trans. Computer-Aided Design*, vol. CAD-4 (3), pp. 166-177 (1985).

19. W. Maly and J. Deszcka, "Yield Estimation Model for VLSI Artwork Evaluation", *Electronics Letters*, vol. 19 (6), pp. 226-227 (1983).

20. A. V. Ferris-Prabhu, "Critical Areas for Rectangular Patterns", IBM TR19.0556 (1981).

21. A. V. Ferris-Prabhu, "Fault Probability and Critical Area for VLSI Yield Predictions", IBM TR19.0562 (1982).

22. A. V. Ferris-Prabhu, "Computation of the Critical Area in Semiconductor Yield Theory", *Proc. European Conference on Electronic Design Automation*, (Univ. of Warwick, U.K.) Publ. 232, pp.171-173 (1984).

23. C. H. Stapper, "Modeling of Defects in Integrated Circuit Photolithographic Patterns", *IBM J. Res. Develop.*, vol. 28 (4), pp. 461-475 (1984).

24. A. V. Ferris-Prabhu, "Modelling the Critical Area in Yield Forecasts", *IEEE J. Solid-State Circuits*, vol. SC-20 (4), pp.874-878 (1985).

25. A. V. Ferris-Prabhu, "Defect Size Variations and their effect on the Critical Area of VLSI Devices, *IEEE J. Solid-State Circuits*, vol. SC-20 (4), pp. 878-880 (1985).

26. A. V. Ferris-Prabhu, "Spectral Analysis of Spatial Distribution of Radiation Damage in Photovoltaic Devices", *Proc. Int'l Symposium on Lattice Defects in Semiconductors, Tokyo 1966*, pp. 418-423, Tokyo University Press (1968).

27. R. Courant and D. Hilbert, *"Methods of Mathematical Physics"*, Vol. I, Interscience Publishers (1953).

28. F. Hildebrand, *"Methods of Applied Mathematics"*, Second Edition, Prentice-Hall (1965).

29. W. Maly, M. Thomas and J. Chinn, "Double-Bridge Test Structure for the evaluation of Type, Size and Density of Spot Defects", Technical Report CMUCAD-87-2, CMU (1987).

30. A. V. Ferris-Prabhu, "Yield Implications and Scaling Laws for Submicrometer Devices", *IEEE Trans. Semiconductor Manufacturing*, vol. 1 (2), pp. 449-61 (1988).

31. C. Kooperberg, "Circuit Layout and Yield", *IEEE J. Solid-State Circuits*, vol. 23 (4), pp. 887-892 (1988). The defect sensitivity function g_{ij} and the term P_{ij} introduced in this paper have the same meaning as the kernel $K(x,w)$ and the fault probability $\Phi(w)$ introduced in references 15, 24, and 25 above.

ON THE PROBABILITY OF FAULT OCCURRENCE

Sharad C. Seth and **Vishwani D. Agrawal***

University of Nebraska
Department of Computer Science
Lincoln, NE 68588

***AT&T Bell Laboratories**
Computer Systems Research Department
Murray Hill, NJ 07974

INTRODUCTION

In this paper we introduce the probability of fault occurrence. For a modeled fault (e.g., a stuck fault) this is the probability with which the fault will occur on a chip. The occurrence of a fault is only observable as fault indication by a test capable of detecting it. We determine the probability of fault occurrence from chip test data. No attempt is made to find these probabilities for individual faults but they are determined only as a distribution for all faults. Combined with detection probabilities (related to the conventional fault coverage), fault occurrence probabilities provide a revised coverage requirement versus chip quality relation.

In the previous work [1,2,3], attempts are made to find a relation between fault coverage and product quality. Product quality is measured as *reject ratio* or the ratio of faulty chips among those passing tests. The reject ratio is found by combining the actual chip (or wafer) data with fault coverage data. For a required reject ratio, it is possible to determine the necessary fault coverage of tests. This coverage, in general, depends on certain processing parameters that also determine the yield.

The work in the papers cited above has been used by several workers to analyze experimental data [4,5,6]. However, there are difficulties with these analyses. For generally acceptable reject ratios, e.g., 1 in 10,000, most computed fault coverage requirements turn out to be almost 100 percent. Such coverage is not considered possible by the designers of large VLSI chips. There are several reasons for this. First, large circuits have some redundant faults and algorithms to find redundancies are very complex. Second, the alternative approach of exhaustive testing (where redundant faults may be ignored) is applicable only to the special class of circuits that can be partitioned into small combinational blocks with access to inputs and outputs. A third reason is the arbitrariness in measuring fault coverage. Different fault simulators make different assumptions about race faults, oscillation faults, hyperactive faults, etc.

The detection probability of a fault is really a conditional probability. It is the probability of detection by an input vector, given the fact that the fault is present. We have studied this probability in a recent paper [7]. In the present work, we also consider the fault occurrence probability. It is defined for each fault as the probability of its being found on a chip. Just as all faults are not equally detectable, they also do not occur with equal probability. The product of these two probabilities is the absolute failure probability of a chip by a test vector. In other words, a more realistic coverage requirement can be obtained if the faults were weighted by their occurrence probability. For example, a fault that never occurs will have a zero weight and is not required to be covered.

We analyze wafer test data to illustrate our analysis. The data on the measured fraction of failing chips versus the number of vectors is used to empirically determine a failure probability density function. From this, true yield and reject ratio are easily estimated without any fault coverage analysis.

NOTATION

c	Total number of chips tested
N	Total number of test vectors applied
c_i	Number of chips that fail exactly at vector number i
y	True yield, i.e., fraction of good chips
y_n	Estimated yield of chips after application of n vectors
r_n	Reject ratio after application of n vector, $(y_n - y)/y_n$

THEORY

We assume that a chip failure on an applied test vector is a random event with the associated density function $f(x)$. Thus, $f(x)dx$ is the fraction of chips that fail on a vector with a probability between x and $x+dx$. Since only a fraction $1-y$ of the total chips can fail, we have

$$f(x) = y\delta(x) + p(x)$$

where, $\delta(x)$ is the Kronecker delta function and $p(x)$ is a partial density function. Clearly,

$$x = Prob.\,(\textit{A fault has occurred } \text{and } \textit{is tested by the vector})$$

Since $f(x)$ is a density function,

$$\int_0^1 f(x)dx = y + \int_0^1 p(x)dx = 1$$

Therefore,

$$\int_0^1 p(x)dx = 1 - y$$

Now suppose after application of n test vectors, a certain fraction of the chips has not failed. The expected value of this fraction is the yield of chips after n vectors and is denoted by y_n. Clearly,

$$y_n = \int_0^1 (1-x)^n f(x)dx = y + \int_0^1 (1-x)^n p(x)dx \quad (1)$$

It is easily verified that $y_0 = 1$ and $y_\infty = y$, that is, before the testing starts the yield is 100% and if testing continues indefinitely, only the good chips would contribute to the yield.

Our aim is to estimate $f(x)$ from the chip failure data obtained while testing a sample of c chips by a test sequence of N vectors. As these vectors are applied, we record the number of chips that fail for the first time at each vector. For vector number i we will denote the number of such chips by c_i. Using Bayes theorem [8] with a uniform *a priori* failure probability, we have

$$f(x) = y\delta(x) + \left(1 - y - \frac{1}{c}\sum_{i=1}^{N} c_i\right) \frac{(1-x)^N}{\int_0^1 (1-x)^N dx} + \frac{1}{c}\sum_{i=1}^{N} c_i \frac{x(1-x)^{i-1}}{\int_0^1 x(1-x)^{i-1} dx}$$

The derivation of the above equation is similar to the one described in a recent paper [7]. The three terms on the right hand side represent, respectively, the contributions to $f(x)$ due to the fault-free chips, the faulty chips not rejected by any of the N vectors, and the remaining chips that are grouped according to vectors at which they failed. On evaluating the integrals on the right hand side we can rewrite the above equation as

$$f(x) = y\delta(x) + \left(1 - y - \frac{1}{c}\sum_{i=1}^{N} c_i\right)(N+1)(1-x)^N + \frac{1}{c}\sum_{i=1}^{N} c_i i(i+1)x(1-x)^{i-1} \tag{2}$$

Substituting this expression for $f(x)$ in Equation (1), we find the yield after n vectors as

$$y_n = y + \left(1 - y - \frac{1}{c}\sum_{i=1}^{N} c_i\right)\frac{N+1}{N+n+1} + \frac{1}{c}\sum_{i=1}^{N} c_i \frac{i(i+1)}{(n+i)(n+i+1)} \tag{3}$$

In the above expressions, the yield y is still an unknown parameter. It can be evaluated by equating the estimated yield after N vectors to the measured yield. Thus

$$y_N = 1 - \frac{1}{c}\sum_{i=1}^{N} c_i$$

Substituting for y_N from Equation (3) and solving for y, we get

$$y = 1 - \frac{1}{c}\sum_{i=1}^{N} c_i - \frac{2N+1}{N}\cdot\frac{1}{c}\sum_{i=1}^{N} c_i \frac{i(i+1)}{(N+i)(N+i+1)} \tag{4}$$

From Equations (3) and (4) it is now possible to compute the reject ratio as follows:

$$r_n = \frac{y_n - y}{y_n} \tag{5}$$

EXPERIMENT

To illustrate the application of the above analysis, we will use the experimental test data of a VLSI chip from a previous paper [2]. The measured yield of this chip for a test sequence of 3062 vectors is 0.469062. Of the 1002 tested chips, 470 did not fail after all vectors had been applied. The measured yield as a function of vectors is shown in Fig. 1. The true yield, as computed from Equation (4), was 0.46082. Figure 1 also shows the estimated yield y_n, as computed from Equation (3), for three different values of true yield. Only for the yield given by Equation (4), y_n closely matches the experimental data. This shows the sensitivity of our expression to the value of the true yield y and hence its suitability for estimating yield.

Reject ratio, as computed from Equation (5), is shown in Fig. 2. At the end of 3062 vectors, the computed reject ratio is about 0.017 or 17 faulty chips per thousand. The fault coverage of these vectors was about 90 percent [2]. Figure 2 predicts a reject ratio of 0.0001 or one faulty chip per 10,000 for about 500,000 vectors. Notice that this computation does not use fault coverage but only involves the use of the probabilities of fault occurrence and detection that are obtained from the actual measured data.

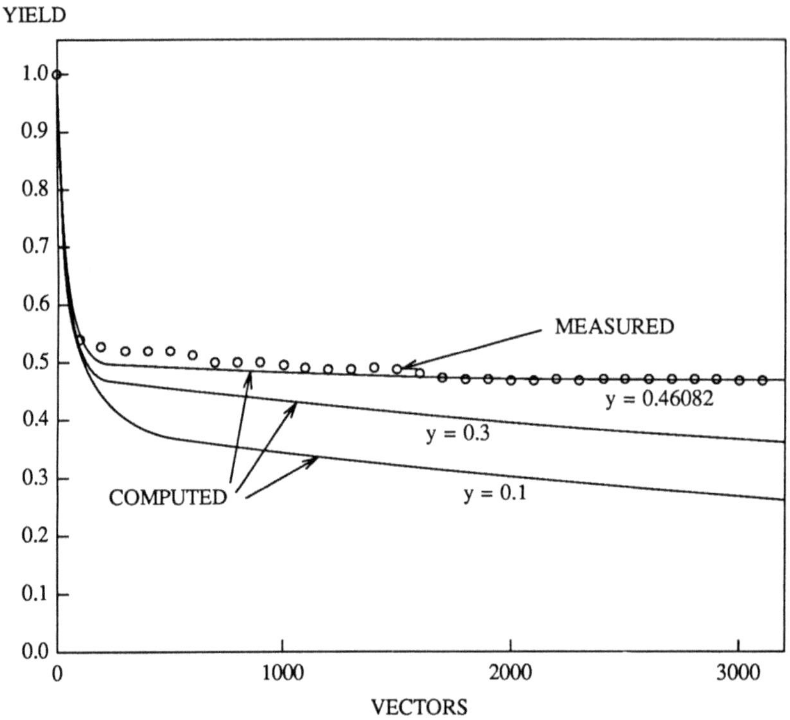

Fig. 1 Yield determination from measured test data.

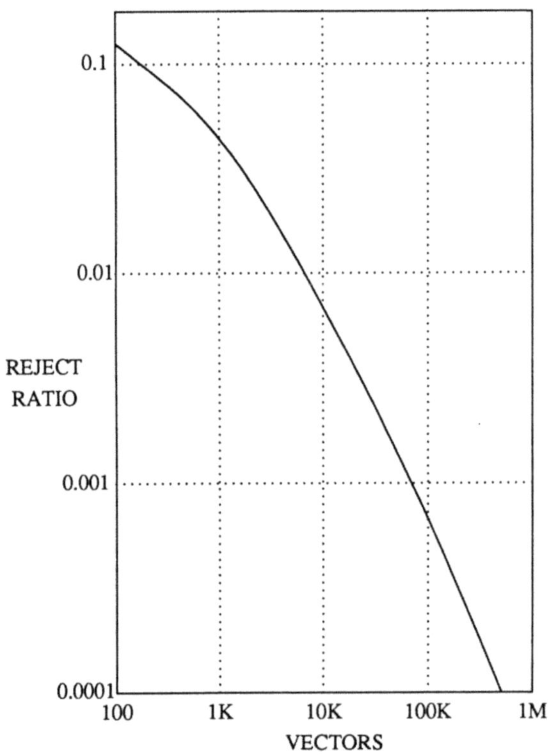

Fig. 2 Computed reject ratio.

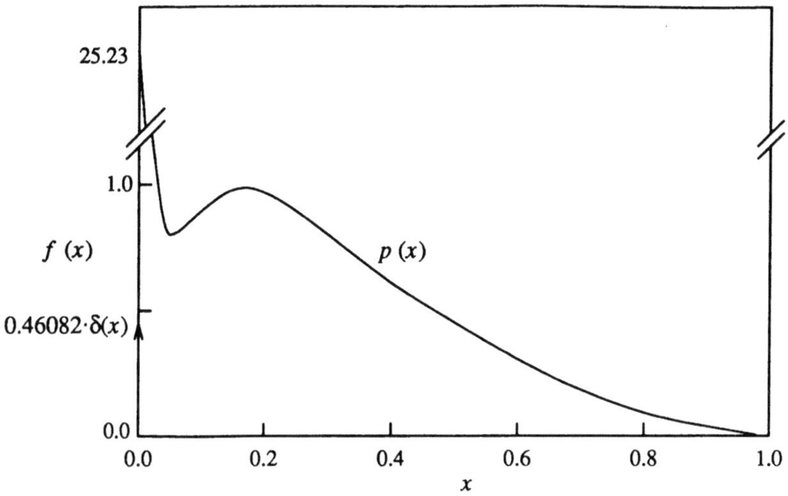

Fig. 3 Failure probability density function.

Figure 3 shows the failure probability density function $f(x)$ as computed from Equation (2). The delta function at the origin corresponds to the estimated true yield or good chips. The remaining part, i.e., $p(x)$, is the density of faulty chips. As explained earlier, this is the joint probability of a fault occurring and of its being detected by a vector. Evidently, $p(x)$ has two parts. One is a sharp peak near the origin and the other is a function of relatively lower magnitude. The peak corresponds to those faulty chips that survived all 3,062 vectors. This portion is mainly responsible for field rejects.

It can be conjectured that the second portion of $p(x)$ corresponds to failure modes similar to stuck type faults. Since the vectors were generated for such faults, the detection probabilities are higher. The faults in the peak near the origin might correspond to other failure modes like delay faults, stuck open faults, etc., for which the tests were not specifically targeted. Further investigation should lead to better insight.

CONCLUSION

We have given an entirely new method of asessing VLSI test quality. The tested product quality is directly evaluated from test data. This eliminates the need for complex fault models and fault simulation. Results are more realistic than coverage evaluation because the probability of fault occurrence is implicit in the analysis. Further investigation may lead to better methods of test generation. For example, tests are required to detect the low detection probability faults that have non-zero probability of occurrence rather than detecting stuck type faults that may never occur.

REFERENCES

[1] V. D. Agrawal, S. C. Seth, and P. Agrawal, "Fault Coverage Requirement in Production Testing of LSI Circuits," *IEEE Journal of Solid-State Circuits*, Vol. SC-27, pp. 57-61, Feb. 1982.

[2] S. C. Seth and V. D. Agrawal, "Characterizing the LSI Yield from Wafer Test Data," *IEEE Trans. on CAD*, Vol. CAD-3, pp. 123-126, April 1984.

[3] T. W. Williams and N. C. Brown, "Defect Level as a Function of Fault Coverage," *IEEE Trans. Computers*, Vol. C-30, pp. 987-988, December 1981.

[4] W. R. Mann, Private communication.

[5] P. T. Wagner, Private communication.

[6] B. W. Woodhall, B. D. Newman, and A. G. Sammuli, "Empirical Results on Undetected CMOS Stuck-Open Failures," *Proc. Int. Test Conf.*, Washington, D.C., pp. 166-170, September 1987.

[7] V. D. Agrawal, H. Farhat, and S. Seth, "Test Generation by Fault Sampling," *Proc. Int. Conf. Comp. Des. (ICCD-88)*, Rye Brook, NY, October 1988.

[8] A. Papoulis, *Probability, Random Variables, and Stochastic Processes*, McGraw-Hill, New York, 1965.

A NEW YIELD FORMULA FOR FAULT-TOLERANT LARGE-AREA DEVICES

C. Thibeault, Y. Savaria[1], and **J.-L. Houle**

Ecole Polytechnique de Montreal
Department of Electrical Engineering
Montreal, Quebec H3C 3A7, Canada

ABSTRACT

A new yield formula is proposed for large-area devices structured as two-level hierarchies [1]. A modified figure of merit is also proposed. Based on these formulas, examples are developed to demonstrate how to select the best fault-tolerance strategy in terms of yield, area, and processing power.

INTRODUCTION

Wafer Scale Integration (WSI) offers a great potential in terms of high chip density and high processing power. However, without some form of yield enhancement through redundant architectures, the yield would be essentially zero. There is clearly a trade-off between the yield of a WSI device and its throughput. The purpose of this work is to study this trade-off. The work is based on our experience acquired through the design of self-reconfigurable VLSI chips [1,2]. A new formula is proposed to evaluate the yield of fault-tolerant integrated circuits (ICs). A method for selecting the most cost-effective strategy is also developed.

PREVIOUS FORMULAS

Some formulas [3,4] have been developed to predict the yield of fault-tolerant regular arrays. For example, the formula proposed by Koren and Pradhan [5] gives the yield of circuits made of identical elements, where some of the elements are spares which can replace any faulty one. However, their formula does not take into account the so called "kill area", where a defect cannot be tolerated. Unfortunately, it turns out that in our chips [1,2] the kill area occupies a significant portion of the total area. Moreover, Koren and Pradhan's formula cannot predict the yield of a hierarchical structure as the one shown in Fig. 1, which is a typical implementation of a SIMD (Single Instruction Multiple Data) architecture with two-level fault-tolerance capability.

In this paper, we consider ICs structured as shown in Fig. 1, made of modules (or rows), where each module contains some replaceable elements, REs, and one control unit, CU, shared by all the REs of a given module. A processing element (PE), which

[1]Yvon Savaria is also appointed as an auxiliary professor with the EE Department of McGill University, Montreal, Canada.

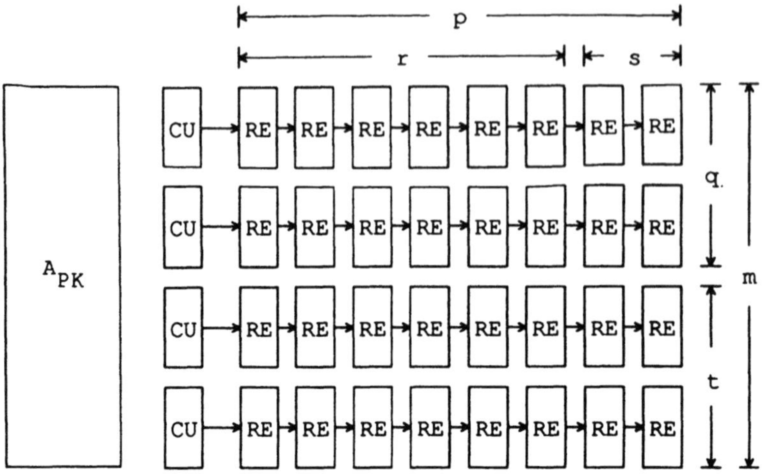

CU : Control Unit;
RE : Replaceable Element (may be one or more processing elements (PEs));
A_{CU} : Control Unit Area (includes any intolerant area in a module);
A_{RE} : Replaceable Element Area;
A_{PK} : Permanent Kill Area;
q : number of modules (spare modules not included);
t : number of spare modules;
m : total number of modules (spare modules included);
r : number of REs per module (spare REs not included);
s : number of spare REs per module;
p : total number of REs per module (spare REs included);

Figure 1: *Typical implementation of a SIMD architecture with fault-tolerance capability*

represents the smallest processing resource, is considered as the smallest practical RE. In Fig. 1, there are m modules containing p REs and one CU. In each module, s of the p REs are spares which can replace any of the remaining r REs. The presence of many control lines and other practical limitations to the reconfiguration of the communication network limit RE replacements within a given module. Note that RE replacements constitute the first level of the fault-tolerance hierarchy. The second level consists of the addition of spare modules in the circuit. Thus in Fig. 1, t of the m modules are spares which can replace any of the q others modules. This means that one or more complete modules can be defective, but the presence of spare modules allows to tolerate them. Note that a portion of the kill area at the first level of the hierarchy becomes part of the fault-tolerant area at the second level. The other portion of the kill area, called the "permanent kill area", is represented by the A_{PK} block.

A NEW YIELD FORMULA

A new yield formula has been developed for structures as the one shown in Fig. 1. The method used to develop our yield formula is similar to the one used by Koren and Pradhan. Thus, the yield of a fault-tolerant IC can be expressed as

$$Y_r = \sum_{x=0}^{\infty} P\{x\} P_{tx} \ , \tag{1}$$

where

$P\{x\}$: probability of having x defects[2] in the total area of the IC;
P_{tx} : probability of tolerating these x defects.

[2]At this stage of development, only point-defects are considered.

As Koren and Pradhan, we use *generalized negative binomial statistics* [6] for $P\{x\}$, but any other type of statistics can also be used. *Combinatorial statistics*, such as *Maxwell-Boltzmann* or *Bose-Einstein* statistics, are used to compute P_{tx}. For that computation, the circuit is modeled as an array of equivalent-area elements, EEs, where the area of each element is A_{RE}. Consequently, the A_{PK} block and all the CUs are divided into EEs, and each RE is treated as an EE. In the chosen notation, the A_{PK} block contains v EEs and each CU is composed of u EEs. Moreover, there are z EEs in the whole circuit area, where $z = v + m \times (p + u)$. Using combinatorial statistics and the previous circuit model, P_{tx} can be defined as

$$P_{tx} = \frac{C_x}{D_x}\,, \tag{2}$$

where the following definitions apply:
C_x : number of distinct patterns where x defects distributed over the IC area are tolerated (i.e. the circuit is functional in spite of these defects);
D_x : number of distinct patterns where x defects are distributed over the IC area.

With all these previous considerations and since it is obvious that $P_{tx} = 1$ when $x = 0$, (1) becomes

$$Y_r = P\{x{=}0\} + \sum_{x=1}^{\infty} P\{x\}\frac{C_x}{D_x}\,. \tag{3}$$

The term C_x is developed by varying k, the number of affected (but not necessarily defective) modules, from 1 to m. Since the development of C_x for $k \leq t$ is quite different from the case where $t + 1 \leq k \leq m$, it is done separately for both cases, identified by $\alpha = 1$ and $\alpha = 2$ respectively. Thus, (3) becomes

$$Y_r = P\{x{=}0\} + \sum_{x=1}^{\infty} \frac{P\{x\}}{D_x}\left[\sum_{k=1}^{t}\binom{m}{k} C_{xk1} + \sum_{k=t+1}^{m}\binom{m}{k} C_{xk2}\right], \tag{4}$$

where C_{xk_α} represents the number of distinct patterns where x defects distributed over exactly k modules are tolerated (exactly meaning no empty modules).

Due to space limitation, the derivation of C_{xk1} and C_{xk2} will not be given here. A complete derivation can be found in [8]. After subtitution of C_{xk1} and C_{xk2} in (4), (5) is obtained.

$$\begin{aligned} Y_r \;=\; & P\{x{=}0\} + \sum_{x=1}^{\infty} \frac{P\{x\}}{D_x}\left\{\sum_{k=1}^{t}\binom{m}{k}\sum_{i_1=1}^{p+u}\cdots\sum_{i_k=1}^{p+u} Q_{xy_1}\prod_{j=1}^{k}\binom{p+u}{i_j} + \right. \\ & \sum_{k=t+1}^{m}\binom{m}{k}\sum_{g=0}^{t}\binom{k}{g}\sum_{h=0}^{g}\binom{g}{h}\sum_{i_1=1}^{u}\cdots\sum_{i_h=1}^{u}\;\sum_{i_{h+1}=s+1}^{p}\cdots\sum_{i_g=s+1}^{p}\;\sum_{i_{g+1}=1}^{s}\cdots \\ & \left.\cdots\sum_{i_k=1}^{s}\sum_{f_1=0}^{p}\cdots\sum_{f_h=0}^{p} Q_{xy_2}\prod_{j=1}^{h}\binom{u}{i_j}\binom{p}{f_j}\prod_{j=h+1}^{k}\binom{p}{i_j}\right\}, \end{aligned} \tag{5}$$

where

$$y_1 = \sum_{w=1}^{k} i_w \,, \tag{6}$$

$$y_2 = \sum_{w=1}^{k} i_w + \sum_{w=1}^{h} f_w \,. \tag{7}$$

The following is the list of new symbols and definitions for equations 5, 6, and 7:

f_w : number of defective REs in the w-th CU-defective module;
g : number of defective modules;
h : number of CU-defective modules;
i_w : number of defective EEs or REs in the w-th module;
Q_{xy_α} : number of distinct patterns for the distribution of x defects into exactly y_α EEs;
y_α : number of affected EEs and REs.

The formula was developed in two versions : first using *Maxwell-Boltzmann statistics*, and second using *Bose-Einstein statistics.* Q_{xy_α} and D_x depend on the type of combinatorial statistics used.

- For *Bose-Einstein* statistics,

$$D_x = \binom{x+z-1}{x} \,, \tag{8}$$

$$Q_{xy_\alpha} = \binom{x-1}{y_\alpha - 1} \,. \tag{9}$$

- For *Maxwell-Boltzmann* statistics,

$$D_x = z^x \,, \tag{10}$$

$$Q_{xy_\alpha} = \sum_{b=1}^{y_\alpha} (-1)^{y_\alpha - b}\, b^x \binom{y_\alpha}{b} \,. \tag{11}$$

EVALUATION OF FAULT-TOLERANCE : A NEW FIGURE OF MERIT

In order to evaluate fault-tolerance strategies, one can use a *figure of merit*, which quantifies the effect of the inclusion of such strategies in an IC. Mangir [7] proposed one, which is expressed as

$$f_m = \frac{Y_r \times A_o}{Y_o \times A_r} \,, \tag{12}$$

where

f_m : Mangir's figure of merit;
Y_r : yield of fault-tolerant IC;
Y_o : yield of non fault-tolerant IC;
A_r : area of fault-tolerant IC;

A_o : area of non fault-tolerant IC.

This figure of merit takes into account the variations of yield and area, but not the variations of IC functionality (number of working PEs on a chip). In some applications (e.g. when the total number of PEs is fixed), inclusion of fault-tolerance implies a variable functionality, thus, we have defined a new figure of merit, which is a generalization of f_m,

$$f_{m2} = \frac{Y_r \times A_o \times f_r}{Y_o \times A_r \times f_o} , \tag{13}$$

where

f_{m2} : new figure of merit;
f_r : functionality of fault-tolerant IC;
f_o : functionality of non fault-tolerant IC.

It is appropriate to include fault-tolerance if $f_{m2} > 1$, and the best fault-tolerant strategy is the one that maximizes f_{m2}. This maximum is found by varying the number of spare REs and spare modules. In addition, there is a third possible way to optimize the architecture of a fault-tolerant IC. It consists of varying the number of PEs per RE. As mentioned earlier, PEs represent the smallest possible REs. In some cases, it may be appropriate to regroup PEs to form bigger REs because it reduces the area devoted to reconfiguration circuitry, A_{RC}, which is generally part of the kill area. This PE regrouping can be done "vertically" i.e. the number of REs per module remains the same but the number of modules decreases. Consequently, the size of CUs and modules increases in the same proportion as the size of REs. For example, a circuit containing initially 16 rows of PEs can be treated in terms of fault-tolerance as a circuit with 16, 8, 4, 2, or even only one module.

EVALUATION OF FAULT-TOLERANCE : EXAMPLES

Examples are now presented to demonstrate how Eq. 5 and 13 can determine the most cost effective fault-tolerant structure. In these examples, the figure of merit f_{m2} of an array as in Fig. 1 is evaluated. The numerical values for the areas of the various components are derived from a chip that we have designed and implemented in a *3-micron double-metal CMOS technology* [2]. This 12 PE chip was structured as one module of six REs where two of the REs were spares. An empirical equation, based on the actual layout of this chip, is used to evaluate A_{RC} (in cm^2) as a function of p, m, and t:

$$A_{RC} = 0.001p \times (m + t) \ . \tag{14}$$

We also use *generalized negative binomial statistics* for Y_o. These statistics are function of two new parameters: c, the *clustering parameter*, quantifying the defect tendency to be regrouped, and d, the *expected value of a defect density function.*

The inclusion of fault-tolerance capability in an IC depends on some factors, such as the type of architecture, the design constraints, etc . We will concentrate here on two possible approaches of fault-tolerance inclusion: a fixed number of working PEs and a fixed number of PEs.

Fixed functionality

This approach is most valuable when the throughput is limited to fixed values (eg. RAM chips) and when there is some area available for the addition of some extra spares (REs or modules). Since the functionality is fixed, f_r equals f_o, so after simplification f_{m2} equals f_m. First, the functional kernel of the IC is built with the number of PEs

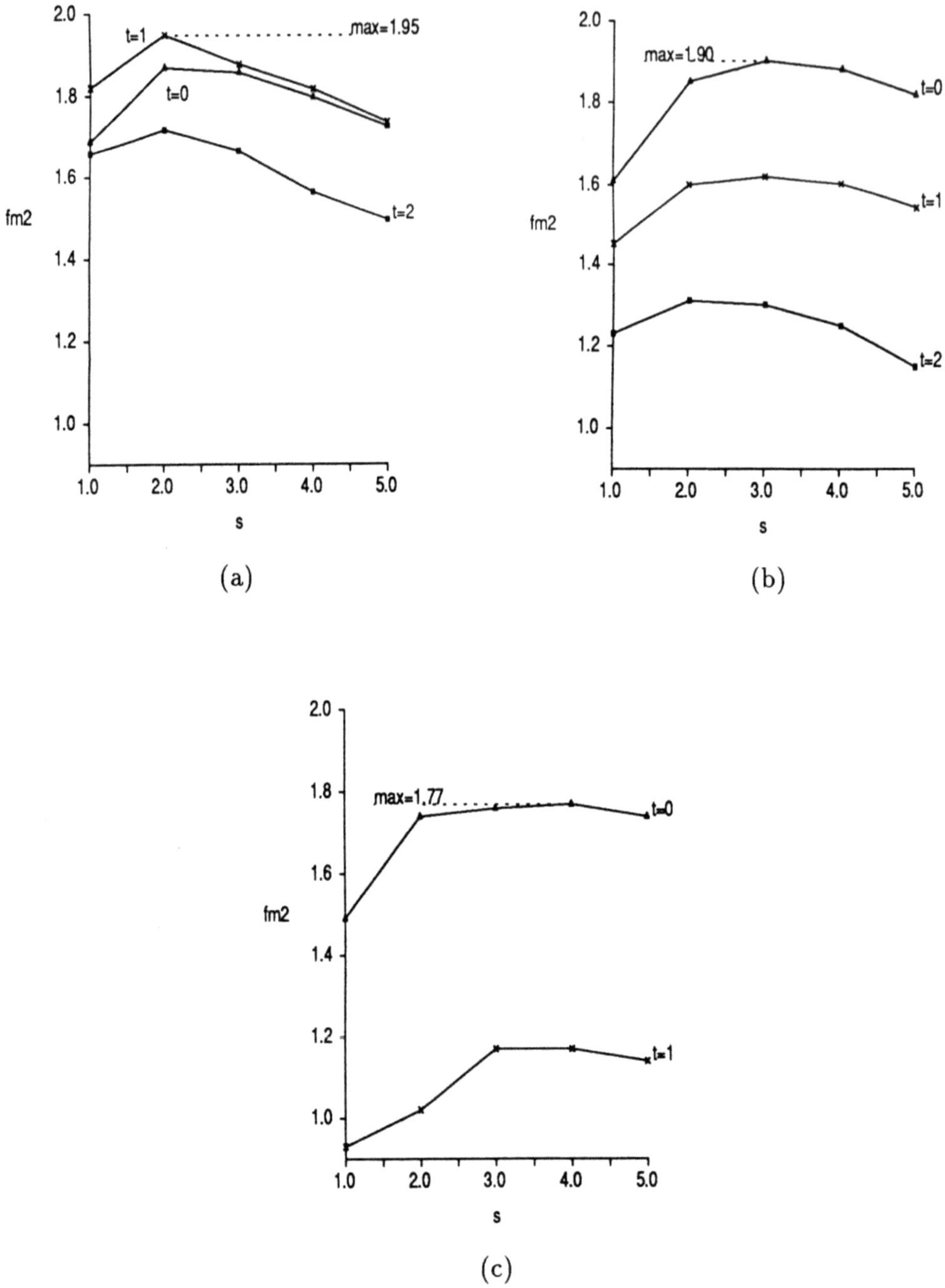

Figure 2: *Figure of merit (f_{m2}) versus number of spare REs (s) and spare modules (t). Fixed functionality.* (a) $q = 4$, $A_{CU} = 0.1 + 0.001p\ cm^2$, $A_{RE} = 0.03\ cm^2$; (b) $q = 2$, $A_{CU} = 0.2 + 0.001p\ cm^2$, $A_{RE} = 0.06\ cm^2$; (c) $q = 1$, $A_{CU} = 0.4 + 0.001p\ cm^2$, $A_{RE} = 0.12\ cm^2$.

required to obtain the desired functionality. As an example, we chose a functional kernel of 4 rows of 16 PEs, where each PE has an area of $0.03cm^2$. Each row also contains a CU of $0.1cm^2$. In this case, the optimum value of f_{m2} is found by adding some extra spare REs or spare modules to the functional kernel, and also by vertical regrouping as explained previously. In the present example, *Bose-Einstein statistics* are applied, c and d are set to one, and A_{RC} is separated into various components, one half goes into A_{PK} and the other half is further added to A_{CU}. The results for this example are shown in Fig. 2, where Fig. 2 (a) represents the starting step of the vertical regrouping (there is only one PE per RE). In terms of reconfiguration, the functional kernel contains 4 modules ($q = 4$), where each module is composed of 16 REs ($r = 16$) and one CU. In Fig. 2 (b) , PEs and CUs are regrouped to reorganize the functional kernel into two modules ($q = 2$). There are still 16 REs per module but REs and CUs areas are now doubled. Similarly, Fig. 2 (c) represents the case where the number of functional modules, q, is 1. We observe that the optimum f_{m2} (=1.95) corresponds to Fig. 2 (a), when $q = 4$, $t = 1$, and $s = 2$. Therefore in this example, it is not worth regrouping PEs to form larger REs, but it is advantageous to have spare modules.

Fixed number of PEs

This approach is good when the final throughput is not limited to precise values and when the available area is limited. In this context, the IC area is filled with the maximum number of PEs to form a fixed kernel (some space is left empty for the reconfiguration circuitry). The optimum f_{m2} is found by vertical regrouping and by varying the number of spares (REs and modules), which implies a variation of the functionality since the number of PEs is kept constant. In this example, *Bose-Einstein statistics* are also applied, c and d are once again set to one, and A_{RC} is separated as in the previous example. Fig. 3 shows some results for this approach. Here, the fixed kernel is composed of 4 rows containing 16 PEs and one CU, where each PE has an area of $0.03cm^2$ and where each CU area is $0.1cm^2$. Fig. 3 (a) is the starting step (PEs=REs), where the fixed kernel contains 4 modules ($m = 4$) of 16 REs ($p = 16$). In Fig. 3 (b), the fixed kernel is now composed of 2 modules ($m = 2$). There are still 16 REs and one CU per module, but each RE and the CU are now twice as large. Similarly, Fig. 3 (c) represents the case where the number of total modules, m, is 1 . We observe that the optimum f_{m2} (=1.87) corresponds to Fig. 3 (a), when $m = 4$, $t = 1$, and $s = 2$. Once again, regrouping REs does not pay in this situation and it is appropriate to have spare modules.

Discussion

One may observe that the figure of merit for both cases was not very high, however it is not surprising since the yield with no fault-tolerance added was approximately 30 %. Much larger figure of merit could be observed for very large ICs where the initial yield could be lower than a few percent. Note also that if the maximum figure of merit for the various situations are very similar, interestingly these maximums are not found for the same levels of fault-tolerance and significantly lower figure of merit can result if an inappropriate level of fault-tolerance is implemented. Moreover, even though it was not found advantageous to merge PEs to form larger REs, parameters can indeed be selected to demonstrate that it can be the case.

COMPARISON BETWEEN BOSE-EINSTEIN AND MAXWELL-BOLTZMANN

The two examples developed with *Bose-Einstein statistics* were redone with *Maxwell-Boltzmann statistics* and the comparison of the results is shown in Fig. 4 and 5. In each example, we can observe that both types of statistics give relatively close results and that *Bose-Einstein statistics* curves are slightly higher than those of *Maxwell-Boltzmann statistics*, which was not obvious *a priori*. Also, in each case, the decision made on the choice of the fault-tolerance strategy remains the same. However more examples have to be studied to verify to what extent both types of statistics always lead to the same choice of fault-tolerance strategy.

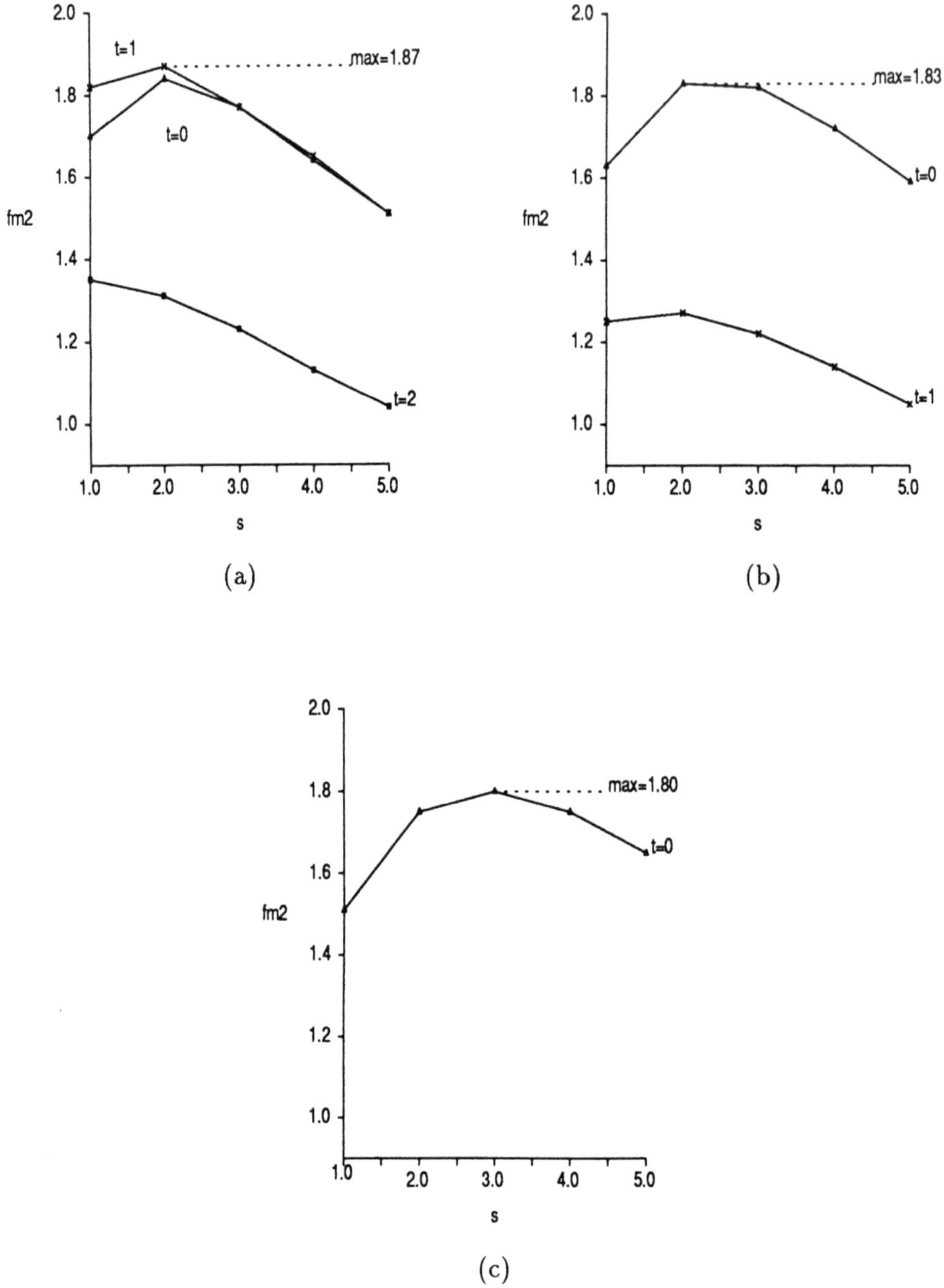

Figure 3: *Figure of merit (f_{m2}) versus number of spare REs (s) and spare modules (t). Fixed number of PEs.* (a) $m = 4$, $A_{CU} = 0.12\ cm^2$, $A_{RE} = 0.03\ cm^2$; (b) $m = 2$, $A_{CU} = 0.22\ cm^2$, $A_{RE} = 0.06\ cm^2$; (c) $m = 1$, $A_{CU} = 0.42\ cm^2$, $A_{RE} = 0.12\ cm^2$.

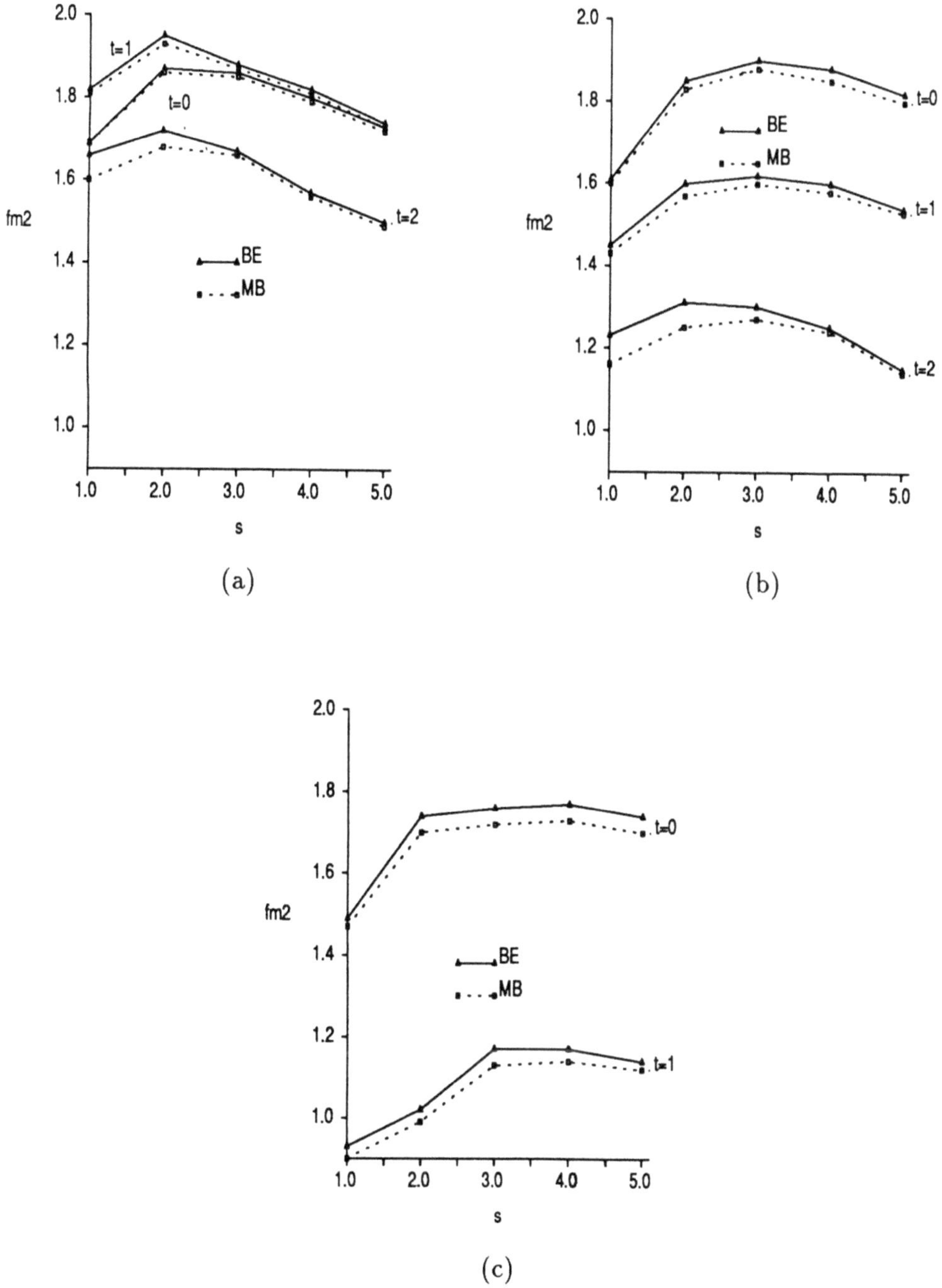

Figure 4: *Figure of merit (f_{m2}) versus number of spare REs (s) and spare modules (t). Comparison between Bose-Einstein statistics and Maxwell-Boltzmann statistics with fixed functionality. (a) $q = 4$; (b) $q = 2$; (c) $q = 1$.*

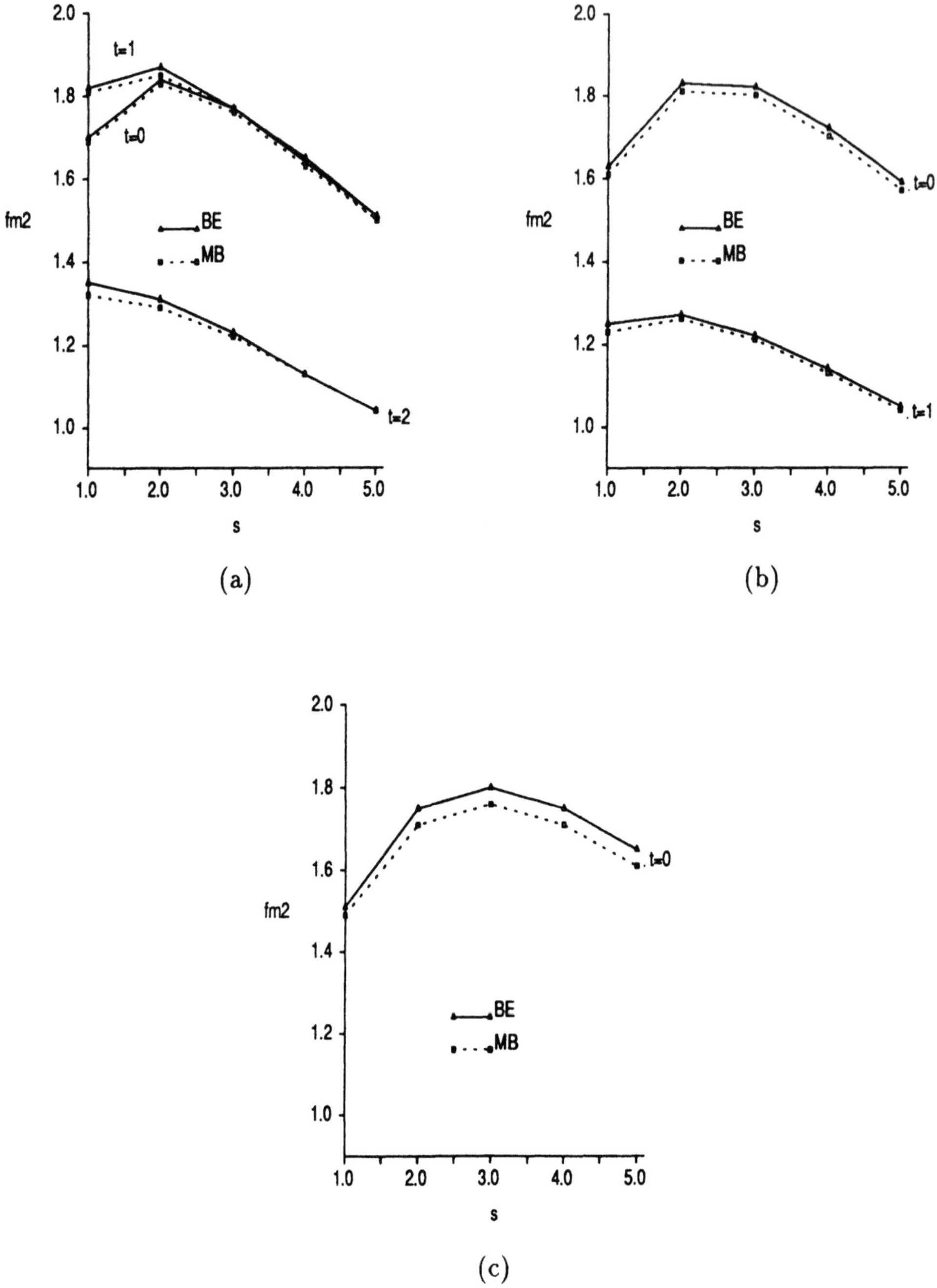

Figure 5: *Figure of merit (f_{m2}) versus number of spare REs (s) and spare modules (t). Comparison between Bose-Einstein statistics and Maxwell-Boltzmann statistics with fixed number of PEs. (a) $m = 4$; (b) $m = 2$; (c) $m = 1$.*

CONCLUSION

A new yield formula for fault-tolerant large-area devices has been proposed. This formula evaluates yield for circuits composed of many modules, where spare REs can replace faulty ones within a given module. Moreover, the formula takes into account the kill area and even the presence of spare modules in an architecture with two-level hierarchies. A new figure of merit has also been proposed. These two formulas allow to select the best fault-tolerance strategy in terms of yield, area, and processing power. Therefore, one can choose the optimum number of spare REs and spare modules. One can also determine if it is appropriate to regroup PEs to form larger REs.

REFERENCES

[1] Cyr C., Audet D., Savaria Y., and Houle J.-L., "A Novel Self-Testing and Reconfiguration Scheme for Yield Improvement of Two-Dimensional Arrays", *ICCD '87*, pp. 494-500, October 1987.

[2] Thibeault C., Savaria Y., Lagarde C., Repeta M., and Rocque P., "IMAGE2M, a VLSI Implementation of a Multi-Processor Chip for Morphological Image Processing", *Proc. 5th Canadian Conference on VLSI*, pp. 235-240, October 1987.

[3] Harden J.C. and Strader N.R., "Architectural Optimization for WSI", *IEEE Trans. on Computers*, vol. 37, no. 1, January 1988.

[4] Ketchen M.B., "Point Defect Yield Model for WSI", *IEEE Circuits and Devices Magazine*, vol. 1, no. 4, pp. 24-34, July 1985.

[5] Koren I. and Pradhan D., "Modeling the Effect of Redundancy on Yield and Performance of VLSI Systems", *IEEE Trans. on Computers*, vol. C-36, no. 3, pp. 344-355, March 1987.

[6] Stapper C.H., "Defect Density Distribution for LSI Yield Calculations", *IEEE Trans. Electron Devices*, vol. ED-20, pp. 655-657, July 1973.

[7] Mangir T., "Sources of Failures and Yield Improvement for VLSI and Restructurable Interconnects for RVLSI and WSI : Part I - Sources of Failures and Yield Improvement for VLSI", *Proc. IEEE*, vol. 72, no. 6, pp. 690-708, June 1984.

[8] Thibeault C., Savaria Y., and Houle J.-L., "Yield formula for two-level hierarchical fault-tolerant integrated circuits", *E.E. Department report*, Ecole Polytechnique de Montreal, August 1988.

ACKNOWLEDGMENTS

This work has been partly supported by sholarships to one of the authors from the Government of Quebec (FCAC funds) and the Computer Research Institute of Montreal, by grants OGP0006574 and OGP0009338 from the Natural Sciences and Engineering Council of Canada, and by grant EQ 2806 from FCAC.

APPENDIX : LIST OF SYMBOLS

A_{CU} : Control Unit Area;
A_{PK} : Permanent Kill Area;
A_{RC} : Reconfiguration Circuitry Area;
A_{RE} : Replaceable Element Area;
A_o : area of non fault-tolerant IC;
A_r : area of fault-tolerant IC;
CU : Control Unit;
C_x : number of distinct patterns where x defects distributed over the whole circuit area are tolerated (i.e. the circuit is functional in spite of these defects);
C_{xk_α} : number of distinct patterns where x defects distributed over exactly k modules are tolerated (exactly meaning no empty modules);
D_x : number of distinct patterns where x defects are distributed over the whole circuit area;

EE : Equivalent-area Element;
P_{tx} : probability of tolerating these x defects;
$P\{x\}$: probability of having x defects in the total area of the IC;
Q_{xy_α} : number of distinct patterns for the distribution of x defects into exactly y_α EEs;
RE : Replaceable Element (may be one or more processing elements (PEs));
Y_o : yield of non fault-tolerant IC;
Y_r : yield of fault-tolerant IC;
c : clustering parameter;
d : expected value of a defect density function;
f_m : Mangir's figure of merit;
f_{m2} : new figure of merit;
f_o : functionality of non fault-tolerant IC;
f_r : functionality of fault-tolerant IC;
f_w : number of defective REs in the w-th CU-defective module;
g : number of defective modules;
h : number of CU-defective modules;
i_w : number of defective EEs or REs in the w-th module;
k : number of affected (but not necessarily defective) modules;
m : total number of modules (spare modules included);
p : total number of REs per module (spare REs included);
q : number of modules (spare modules not included);
r : number of REs per module (spare REs not included);
s : number of spare REs per module;
t : number of spare modules;
u : number of EEs included in a CU;
v : number of EEs included in the A_{PK} block;
x : number of defects in an IC;
y_α : number of affected EEs and REs;
z : number of EEs included in the whole IC area;
α : mode subscript : $\alpha = 1$ for $k \leq t$; $\alpha = 2$ for $t + 1 \leq k \leq m$.

4 DEFECT-TOLERANT DESIGNS

DEFECT TOLERANT INTERCONNECTS FOR VLSI

Michael C. Howells, Robert Aitken, and **Vinod K. Agarwal**

McGill University
Department of Electrical Engineering
Montreal, Quebec H3A 2A7, Canada

INTRODUCTION

With the increasing size and complexity of VLSI circuits, the need has arisen for techniques to increase the testability and reconfigurability of VLSI circuits in order to control the costs associated with testing and to ensure economically viable device yields. In order to increase the testability of complex digital logic, techniques such as scan-design[1], built-in self-test based on linear feedback shift registers[2], and boundary-scan[3] have been developed and implemented. For yield and reliability enhancement, many different reconfiguration techniques have been proposed for circuitry with a high degree of regularity, such as memory[4], cellular arrays[5] and multi-processor architectures[6,7,8].

While most of this effort has focused on the logic blocks, surprisingly little of the research in this area has looked at the large structured interconnects (busses, communication channels, interconnection lattices) that often account for more than half the area of a VLSI chip. As integrated circuit pin counts increase, wider data and address paths evolve, and the number of processing elements which need to be interconnected grows, the interconnect architecture also grows in size and complexity.

This paper will address the issue of increasing the testability and reconfigurability of structured interconnects with the intent of developing a method of increasing the defect tolerance of the interconnects through testing and reconfiguration. The approach is to insert additional logic and spares into the interconnect in order to make it more easily testable and reconfigurable, as is done successfully with logic blocks and multiprocessor architectures. The method is intended to be a structured design approach which can be integrated into a CAD tool for the design environment. The paper will look at the application of the technique to multi-processor arrays with global busses.

The remainder of this paper will be divided as follows: Section two discusses the problems related to testing and reconfiguring interconnects. Section three describes the hardware approach being proposed. Section four discusses testing, diagnosis and reconfiguration using the approach. Section five gives an evaluation of the yield enhancement and area costs of the scheme.

TESTING AND RECONFIGURING INTERCONNECT

Although the processing steps necessary for interconnect are less complex than those for the processing elements, the interconnect is nonetheless a likely candidate for yield problems. The fact that a structured interconnect often occupies a large area of the chip, and often provides some global functionality, makes it a likely, single point of failure for the entire chip. In addition, it is likely that an interconnect that is spread all over the chip area will traverse through regions of higher defect density.

Similarly, although the functionality of the interconnect is very simple, it is a distributed functionality related to the number of end-points it connects and the number of branch-points it contains. If, for instance, the fault-model being used requires that stuck-opens be tested for in the interconnect, then all the end-points of the interconnect must be accessible and observable even though they are widely distributed. From the point of view of reconfiguration, the number of endpoints increases the hardware necessary for implementing reconfiguration whether by programmable switches, fusible links, or laser/e-beam personalization.

One can try to avoid the complexity of testing and reconfiguring the interconnect by using a fault masking scheme such as coding[9], triple modular redundancy[10], or transmission time redundancy[11,12]. Unfortunately these schemes require prohibitively large quantities of redundant hardware, for all but the simplest structured interconnects.

If one is dealing with a multi-processor array into which a scheme for testing and reconfiguring the processors has been designed, then the interconnects (or some portion of them) could be considered part of the processor. In such a case defects in the interconnects are indistinguishable from defects in the processor, and reconfiguration involves bypassing and possibly replacing both as a unit. While this approach has its uses, particularly for systolic arrays, it cannot handle interconnects that have more than a few endpoints.

PROPOSED APPROACH

This section describes the approach being proposed for introducing testability and reconfigurability into structured interconnects in a multi-processor array. The term multi-processor array is used here to describe any chip comprised of a number of quite large, functionally identical processing elements (PEs). The approach was developed as part of a study of a commercial chip comprising an array of 32 PEs connected together by a common, global bus of 40 bits, implemented in 1.2μ CMOS. Nonetheless, the ideas presented are extensible to other types of interconnect and processor architectures. It is assumed that there is at least one periphery cell (PC) on the chip which interfaces the array to the external pins on the chip package.

Initial Definitions

Definition 1: An endpoint on a processing element or periphery cell is one of either an input pin, an output pin or an input/output pin.

Definition 2: A site is the collection of all endpoints which belong to a single processing element.

Definition 3: A line is a physical wire that connects an endpoint on a peripheral cell to an endpoint on one or more processing elements.

Definition 4: An interconnect is a group of one or more lines.

Definition 5: A path is the minimal portion of a line necessary to connect an endpoint on a periphery cell and an endpoint on a processing element.

Definition 6: A line pair is a pair of lines whose endpoints are adjacent to each other at all sites and whose lines are tested and reconfigured as a unit.

The Fault Model, Testability And Reconfiguration Strategy

It is desired to test for the following fault categories in the interconnect:

1) All types of AND or OR bridges between one or more lines.
2) Stuck-at faults, both zero and one on one or more lines.
3) Stuck-opens on one or more paths anywhere.

The proposed testability scheme is to provide a mechanism that allows test vectors to be applied from the periphery cell down the set of paths going to each individual site for all sites in turn. The test results are collected via the same sets of paths to which the test vectors were applied. Thus that portion of the interconnect going to each individual site is independently testable. This implies that each site requires some selection logic to enable its paths to be selected for testing. The test pattern generator and test result diagnosis hardware may be either built-in or off-chip.

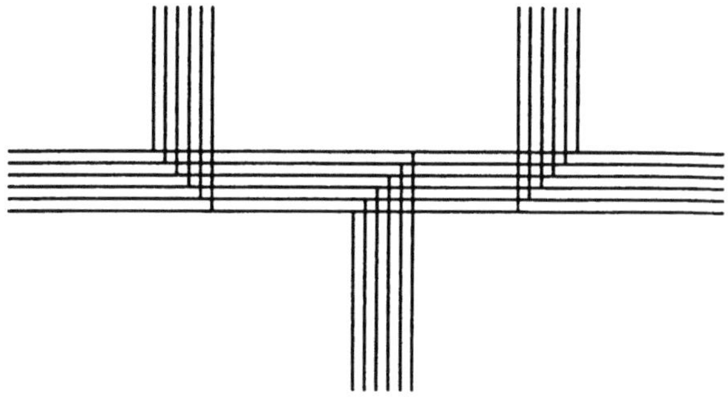

Figure 1 Global Bus Architecture

The proposed approach to reconfiguration is to provide spare lines for replacing faulty lines. A faulty line is disconnected from its endpoints and a spare line connected to the endpoints of the faulty line. The connections are made using soft-configurable switches controlled by status latches. The status latches are programmable from the periphery cell through the interconnect itself.

Implementing The Approach

The implementation presented here focuses on an interconnect architecture comprising a globally shared bus, which was found on the commercial chip previously mentioned. While not all multiprocessor architectures are based on globally shared busses, many do have some global busses, particularly for control signals and host access.

An example of a global interconnect architecture is shown in figure 1. The processing elements would be located at four of the five endpoints of the interconnect and the periphery cell at the fifth.

The overall block diagram for the testability and reconfigurabilty approach as applied to this interconnect architecture is shown in figure 2. The approach introduces a number of new elements into the architecture: these being the TL and EL logic and switching blocks, the SL site control logic blocks, the TC test and program scan chain, the SC serial control chain and, of course, a few spare lines.

Figure 2 shows six lines with two spare lines. In general the bus lines are considered to be grouped into line pairs. and spares are considered to be allocated as spare line pairs. As will be discussed later, this pairing also applies to testing and replacement, and is done to increase the ease of testing. All the lines in the bus are considered as bi-directional in this implementation. This is the worst-case implementation, having the most redundant hardware. More optimal realizations could be achieved by matching input and output lines as line pairs. The minimum size reconfigurable interconnect

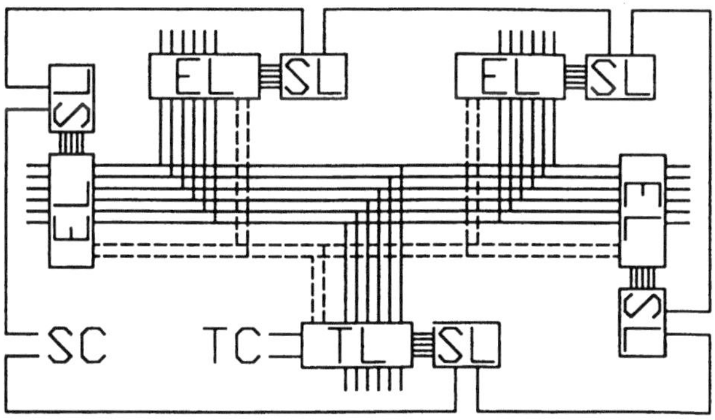

Figure 2 Testable and Reconfigurable Global Bus Architecture

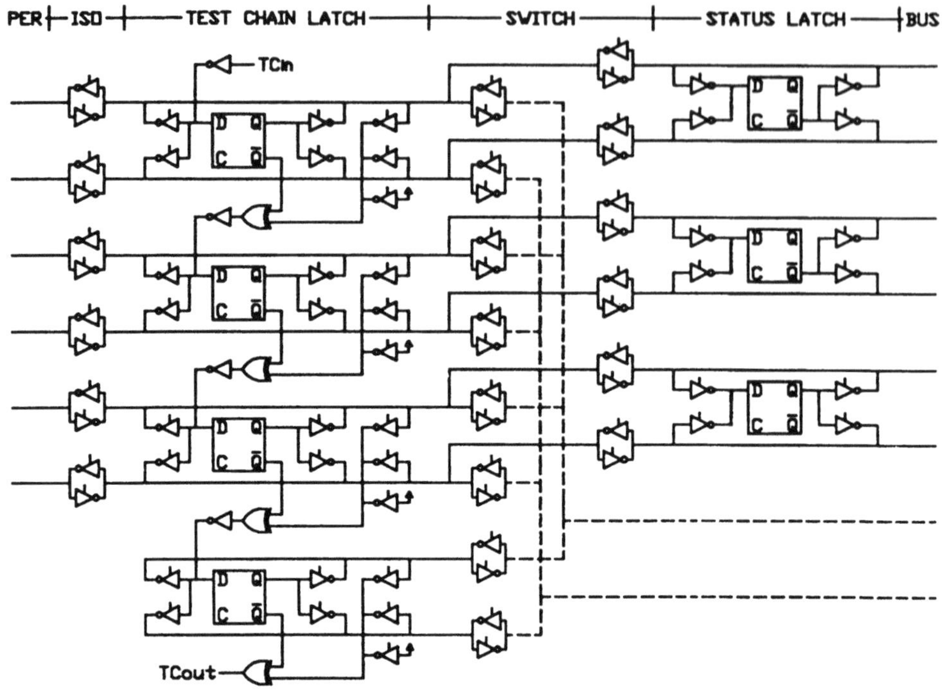

Figure 3 Testpoint Logic for 6 lines with 2 spares

is three line pairs with one spare line pair (as shown). The reconfiguration scheme only permits one spare line pair per set of line pairs, in order to keep the switch complexity low. Thus, to provide a larger number of spare lines to a much larger set of lines, one must break the set into smaller subsets each with one spare.

All the switching and multiplexing elements required in the logic blocks are implemented using a 4 transistor tristate CMOS inverter. This device has the advantage of being completely testable for the stuck-at and stuck-open fault models. A transmission gate implemented from these unidirectional elements therefore requires 8 transistors, as opposed to 2 in the conventional, untestable CMOS transmission gate.

The Test and Program Scan Chain (TC) is the scan chain through which all the test vectors are applied, the test results gathered and the switches configured. This scan chain interfaces the interconnect periphery cell with the test, diagnosis and reconfiguration unit. The test, diagnosis and reconfiguration unit could be off chip or built-in and its implementation will not be discussed or analyzed in this paper.

The Serial Control Chain (SC) is the control chain through which all the control signals for each site are shifted. A serial approach was used to reduce the number of additional interconnects required for control purposes, as these interconnects are not reconfigurable. This control chain also interfaces with the test, diagnosis and reconfiguration unit from which the control signals originate.

At the PC site is placed the Testpoint Logic (TL). This block contains the logic necessary to apply the test vectors to the line pairs, collect the test results coming from the line pairs and switch in a spare line pair to replace a faulty line pair. The Testpoint Logic for a bus having three line pairs with one spare line pair is shown in figure 3.

There are four major sections to the Testpoint Logic. The Test Chain Latch on each line pair (including the spare) implements the logic for the Test and Program Scan Chain. It can serve simply as a shift register, or can apply data to the lines and collect and compare the results. Depending on which inverters are enabled, the test data can be applied to the line pair in either direction. The Status Latches can also be programmed with the data in the Test Chain Latches. The Iso on each line allows the interconnect to be isolated from the periphery cell during testing, thereby assuring that the test is independent of signal values from the periphery cell. The Switch units on each line allow a defective line pair to be replaced by the spare line pair. The Switches also serve as bus drivers.

On each PE site is placed the Endpoint Logic (EL). This block contains the logic necessary to switch in a spare line pair to replace a faulty line pair, to loop back the test bit coming in on one line of the line pair back on the second line and to selectively connect the Status Latch into the test path. The Endpoint Logic for a bus having three line pairs with one spare line pair is shown in figure 4.

There are four major sections in the Endpoint Logic. The Loop switch groups two adjacent lines into one line pair. It inverts the test data coming in on one line and sends it back down the other line in the pair. The Loop can be set in either direction so as to allow the Switch units to be tested in either direction. The Iso on each line is used to isolate the processing element from the interconnect during testing. The Switches and Status Latches serve the same purpose as in the Testpoint Logic and their settings and programmed values mirror those of the Testpoint Logic. The Status Latches in the Endpoint Logic are programmed from the Testpoint Logic.

At each site is placed the Site Logic (SL). This block contains the shift register latches used in the Serial Control Chain. The control information for the Endpoint Logic or the Testpoint Logic at the site is shifted serially into the Site Logic, where it is decoded to produce all the necessary control signals. The Switch setting for line pair i is determined by the Site Logic using the function

$$Setting_i = Status_{(i+1)\ mod\ j} \cdot \overline{Status_{(i+2)\ mod\ j}},$$

where j is the number of line pairs in the group (not including the spare) and the Status signals are from the Status Latches at the site. This function assures that any single line pair and its Status Latches can fail and the Switch setting will be correct. More than one failure cannot be handled in a single group in any case.

The application of the approach could be automated in a schematic capture CAD tool, provided the designer follows the basic rule of assuring that the interconnect architecture can be formed into valid line pairs. Once the interconnect and its interface to the PEs and PC are captured, one could invoke a tool which replaces the receiver/driver blocks in the original interconnect interface with the Testpoint Logic or the Endpoint Logic block and the Site Logic control block. In addition, the Test and Program Scan Chain and the Serial Control Chain must be globally routed.

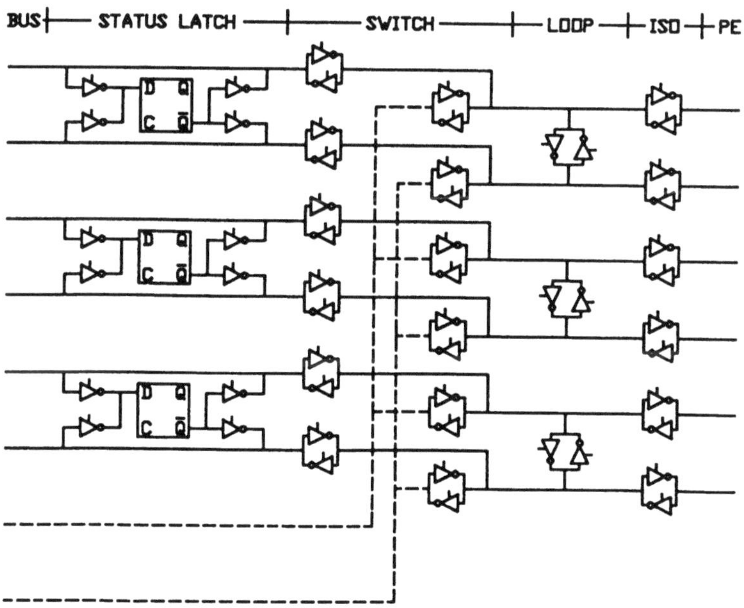

Figure 4 Endpoint Logic for 6 lines with 2 spares

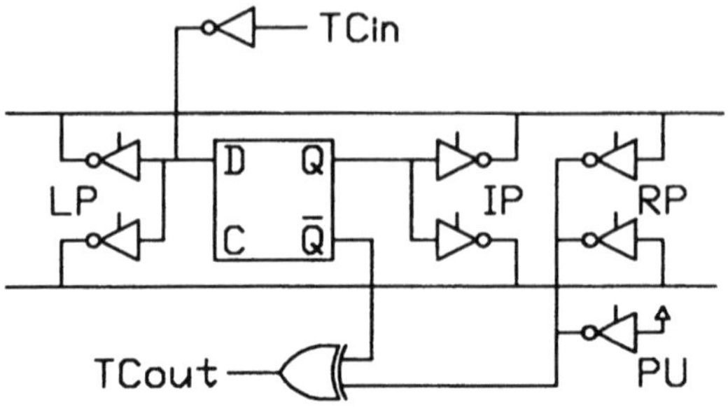

Figure 5 Details of the Test Chain Latch

TESTING, DIAGNOSIS AND RECONFIGURATION

Testing is performed in a systematic fashion, with each step designed to check for a particular class of fault. This technique facilitates diagnosis. Once a diagnosis has been made, the circuit is either reconfigured, if possible, or declared unusable.

The basic idea of the testing scheme is quite simple: A signal is sent along one path to a given site, then returned along another path to its starting point, where it is compared with its original value. This scheme has been expanded to permit testing of all the line pairs as well as the Switches, Status Latches and Loop circuitry. The values placed on the various line pairs allow the detection of stuck-at, bridging and stuck-open faults on the interconnect paths.

Testing proceeds as follows: First the interconnect and its associated test circuitry are isolated from the PEs and PC (using the Iso). Next, the Serial Control Chain is tested in order to determine that control signals may be shifted around the chip. The Site Logic is only tested functionally, not structurally, during the course of the test procedure. Next, the Test and Program Scan Chain is tested with TCin as input and TCout as output. The Test and Program Scan Chain is created using the pull-up inverter (PU) in the Test Chain Latch circuitry (see figure 5) which creates a shift register out of the latches. If these tests fail, the circuit must be discarded. In the next step, the Status Latches of the Testpoint Logic are checked. A modified walking 1s test[13] is used for this purpose. The test consists of vectors with a single 1 which migrates from one position to another. In order to force both a 0-1 and 1-0 transition on each bit, the all 0s pattern is both prepended and appended to the test set. An example test set is shown in the first column of table 1.

Any stuck-at fault in a bit will clearly be detected by such a test set. Columns 2 and 3 of Table 1 show bit 2 stuck-at 0 and 1. Detection and diagnosis of bridging faults is also simple, regardless of whether the bridging is of the AND or OR variety. An AND bridge between bits 2 and 4 will change the initial test set into that shown in column 4, while an OR bridge in the same location will result in column 5 being

Table 1 Walking 1s Test Vectors with a Variety of Errors

Test Set	s-a-0	s-a-1	AND Br.	OR Br.	stuck-open
0000	0000	0100	0000	0000	0X00
1000	1000	1100	1000	1000	1X00
0100	0000	0100	0000	0101	0X00
0010	0010	0110	0010	0010	0X10
0001	0001	0101	0000	0101	0X01
0000	0000	0100	0000	0000	0X00

received. The fact that both transitions occur on each bit also permits detection of a stuck-open line, which will manifest itself as a stuck-at-X fault, as shown in column 6.

Testing of the Status Latches in the Testpoint Logic proceeds as follows: Each pattern in the test set is clocked into both the Test Chain Latch and (using the upper LP switch in figure 5) its corresponding Status Latch. The signals then travel (using the lower RP switch) into the XOR gate, where the two values are compared. The result of this comparison is clocked into the subsequent Test Chain Latch, and then scanned out. The test controller circuitry analyzes these results, diagnoses any faults, and then sends the next pattern. Because of the bidirectional nature of the design, both inputs to the Status Latch must be checked independently (using the opposite switches to above). A single Status Latch failure here requires the replacement of its line pair by the spare line pair. Because the faulty Status Latch is not used in setting its own Switches, this replacement is possible.

It is at this point that the interconnect itself is first tested (although only partially), as testing moves from the Testpoint Logic to the Endpoint Logic. The Status Latches in the Endpoint Logic are tested first. Testing of the interconnect must be performed for each site individually, hence only those Status Latches at the same site are tested together. Testing proceeds as described for the Testpoint Logic. As with the Testpoint Logic, a faulty Status Latch in the Endpoint Logic is treated as a faulty line pair.

Once the Status Latches have been checked, testing of the interconnect is completed by testing the line pairs themselves. This requires the use of the IP switches in the Test Chain Latch and the Loop at the site. Again, the modified walking 1s test is suitable, particularly if the test, diagnosis and control unit is built-in. If external testing is used, more pattern-efficient test sets, such as the well-known $O(log\, n)$ set[14], may be used. Both methods will detect all single and multiple stuck-at, stuck-open and bridging faults among the line pairs. Bridging faults between lines in a line pair are not covered by either method. However, the fact that the Loop inverts the signal on the line pair permits the detection of this type of bridging.

Of course, there may be some faults which are not detected by these means, such as multiple faults involving several components: latches, line pairs, control logic. Further research is necessary both to determine the existence and to enable the detection of these faults.

Once faults have been identified, reconfiguration is carried out if possible. The Switches permit a single line pair to be replaced by a spare line pair. The control logic requires the agreement of two Status Latches before a Switch can be set to use the spare line pair. Thus, a single faulty Status Latch cannot activate a spare incorrectly, or, since no Status Latch controls its own Switches, prevent itself from being bypassed. Once replacement has been performed, additional testing is required to determine that the modified network is still operating. This is simply a re-test of the line pairs to verify switch operation and the functionality of the spare pair. Once the spare of a group spare has been activated, additional errors result in an unusable circuit.

YIELD AND REDUNDANCY EVALUATION

Area Overheads

The extra hardware used to implement the proposed approach is, for the most part, a requisite for any interconnect test and reconfiguration scheme. Each line pair (including the spare line pairs) has had its pair of drivers augmented by the additional logic found in the Endpoint Logic or Testpoint Logic. Each site has had its hardware increased by a Site Logic block. Globally the entire interconnect has been increased by the Serial Control Chain and Test and Program Scan Chain, as well as the spare line pairs.

For any soft reconfiguration scheme for interconnects, switches and latches are compulsory. For k switches at least $log_2 k$ latches are required to store the setting, and if $log_2 k$ latches are used then a $log_2 k$ to k decoder is required. In the proposed scheme there are k latches for k pairs of switches. There is little difference in the cost of the latches and the decoder. Some area could have been saved by designing a hard reconfigurable switch, such as a electrical or laser fuse, but this would have resulted in a

Table 2 Detailed Hardware Overheads

Element	Size	Reconfigurable
line pair	2 I	yes
spare line pair	2 I	yes
Status Latch	40 T per line	yes
Switch, Loop, ISO	60 T per line	no
Switch, Test Chain Latch, ISO	87 T per line	no
SL	400 T	no
SC	3 I	no
TC	0 I	no

less flexible approach. The number of switches has been minimized by having only one spare line pair per group of line pairs. Had more spares been permitted per group then the switch cost would have increased. Obviously the spares themselves are required for any reconfiguration scheme. In the proposed scheme, the number of spares used may be optimized by performing a yield/area analysis to determine optimal group size.

For any testability scheme, additional control signals and control logic are needed to select the various test modes. In addition, extra test logic is needed to ensure that the circuit being tested is controllable and observable. In the proposed scheme, the control signals have been implemented using the Serial Control Chain and the Site Logic at each site. The Serial Control Chain is the minimum size possible, comprising only 3 lines: clock, data and enable. It was verified by a yield/area analysis that the serial approach was generally more economical than running a group of global control signals to all sites.

Table 2 outlines the break down of the hardware cost in terms of transistors or interconnects. It also indicates whether or not each type of redundant hardware is reconfigurable. The transistor counts are per site (those marked "per line" are therefore per line per site).

One interconnect unit I is equivalent to the area of one line. The vast majority of the interconnect area is reconfigurable with only the Serial Control Chain being kill area. The Test and Program Scan Chain is assumed to be very short, running only between the pins and the Testpoint Logic. One transistor unit T is equivalent to the area that one transistor typically occupies in a logic block, and will always be much less than one I. For small bus bandwidths the Site Logic will account for the majority of the transistor kill area.

Yield Analysis

To evaluate the yield of the scheme two different yield models were used; the first of these being based on the simple Poisson model, and the second being based on a negative binomial model[15].

In the case of an interconnect architecture in the form of a global bus, the Poisson model would seem to be quite good. There is little doubt that clustering of defects into areas of the chip not containing interconnect will lead to increased yield of the interconnect, but the fact that the interconnect will traverse many different regions of the chip, make it more likely that the interconnect will encounter a region of high defect density, thereby decreasing the interconnect yield. In different regions of the chip, each line has an equal probability of being defective, independently of whether or not any other lines are defective in other regions of the chip. Thus the defect distribution, for lines running all over the chip, would appear uniform, as the locations of defects would be far enough apart to be independent of clustering. It is still possible for a cluster of adjacent lines in the bus to be defective in one location. Thus the Poisson model is not entirely accurate and a negative binomial model may be appropriate. In what follows both models will be used.

Huang and Daughton[16] develop a yield equation based on the Poisson model for modules containing N elements of which M good elements are required. Each element takes area A and the defect density is D. Their equation for the yield of such a module can be expressed as

$$Y_{pm} = \sum_{j=0}^{N-M} \binom{N}{j} exp^{-AD(N-j)}(1 - exp^{-AD})^j$$

Harden and Strader[17] develop a yield equation based on the negative binomial model for modules containing N elements of which M good elements are required. Each element takes area A and the average defect density of the module is given by D. The clustering parameter α is also included. If the assumption is made that each element is identical and closely spaced, allowing no spatial fluctuation of the number of defects in the module, then their equation for the yield of such a module can be expressed as

$$Y_{cm} = \sum_{j=0}^{N-M} (-1)^{N-M-j} \binom{N}{N-j} \binom{N-1-j}{M-1} (1 + ((N-j)AD/N\alpha))^{-\alpha}$$

These two equations were used as the basis of a computer program to evaluate the yield of the proposed scheme for globally bussed interconnect architectures having various parameters. The parameters varied included the number of line pairs in the interconnect, the number of groups (each with one spare line pair) into which the interconnect was partitioned, the number of sites on the interconnect, the area of a single equivalent line I, the area of a single equivalent transistor T, the defect density and the clustering parameter.

Each group having one spare was considered as one module. The area E of the redundant logic or lines which were not reconfigurable was computed. The yield of this area was computed using either the basic Poisson yield equation or the basic negative binomial yield equation, depending on the model used to compute the module yield. This yield was simply multiplied by the module yield to give the final group yield (assuming the yield of the unreconfigurable area is independent of the module yield). The Poisson and negative binomial group yield equations are therefore given by

$$Y_{pg} = Y_{pm} \cdot exp^{-ED} \qquad and \qquad Y_{cg} = Y_{cm} \cdot (1 + (ED/\alpha))^{-\alpha} \ .$$

The final yield for the entire interconnect was taken to be the product of all g group yields involved. This assumes that the yields of the groups are independent of each other, which may not be entirely correct. The Poisson and negative binomial interconnect yield equations are therefore given by

$$Y_p = (Y_{pg})^g \qquad and \qquad Y_c = (Y_{cg})^g \ .$$

The area of one equivalent transistor T in the logic was taken to be $150\mu^2$ for a 1μ process technology. The size of the each cell in the National Security Agency CMOS3 cell library [18] was divided by the number of transistors in the corresponding cell. This figure was then averaged and scaled to 1μ, and was found to be approximately $150\mu^2$.

The area of one line I was taken as the area of both a wire and its required separation. Assuming a 1μ line width on a chip of $1cm$ on a side, then the minimum area of one line in a bus running the width of the chip is $2 \cdot 10^4\ \mu^2$. For complex arrays, the bus may have to run multiple widths of the chip. Alternately the designer may want to widen the line. Thus larger line areas were also be considered in the analysis.

Two graphs are shown below, figure 6 and figure 7, which give the yield and area of the total interconnect versus the area of a line. The graphs cover two cases: the first when there are only 5 sites and a bus bandwidth of 256, the second when there are 17 sites and the bus bandwidth is 64. In the second graph, the area of a line has been doubled to reflect the fact that the bus lines for the larger array size will probably need to be longer. Note that, for these graphs, the Poisson and negative binomial model give similar yield values. There were some differences in the yields, for the interconnect with spares, at lower alpha values and higher defect density values.

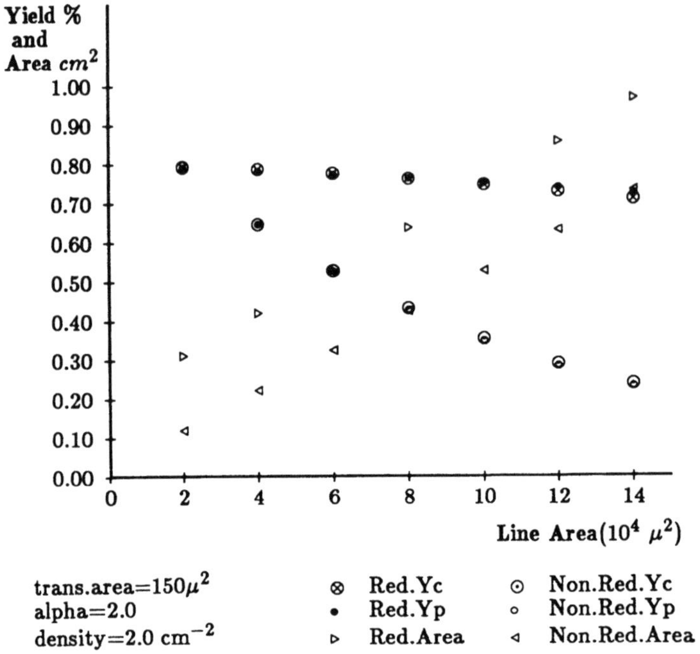

Figure 6 Yield and Area Graph for m=256 and s=5

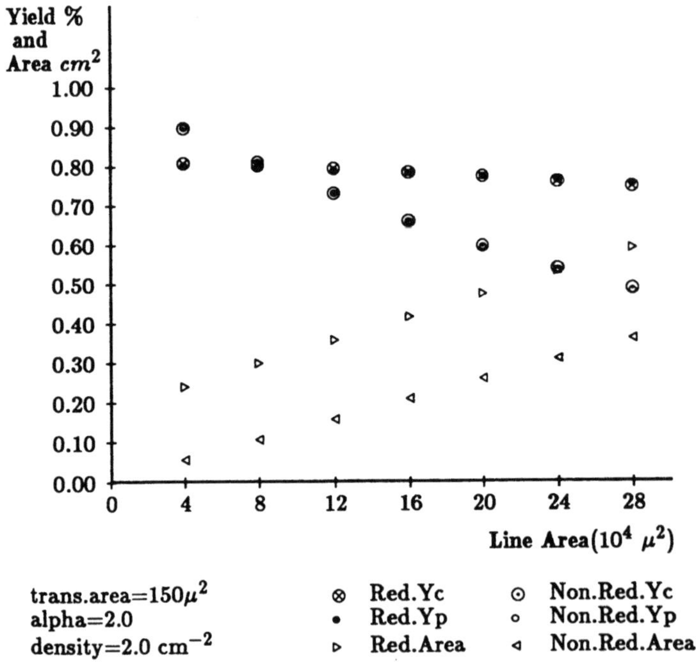

Figure 7 Yield and Area Graph for m=64 and s=17

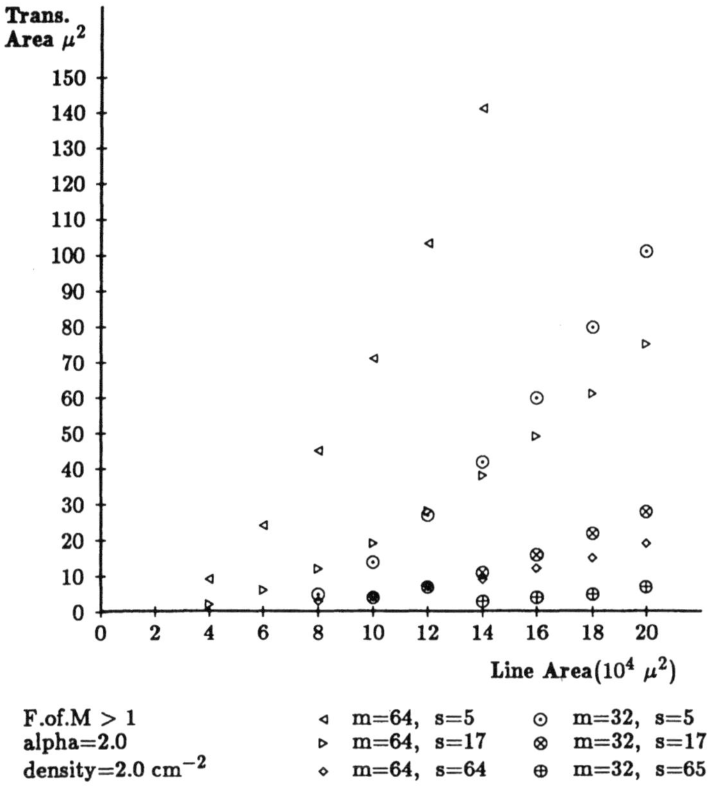

Figure 8 Transistor Area versus Line Area

As can be seen from the graphs, the proposed approach gives favorable yield enhancement at low area cost for low site counts and high bus bandwidths, even when the line area is small. For higher site counts and lower bus bandwidths the line area needs to be larger, for T fixed at $150\mu^2$, before a reasonable yield enhancement at a reasonable area cost is achieved. The approach does better for larger bus bandwidths because the area of the reconfigurable line pairs far outweighs the area of the unreconfigurable logic and unreconfigurable control chain. Similarly, extra sites rapidly increase the extra kill area.

A computer program was written to study how the equivalent transistor area T relates to the line area required to achieve a reasonable yield enhancement at a reasonable area cost. The measure of success was based on the crude figure of merit equation

$$F\ of\ M = (Y_{red} * A_{non-red})/(Y_{non-red} * A_{red}) \ .$$

The program searched for the maximum area of an equivalent transistor for which the $F\ of\ M$ was greater than 1 given a specific line area, bus bandwidth and number of sites.

The graph given in figure 8, below, shows the results of this analysis for various busses of different size connecting various numbers of PEs. As can be seen from the graph, the transistor size needs to be significantly smaller than $150\mu^2$ for busses with low bandwidth and high site counts. It is nonetheless conceivable, that as process technology continues to shrink below 1 μ, the size of the transistor area will decrease with respect to the size of the line area. There may also be some implementations of multiprocessor architectures which have larger line area. The case-study chip previously mentioned was found to have a line area of $100 \cdot 10^4\ \mu^2$ as it had a very complex bus routing.

CONCLUSIONS

This paper presents an approach for increasing the testability and yield of structured interconnects. An implementation of the approach for globally shared busses is given. Using the approach, all stuck-at, AND/OR bridges and stuck-opens in the interconnect and many faults in the related circuitry can be detected and diagnosed. Up to one faulty line pair per group can be reconfigured, and the interconnect can be configured into many groups. The yield and redundancy analysis shows that the approach works well for low site counts and high bus bandwidth, and, depending on the size of an equivalent transistor T versus the line area I, should work well for larger site counts and lower bus bandwidths.

REFERENCES

1. E. B. Eichelberger and T. W. Williams, A Logic Design Structure for LSI Testability, Proc. IEEE Design Automation Conf., 462:468 (1977).
2. R. David, Feedback Shift Register Testing, Proc. Fault-Tolerant Computing Symp. 103:107 (1978).
3. C. Maunder and F. Beenker, Boundary-Scan: A Framework for Structured Design-for-Test, Proc. Intl. Test Conf., 714:723 (1987).
4. W. R. Moore, A Review of Fault-Tolerant Techniques for the Enhancement of Integrated Circuit Yield, Proc. of the IEEE, 684:698 (May 1986).
5. F. B. Manning, An Approach to Highly Integrated, Computer-Maintained Cellular Arrays, IEEE Trans. on Comp. Vol. 26, 536:552 (Jun. 1977).
6. A. Rosenberg, The Diogenes Approach to Testable Fault-Tolerant Arrays of Processors, IEEE Trans. on Comp., 902:910 (Oct. 1983).
7. M. Sami and R. Stefanelli, Reconfigurable Architectures for VLSI Processing Arrays Proc. of the IEEE, 712:722 (May 1986).
8. L. Snyder, Introduction to the Configurable Highly Parallel Computer, COMPUTER 47:56 (Jan. 1982).
9. W. W. Peterson and E. J. Weldon, "Error Correcting Codes," 2nd ed., The MIT Press, Cambridge, Mass. (1972).
10. D. P. Siewiorek and R. S. Swarz, "The Theory and Practice of Reliable System Design," Digital Press, Bedford, Mass. (1982).
11. D. P. Agrawal and V. K. Agarwal, On-Line Bus Fault Diagnosis in Microprocessor Systems, J. Digital Sys., 337:391 (1980).
12. L. H. Pollard and J. H. Patel, Correction of Errors in Data Transmission using Time Redundancy, Proc. Fault-Tolerant Computing Symp., 314:316 (1983).
13. A. Hassan, J. Rajski and V. K. Agarwal, Testing and Diagnosis of Interconnect using Boundary Scan Architecture, Proc. Intl. Test Conf., to be published (1988).
14. W. H. Kautz, Testing for Faults in Wiring Networks, IEEE Trans. on Comp., 358:3 (Apr. 1974).
15. C. H. Stapper, On Yield, Fault Distributions, and Clustering of Particles, IBM Journal Research and Development, 326:338 (May 1986).
16. J. S. T. Huang and J. M. Daughton, Yield optimization in wafer scale circuits with hierarchical redundancies, INTEGRATION, the VLSI journal, 43:51 (Mar. 1986).
17. J. C. Harden and N. R. Strader II., Architectural Yield Optimization for VLSI IEEE Trans. on Comp., 88:110 (Jan. 1988).
18. D. V. Heinbuch, "CMOS3 Cell Library", Addison-Wesley, Reading, Mass. (1988)

COMBINING ARCHITECTURE AND ALGORITHM FOR YIELD ENHANCEMENT AND FAULT TOLERANCE

R.J. Cosentino, B.L. Johnson, and **J.J. Vaccaro**

The MITRE Corporation
Bedford, MA 01730

INTRODUCTION

IC manufacturers have traditionally met customers' demands for ever increasing chip processing capability by decreasing feature size. The alternative -- significantly increasing the chip area -- has thus far not been economically competitive, owing primarily to the low yield of larger circuit areas [1].

Techniques for raising the yield of large-area circuits can be categorized as either reconfiguration or as defect-masking. In this paper, "defect masking" means the isolation of errors arising from manufacturing defects in a circuit. "Fault masking" means the isolation of errors arising from circuits that become faulty during operation, following production testing. By far the greatest effort to date has been placed on techniques for reconfiguring circuits by physically or electronically rerouting around faulty modules. Drawbacks of reconfiguration are the extensive localized testing and the cost of the additional processing steps. Defect masking is most commonly associated with triple modular redundancy, which requires too much redundancy to compete successfully with conventional VLSI [1]. However, defect masking avoids many of the problems and costs associated with reconfiguration.

This paper reports on a novel defect-masking technique that provides more defect tolerance than triple modular redundancy, at a small fraction of the cost in redundancy. We have combined the area-efficient redundancy of residue number systems (RNS) with the multi-defect pattern coverage of a class of parallel-processing systolic architectures to produce a more cost-effective defect-masking approach to the yield enhancement of large-area signal processing circuits. We have extended this technique to provide efficient, effective fault tolerance for the correction of hard, soft, and intermittent errors in operation.

We first discuss a class of systolic architectures that calculate their output terms in parallel. This well known architectural technique offers significant VLSI design and testability advantages and provides fault correction capability for an extended number of defect patterns. Next we describe the properties of RNSs and redundant residue number systems (RRNS) that provide defect-masking and fault-masking with minimum redundancy. Then we show that the natural combination of RRNSs with a systolic architecture, which calculates its outputs in parallel, forms efficient and effective defect-tolerant and fault-tolerant circuits.

SYSTOLIC ARCHITECTURES

Systolic-architecture design techniques provide high-performance, special-purpose processors for compute-bound problems [2]. These techniques decompose an architecture into a few simple processing elements (cells) that are interconnected in a simple, repetitive manner (an array) to provide high degrees of pipelining and multiprocessing. This regularity allows an entire VLSI architecture, containing hundreds of thousands of transistors, to be implemented by carefully designing a few, low-complexity building blocks [3,4]. Systolic architectures are not universally applicable; however, they are ideally suited for numerous digital signal processing (DSP) algorithms.

In systolic architectures, data is passed from cell to neighboring cell. On each clock cycle, the data within a cell is used by that cell's processing element (PE) to calculate a portion of a particular output term. In some architectures, each output term is constructed from partial results as it passes from cell to cell. In this case, the calculations are pipelined: each PE adds its contribution and passes on the accumulated partial result to the next cell for further processing. In other architectures, only the input data is passed between cells, and each output term is calculated in a separate systolic cell (in-place calculation). In this case, the calculations are in parallel: the partial results are retained within each PE and used to calculate a single cell's output term. In general, both architectures can implement the same function and the internal architectural subtleties are virtually transparent to users.

However, parallel-processing systolic architectures facilitate the identification of defective cells in production testing. The presence of an error at a given output port during a given clock period identifies the PE that produced it. The structure of the input and output data paths resembles a scan design: the PE inputs are controllable and the PE outputs are observable through the cells' data registers [5]. Additionally, for regular arrays with identical PEs, an exhaustive test may be performed in little more than the time required to test a single PE.

Since the outputs from PEs with in-place calculations are passed on to the output ports unmodified, a single error-correction circuit at the output port is required. This allows redundancy on the level of the PEs without a multiplicity of error-correction circuits (voters for triple modular redundancy).

RRNS PROCESSING

RNS processing is a design approach that possesses inherent parallelism [6]. An integer processor with a dynamic range M is realized with N independent processors, each using a different modulus. In addition to parallel processors, a typical system requires input and output converters. The advantage of residue arithmetic is that the original problem is partitioned into a number of smaller, faster, independent, parallel calculations [7]. Additionally, the partitioning provides VLSI system implementation advantages and facilitates system testing. Of course, RNS is not applicable to general purpose computations; but the strengths of RNS --multiplication and addition -- are the most prevalent DSP operations.

The design effort can be eased considerably by combining a systolic architecture with RNS processing. As an illustration, we reduced the design of a one-half million transistor programmable FIR filter to the readily manageable design of a 700-transistor cell. This 128-stage filter, having five 5-bit moduli, was implemented in 1984 with 4-μm nMOS. The 128 x 5 = 640 PEs were fabricated on 20 chips, and each converter was fabricated on a separate chip, so that the system required 23 VLSI chips in all.

RNS also has some unique properties that allow the incorporation of defect tolerance comparable to triple modular redundancy, at a small fraction of the cost in additional chip area. At least three approaches to applying the properties of redundant RNS for fault tolerance have been proposed [8,9,10]. We have found that the approach proposed in [10] provides the most area-efficient correction circuits for use in a fault-tolerant version of the FIR filter in which the clock rate approaches the speed limit determined by the technology. To establish a base for comparing redundancy requirements, we assume that the sensitive chip area required by each of two systems, one employing conventional arithmetic and the other employing residue arithmetic, is the same. Then, if the calculations in a PE employing conventional arithmetic are performed by an N-residue RNS, each of the N-residue PEs requires 1/N the area of the conventional PE. The properties of RNS allow any of the N residues to be, in effect, replaced by a redundant residue with a modulus larger than and relatively prime to the N nonredundant moduli, which are themselves pairwise relatively prime. Thus, we need add only one redundant PE. The area increase due to the redundant PE is 1/2N the area increase required by triple modular redundancy. For the FIR filter described above, which has five channels, the area requirement is one tenth that needed for triple modular redundancy.

The need for adding only one redundant residue PE assumes that the system "knows" which PEs are defective. Those residues are then erasures (errors that are known as such). The identification of the defective PEs in production testing is facilitated by the in-place systolic architecture.

The error-correction circuits for a one-erasure correction system (i.e., a system that tolerates at most one erroneous residue in any (N+1)-tuple of residues appearing simultaneously at the output converter) requires two inputs: one input identifies the erasures; the other is the difference between two independent calculations of the residue with respect to the redundant modulus -- one from the N nonredundant residues, the other from the redundant channel (Figure 1).

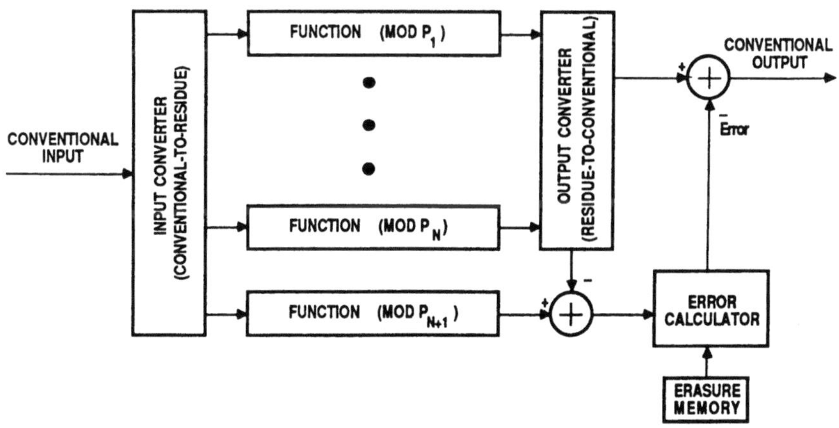

Fig. 1. Redundant Residue Number System Processing.

YIELD ESTIMATES

The combination of RRNS processing and systolic architecture provides multi-defect pattern coverage efficiently. Multi-pattern robustness arises from the localized computations provided by the parallel-processing systolic array. Effectiveness results from the low percentage of redundancy required with RRNS, combined with a single output resolver that is a by-product of the parallel-processing systolic architecture.

The capability of RRNS processing combined with systolic architecture can be illustrated by a monolithic version of the programmable 128-stage FIR filter projected to today's 1.5-μm CMOS technology. For analysis purposes the system architecture is decomposed into four distinct parts that receive separate treatment in providing redundancy and calculating yield: the PEs, the data paths, the input converter, and the output converter. Without redundancy the RNS systolic processing elements account for 80% of the silicon area and represent the dominant yield limiter.

Yield Estimates for a Nonredundant FIR Filter

Reasonable yield estimates may be made by considering the range of experiences of IC manufacturers [11]. They are typically within the bounds of the two curves described by the formula,

$$Y = (1 + N_0/\alpha)^{-\alpha} , \qquad (4.1)$$

where N_0 is the average number of fault-causing defects in the circuit, and the parameter α ranges between 0.5 and 4 (Figure 2). N_0 is the product of the defect density, D_0, of the defects on the wafer and the active area, A, of the circuits. The bottom curve for $\alpha = 4$ provides conservative yield estimates, and will be used in this section. Wafer defect densities are assumed to fall between one and two defects per square centimeter [12].

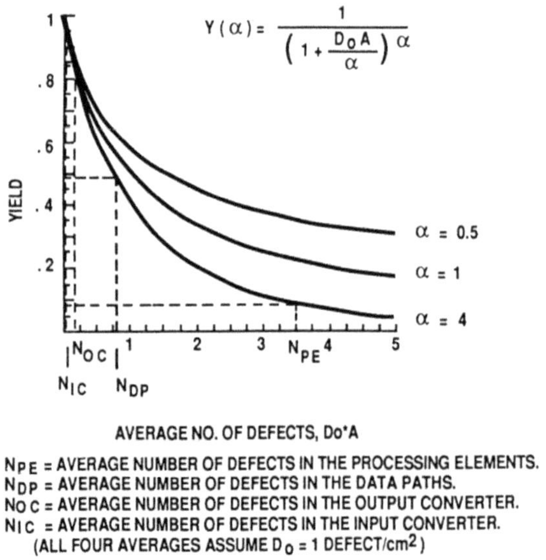

Fig. 2. Limits of Typical Yield Experience of IC Manufacturers ($\alpha = 0.5$, 4). The Middle Curve ($\alpha = 1$) is the Yield for a Bose-Einstein Defect Distribution.

Table 1. Estimated Sensitive Areas of a 1.5-μm CMOS Filter

Part	(a) Nonredundant Filter Area cm^2	%	(b) Filter with One Redundant Channel and Redundant Data Paths Area cm^2	%
Processing Elements	3.52	78	4.22	51
Data Paths	0.79	17	3.8	45
Output Converter	0.18	4	0.26	3
Input Converter	0.05	1	0.06	1
Total	4.54	100	8.34	100

We calculated the sensitive area of each of the four parts based upon a multiple-chip implementation of a 1.5-μm CMOS filter. For a density of one defect/cm^2, the curve gives an impractically low yield of 8% for the PEs alone. The yields of the data paths and of the input and output converters are 48%, 84%, and 95%, respectively. The 3.4% yield of the total circuit is limited primarily by the PEs; therefore, we first consider adding defect tolerance to the PEs.

A Defect-Tolerant FIR Filter

In an RRNS implementation, one redundant channel, together with a single output resolver circuit and a small memory that stores the location of defective PEs, masks errors produced by defective PEs. With a single redundant channel, each localized parallel systolic computation can survive a single error in any residue channel. With this approach an L-stage filter can be completely functional in the face of up to L defective PEs -- one in each stage. Let P_{PE} denote the probability that an individual PE is defect-free. This probability is determined from (4.1), where $\alpha = 4$ and the area of a single PE, A_{PE}, = 0.0055 cm^2. The yield of a redundant set of PEs, Y'_{PE}, may be approximated by[1]

$$Y'_{PE} = [P_{PE}^6 + 6 \cdot P_{PE}^5 (1-P_{PE})]^{128} \qquad (4.2)$$

for a 128-stage filter with six moduli. The first term in (4.2) is the probabilty that all six PEs in any column are defect-free; the second term is the probability that exactly one is defective. For a defect density of one defect/cm^2, Y'_{PE} = 95% -- a yield equal to that of the input converter (Figure 3), which has less than 2% of the area of the PEs (Table 1(a)).

The redundancy in the PEs raises the total yield significantly -- to 30% for one defect per square centimeter (Figure 3). The cost in redundancy is a 20% increase in the data paths and input converter, needed to provide the input residues for the redundant channel, and a 44% increase in the area of

1. Equation (4.2) is an approximation if defect clusters cover two or more PEs. RRNS PEs are disconnected and may be physically separated; the equations hold for separations that are larger than the cluster extents.

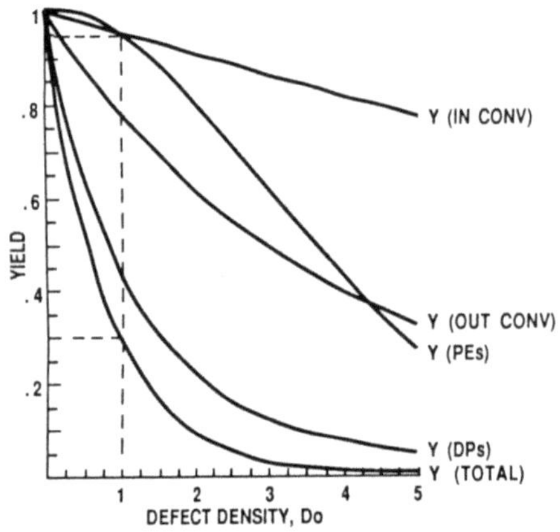

Fig. 3. Yields With One Redundant Residue.

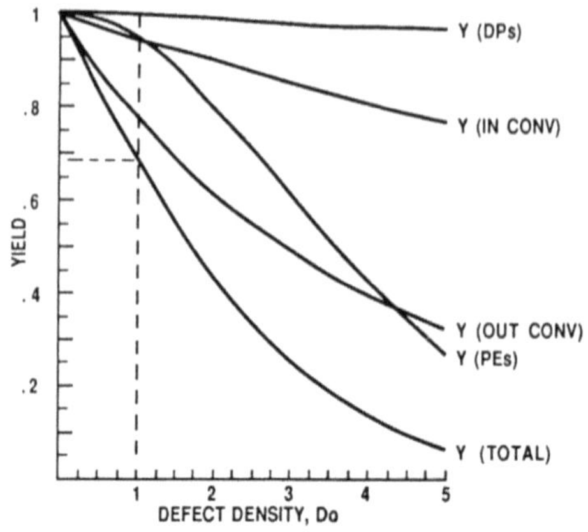

Fig. 4. Yields With One Redundant Residue and Quadruple Redundant Data Paths.

the output converter,[2] for a total increase of 21% over that of the nonredundant filter. Thus, a modest 21% increase for redundancy provides us with a much higher yield, limited primarily by the data paths.

Redundancy added to the data paths in the form of interwoven (quad) logic raises the yield still further[3], so that for a density of one defect/cm^2 the calculated total yield is over 60% (Figure 4) -- now limited by the output converter. The cost of this additional redundancy is a quadrupling of the data path area (Table 1(b)). The increase in area over the nonredundant filter is 86% -- a little less than a doubling of the area. The very high yield of the data paths suggest that data path yield could be traded off against redundant area requirements for the data paths without affecting the total yield significantly.

The RRNS technique provides a large amount of defect tolerance for the greater part of the circuits at low cost. More conventional, relatively higher cost techniques need to be applied only to the remainder of the circuits, resulting in less impact on the overall cost.

The technique is easily extended to more redundant channels. A second redundant channel increases the PE yield of the 128-tap filter to 99.99% for a defect density of 1 defect/cm^2.

A Fault-Tolerant FIR Filter

Two redundant channels provide tolerance to a set of defect patterns similar to that for one redundant channel plus memory -- those patterns having at most one defective PE in a column. However, (4.2) now becomes

$$Y''_{PE} = [P^7_{PE} + 7 \cdot P^6_{PE}(1-P_{PE})]^{128} \qquad (4.3)$$

---

2. The area of the output converter is roughly proportional to the square of the number of channels.

3. To deal with the intricate defect patterns of the data paths, the Bose-Einstein statistics were assumed, which gives the middle yield equation (α=1) in Figure 2.

owing to the presence of an additional channel. The redundancy is double (neglecting the memory) and the yield is down slightly due to the additional area; however, this two redundant channel error-correction circuit has a couple of advantages over the one redundant channel erasure-correction circuit. First, faulty PEs need not be located and identified in memory; a simpler GO/NO-GO production test can be used instead. Secondly, the system can provide error correction for residues from each set of defect-free PEs, providing correction against soft errors, intermittent errors, and hard errors occurring in operation [13].

CONCLUSIONS

We have proposed an efficient means of improving the yield of large digital signal processing functions. The natural combination of redundant residue number systems with an in-place systolic architecture forms an efficient and effective defect-tolerant or fault-tolerant circuit. The in-place systolic array allows fault isolation and requires a single output error-correction circuit. The RRNS technique provides PE defect-masking and fault-masking with a minimum of redundancy.

We compared the defect-masking redundancy needed and the coverage obtained from a FIR filter employing residue arithmetic and a parallel-processing systolic architecture with one using conventional arithmetic and a pipelined architecture. A novel defect-tolerance technique, applicable to the former, required an order-of-magnitude less redundancy than conventional triple modular redundancy applied to the latter.

Because the PEs typically use the largest sensitive chip area by far, conventional defect-coverage techniques applied to the remaining parts of the system have less impact on the percentage increase in area. An area for further investigation is the development of defect-tolerant data paths at less expense than quadruple redundancy. A modest aim would be to raise the yield of the data paths above that of the output converter while keeping the total redundancy below 50%. Then the yield of a very-large-area system (an order-of-magnitude larger than conventional areas) will be essentially that of a VLSI circuit of conventional area, without the need for an excessive amount of redundancy or for extra or costly manufacturing steps.

With the advent of VLSI, which allows entire systems to be fabricated on a chip, designers now have the opportunity to increase the chip performance by improving the system design. The synergistic combination of the systolic architecture and the redundant residue number system described in this paper illustrates how an intelligent choice of architectures and algorithms can be used to more fully exploit the potential of VLSI technology.

ACKNOWLEDGMENT

This work was supported both by the USAF, Electronic Systems Division, AFSC, under contracts F19628-84-C-0001 and F19628-86-C-0001, and by the MITRE-Sponsored Research Program.

REFERENCES

1. R.O. Carlson and C.A. Neugebauer, Future Trends in Wafer Scale Integration, Proc. IEEE, Vol. 74, No. 12 (1986).
2. H.T. Kung, Why Systolic Architectures?, Computer (IEEE) (January 1982).
3. J.J. Vaccaro, B.L. Johnson, and C.L. Nowacki, A Systolic Discrete Fourier Transform Using Residue Number Systems Over the Ring of Gaussian Integers, Proc. 1986 IEEE International Conference on Acoustics, Speech, and Signal Processing (ICASSP) (1986).

4. G.A. Doodlesack, M. Gray, B.L. Johnson, C.L. Nowacki, and J.J. Vaccaro, VLSI Chip Designs for QRNS-Based DFTs, Proc. 1987 International Symposium on Circuits and Systems (ISCS) (1987).
5. R.J. Cosentino, Concurrent Error Correction in Systolic Architectures, IEEE Trans. Computer-Aided Design, Vol. CAD-7, No. 1 (1988).
6. B.L. Johnson, E.A. Palo, R.J. Cosentino, and J.J. Vaccaro, The Residue Number System for VLSI Signal Processing, Proc. SPIE, Advanced Algorithms and Architectures for Signal Processing, Vol. 696 (1986).
7. H.L. Garner, The Residue Number System, IRE Trans. Elect. Comput., Vol. EC-8, No. 6 (1959).
8. F. Barsi and P. Maestrini, Error Correcting Properties of Redundant Residue Number Systems, IEEE Trans. Comput., Vol. C-22, No. 3 (1973).
9. D. Mandelbaum, Error Correction in Residue Arithmetic, IEEE Trans. Comput., Vol. C-21, No. 6 (1972).
10. R.W. Watson, "Error Detection and Correction and Other Residue-Interacting Operations in a Redundant Residue Number System," Ph.D. Dissertation, Dept. of Elec. Engrg., University of California, Berkeley (1965).
11. M.B. Ketchen, Point Defect Yield Model for Wafer Scale Integration, IEEE Circuits and Devices Magazine, Vol. 1, No. 4 (1985).
12. H.K. Dicken, Demystifying ASIC Costs, VLSI Systems Design (1988).
13. R.J. Cosentino, Fault Tolerance in a Systolic Residue Arithmetic Processor Array, IEEE Trans. Comput., Vol. C-37, No. 7 (1988).

DESIGN OF A FAULT-TOLERANT DRAM WITH NEW ON-CHIP ECC

P. Mazumder

University of Michigan
Department of Electrical Engineering and Computer Science
Ann Arbor, MI 48109

INTRODUCTION

Soft errors induced by incident alpha particles are one of the most serious problems at present and in future VLSI dynamic random-access memory (DRAM) development. Alpha particles emitted during radioactive decay of uranium and thorium, contained in minute proportions in packaging materials [1], have been shown to cause 98% failures that occur during normal operation of the DRAM chip [2,3]. On-line hard and medium errors are relatively very low in a well-designed chip. Hard errors are caused by permanent defects such as stuck-at faults, while the causes of medium errors are obscure, and frequently these errors result due to pattern-sensitive leakage currents. Such errors are virtually eliminated by testing the chips rigorously after manufacture. A small fraction of failures are due to transients like the voltage spikes and man-made statics. By introducing suitable filtering circuits these sources of transient errors can be sufficiently suppressed. But the alpha-particle-induced soft errors are becoming more and more critical as the the cell dimension is reducing.

The capacitor value and the cell topography play an important role in deciding how many cells can fail by the incidence of a single alpha particle. In planar implementation of the storage capacitor, the track of alpha particle usually restricts within the boundary of a single cell, and the resulting fault causes a single-event upset. The present three-dimensional DRAM employs a deep trench capacitor that extends from the planarized surface through the n-well into the p^+ substrate. The capacitance is typically 30fF, and the capacitor is highly susceptible to be discharged by noise electrons. The memory-cell soft errors can be principally classified into two types: i) single-cell upset and ii) double-cell upset. In a single-cell upset, an alpha particle strikes on a single capacitor discharging it alone. On the contrary, if the alpha particle strikes on the intervening space between two adjoining capacitors, plasma discharge may occur between both capacitors resulting in charge loss in both. Such soft errors are known as double-cell upsets, and are very common in three-dimensional DRAM. If σ^2 is the effective area of a cell and δ is the interceller distance, then the probability of an alpha particle striking on the intervening space is proportional to δ/σ. Thus a large number of alpha particles may cause double-bit errors if they have sufficient kinetic energy to discharge both the capacitors. The Monte Carlo simulation done by Sai-Halasz *et al.* *[3]* has revealed that as the feature width and the critical charge in storage capacitor decrease, the double-bit soft errors dominate over the single-bit soft errors.

The conventional system-level error-correcting circuits use SEC/DED type Hamming code, and, therefore, they fail to correct the double-cell upsets in three-dimensional DRAM's. Another potential limitation of system-level coding is that in order to retrieve w-bit word from the

This research was partially supported by the following grants: URI-Army's DAAL 03-87-K-0007, NSF's MIP-8808978 and ONR's N00014-85-K-531.

1. **Error latency** is defined as the time interval between the occurrence of a fault and its manifestation as a malfunction.

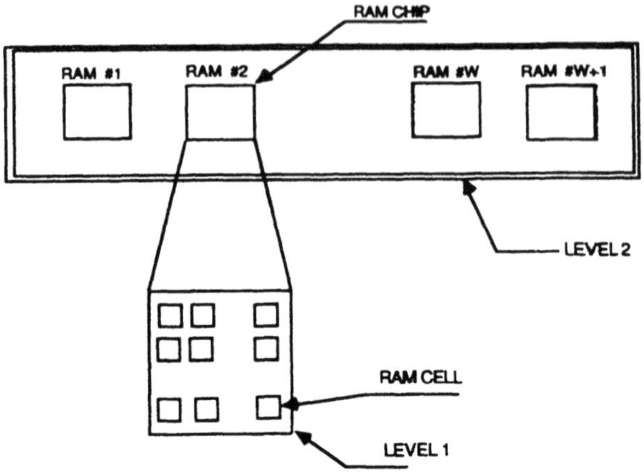

Fig. 1. 2-Level Memory System

memory system, it sensitizes only $w+r$ memory cells, where $w/(w+r)$ is the code rate. This scheme introduces a high error latency[1] and, thereby, very poor uncorrectable-error rate (UER). In order to improve the UER, it is necessary to simultaneously sensitize all the cells on the selected row (word line in the chip) of the different chips. If an n-bit DRAM chip is organized as s identical square subarrays, then as many as $(w+r)\sqrt{sn}$ cells can be sensitized simultaneously in one memory cycle while accessing for a word. This paper discusses four on-chip error-correcting coding techniques that can be employed to improve the reliability of the memory systems.

SINGLE-BIT ERROR-CORRECTING CODES

In this section, three different on-chip single-bit error-correcting codes are described that can be easily integrated within a DRAM chip to improve the soft-error rate. The first code exploits the organization of the memory system, while the other two codes are independent of memory organization and they can automatically correct all single errors.

2-Level Parity Code

A simple on-chip error correction technique which improves the soft-error rate by more than 10^6 times [4] can be obtained by exploiting the hierarchical (2-level) organization of a w-bit n-word memory system shown in Fig. 1. Level-2 array consists of $w+1$ DRAM chips in which the last chip is used to store the parity bit of the word. Each memory chip consists of a two-dimensional array of $\sqrt{n} \times \sqrt{n}$ single transistor memory cells for data storage and an extra $\sqrt{n}$ memory cells one on each column that store the parity bit for the corresponding on-chip word line. Let $b_k(u,v)$ be the content of the cell on u-th row and v-th column within the k-th DRAM chip. The parity bit of u-th row in k-th chip will be given by $P_k{}^1(u,v) = P_k{}^1 = \oplus \sum_{v=0}^{\sqrt{n}} b_k(u,v)$. The level-2 parity bit can be calculated from the selected $w+1$ memory cells as given by $P^2(u,v) = P^2 = \oplus \sum_{k=1}^{w+1} b_k(u,v)$. Fig. 2 illustrates how by using these parity bits single error can be corrected. The level-2 parity bit indicates whether any error has occurred, while the level-1 parity bit indicates which DRAM is faulty, and thereby the error can be easily corrected by complementing the data bit of the faulty DRAM. Table 1 shows the different possible outcomes, and conditions for different types of errors.

Product Code

Let a DRAM chip with $n = (sm^2)$ information or data bit be organized into s subarrays each of size $m_1 \times m_2$. The DRAM is called nonredundant if $m_1 m_2 = m^2$. In a fault-tolerant DRAM chip usually $m_1 m_2 > m^2$. Let $C^k{}_{u,v}$ be the memory cell located at the crosspoint of u-th word line and v-th bit line of the k-th subarray, and $c^k{}_{u,v}$ be the content of the cell $C^k{}_{u,v}$.

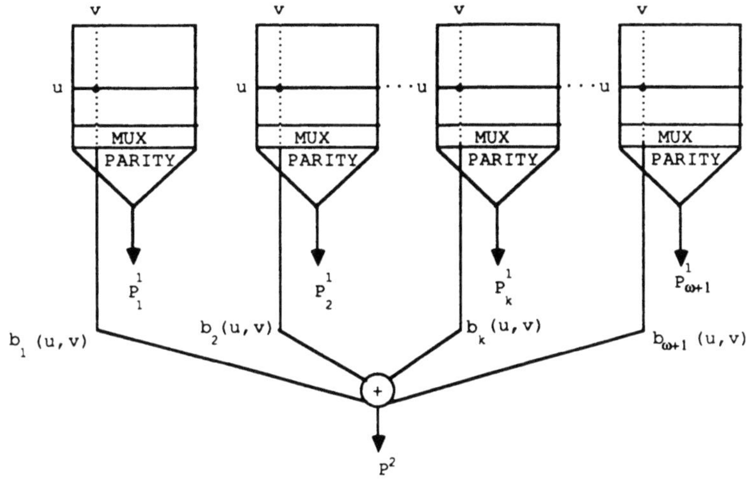

Fig. 2. 2-Level Parity Code

Definition 1: For all $0 \le u \le m_1-1$ and $0 \le k \le s-1$, the string of binary bits $\{c^k_{u,0}, c^k_{u,1}, \cdots, c^k_{u,m_2-1}\}$ forms a product codeword, if m_2 bits are organized into $(p+1)(q+1)$ rectangular array such that if $b(i,j) = c^k_{u,iq+j}$ then for all $0 \le i \le p-1$, $b(i,q) = \oplus \sum_{s=0}^{q-1} b(i,s)$ and for all $0 \le j \le q-1$ $b(p,j) = \oplus \sum_{r=0}^{p-1} b(r,j)$. The Kronecker product of $C_1 \otimes C_2$, where for all i, $C_1 = \{b(i,0), b(i,1), \cdots, b(i,q)\}$ and for all j, $C_2 = \{b(0,j), b(1,j), \cdots, b(p,j)\}$ defines a product code, $PC(pq, p+q+1)$, where $pq = m$ and $m_2 - m = p+q+1$.

The ratio of the number of redundant bits to the total number of bits in the codeword is called the coding efficiency. If $p \approx q$, the coding efficiency for the product code will be maximized to $(2p/(p+1)^2)+1/(p+1)^2 \approx 2/\sqrt{m}$. Thus, in a DRAM with n information bits, altogether $2(sn^3)^{1/4}$ bits are required to correct as many as $\sqrt{sn}$ single-bit/word-line errors. For example, in a 16M-bit DRAM organized into 16 subarrays altogether 1M (i.e., 1,048,576) redundant cells (6.6%) are needed to correct 16,384 single-bit/word-line soft errors.

In order to read a cell $C^k_{u,iq+j}$, the following two equalities are checked by parity generators: $\oplus \sum_{r=0}^{p} b(r,j) = 0$ and $\oplus \sum_{s=0}^{q} b(i,s) = 0$. If both the equalities are not true, then a single-bit error is corrected by complementing $c^k_{u,iq+j}$. If only the first equality is not true, then for $0 \le i \le p-1$ the second equality is tested, and if the second equality is not true for $i=t$, then

Table 1. Different Types of Errors

P^2	P_i^1	P_j^1	Remarks
No Error	No Error	No Error	Error-Free Word
No Error	Error	No Error	Error-Free Word 1-Bit Error Detected in i-th RAM
No Error	Error	Error	Two 1-Bit Errors in the Word Detected
Error	Error	No Error	1-Bit Error in the Word Detected & Corrected
Error	No Error	No Error	1-Bit Error in the Word Detected Two 1-Bit Error in i-th or j-th RAM Detected
Error	Error	Error	1-Bit Error in the Word Detected, Not Corrected 1-Bit Error in i-th and j-th RAM Detected

P^2: Level-2 Parity Bit, P_i^1: Level-1 Parity in i-th Chip, P_j^1: Level-1 Parity in $j(\neq i)$-th Chip

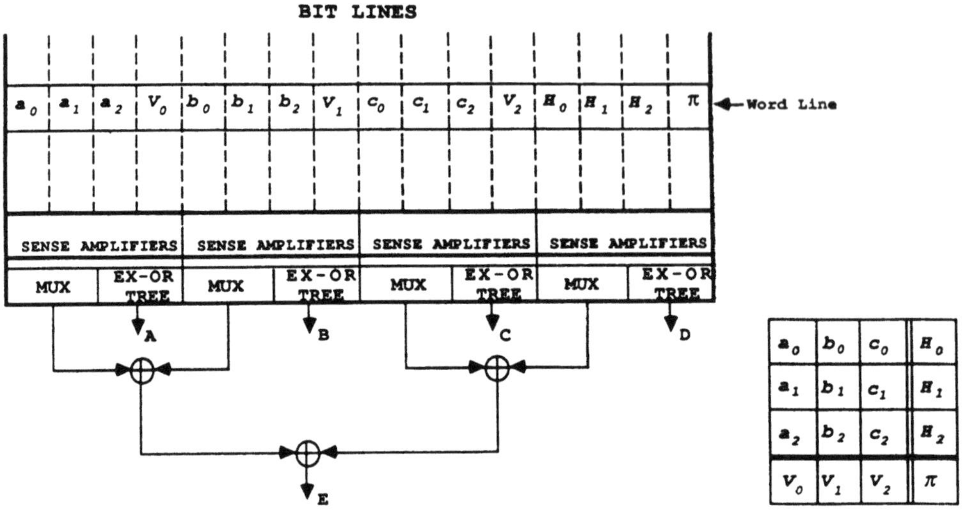

Fig. 3. Product Code (16,9) Organization in a DRAM

$c^k_{u,lq+j}$ is complemented to correct the single-bit error. Similarly, if only the second equality is not true, then for $0 \le j \le q-1$ the first equality is tested, and $c^k_{u,iq+t}$ is complemented if for $j=t$ the first equality is not true. Thus all single bit errors can be corrected. If there are two errors (say, cells $C^k_{u,lq+j}$ and $C^k_{u,iq+r}$) then the above scheme will incorrectly complement the cell $C^k_{u,iq+j}$. Hence, product code is inadequate to correct double-bit memory-cell upset. In order to write data on cell $C^k_{u,iq+j}$, its content $c^k_{u,iq+j}$ is determined by a read operation, and it is compared with the data. If the result of comparison is the same, the content of the cell remains unchanged. If the result of comparison is different, the contents of cells $C^k_{u,iq+j}$, $C^k_{u,pq+j}$, $C^k_{u,iq+q}$ and $C^k_{u,pq+p+q}$ are complemented.

A typical implementation of a product code, PC(16,9) is shown in Fig. 3 where to each word line, having 9 information bits, 7 parity bits are added, so that the overall word-line size is 16 bits. These 16 bits are organized into a logical 4×4 array. The parity bit of the i-th row is stored in the cell denoted by H_i, and the parity bit of the j-th column is stored in the cell denoted by V_j. The overall parity of the word line is stored in the cell designated by Π. Altogether, five parity trees and four multiplexers are used to construct the triplet from which the corrected bit can be read. In general, for a square array with a codeword of m bits, altogether $\sqrt{m}+2$ parity trees each of height $0.5\log m$, and $\sqrt{m}+1$ multiplexers of size $\sqrt{m}+1:1$ are needed.

Cubic Code

The coding efficiency of the product code can be improved by organizing the bits in the form of a three-dimensional array instead of a rectangular array. In order to maximize the coding efficiency the array should be a cube, and hence is called here *cubic code*.

Definition 2: For all $0 \le u \le m_1-1$ and $0 \le k \le s-1$, the string of binary bits $\{c^k_{u,0}, c^k_{u,1}, \cdots, c^k_{u,m_2-1}\}$ forms a cubic codeword if m_2 bits are organized into $p \times q \times r$ three-dimensional array consisting of information bits and a linear array of $p+q+r$ parity bits such that if $b(x,y,z) = c^k_{u,zpq+yp+x}$ and $l(t) = c^k_{u,pqr+t}$ then for all $0 \le z \le r-1$, $l(z+p+q) = \oplus \sum_{r=0}^{p-1} \sum_{s=0}^{q-1} b(r,s,z)$, for all $0 \le y \le q-1$, $l(y+p) = \oplus \sum_{r=0}^{p-1} \sum_{s=0}^{r-1} b(r,y,s)$, and for all $0 \le x \le p-1$, $l(x) = \oplus \sum_{r=0}^{q-1} \sum_{s=0}^{r-1} b(x,r,s)$. The Kronecker product of $C_1 \otimes C_2 \otimes C_3$, where for all x, $C_1 = \{b(x,0,0), b(x,1,0), \cdots, b(x,q-1,r-1), l(x)\}$ and for all y, $C_2 = \{b(0,y,0), b(1,y,0), \cdots, b(p-1,y,r-1), l(y+p)\}$ and for all z, $C_3 = \{b(0,0,z), b(1,0,z), \cdots, b(p-1,q-1,z), l(z+p+q)\}$ defines a cubic code, CC($pqr, p+q+r$), where $pqr = m$ and $m_2 - m = p+q+r$.

The logical organization of the memory cells in a word line is shown in Fig. 4, where corresponding to information bits in each x, y and z-plane, there is a parity bit. Thus altogether there are $p+q+r$ parity bits and the resulting $CC(pqr, p+q+r)$ code has a coding efficiency of $(p+q+r)/(pqr+p+q+r)$. For a cubic array with codeword of $m = p^3$ bits, the coding efficiency is approximately $3/m^{2/3}$. Thus the improvement of efficiency in a cubic code over the product code is by a factor of $0.67m^{1/6}$. For a 16M-bit DRAM chip organized into 16 subarrays altogether 48K (which is less than 0.3%) redundant bits are required. The encoding and decoding logic for such a code is similar to a product code. In general, for a cubic array of an m-bit codeword, $m^{1/3}+1$ parity trees each of height $0.67\log m$, and $m^{1/3}+1$ multiplexers are needed.

DOUBLE-BIT ERROR-CORRECTING CODES

The main limitation of the 2-level parity, product, and cubic codes is that they fail to correct double-bit errors, and in a three-dimensional DRAM where the double-bit soft errors are relatively common these codes are inadequate for on-chip ECC applications. The conventional double-bit error-correcting codes such as Bose-Chaudhury-Hocquenghem (BCH) [5], Reed-Solomon [6], Golay [5], etc., cannot be readily applied to correct double-bit errors in DRAM. These codes are frequently used in digital communications to correct t-bit ($t \geq 1$) errors. The encoding and decoding circuits of these codes employ multi-bit linear feedback shift register (LFSR) which if used in DRAM will introduce very high access delay. By concatenating *finite projective geometry codes* (PGC) [5], a linear code can be constructed which can correct double-bit soft errors in a memory chip [7]. The PGC suffers from many practical limitations. It needs an n-bit memory to be organized into $n^{1/3}$ subarrays (for example, a 16M-bit DRAM needs to be partitioned into 256 subarrays). It needs a complex decoder which computes address arithmetic over Galois Field, $GF(n^{1/3})$. It cannot correct sense amplifier faults and it can correct at most two errors in the chip as opposed to $\sqrt{sn}$ errors in product code. Even though its code rate is better than the product code, it needs much more layout area and it uses a software algorithm to correct double-bit errors. A new code is proposed that can be easily implemented within the framework of the high-density memories. One of the constraints on the code is that its encoding and decoding circuit should be compatible to the low (3λ) interceller pitch width. This new coding scheme is called here *augmented product code,* and is constructed by adding a set of diagonal parity bits to the conventional product code which uses the vertical and horizontal parity bits.

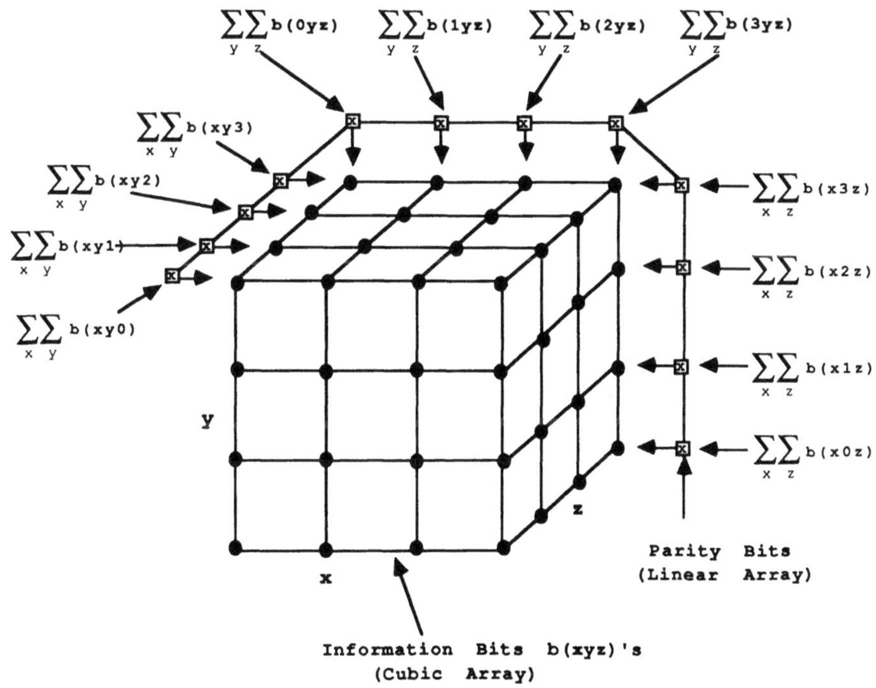

Fig. 4. Cubic Code

Table 2. All Possible Patterns in Coding

Π	$X(i)$	$Y(j)$	$D(t)$	Remarks
No Error	No Error	No Error	No Error	No Error $b(i,j)$ Error-Free
Error	Error	Error	Error	Single Error, $b(i,j)$ Erroneous
Error	Error	No Error	No Error	Single Error, $b(i,j)$ Error-free, $b(i,\bar{j})$ Erroneous
Error	No Error	Error	No Error	Single Error, $b(i,j)$ Error-free, $b(\bar{i},j)$ Erroneous
Error	No Error	No Error	Error	Single Error, $b(i,j)$ Error-free, $b(\bar{i},\bar{j})$ Erroneous
Error	No Error	No Error	No Error	Single Error, $b(i,j)$ Error-free, $b(\bar{i},\bar{j})$ Erroneous
No Error	No Error	Error	Error	Double Error, $b(i,j)$ & $b(\bar{i},j)$ are Erroneous
No Error	Error	No Error	Error	Double Error, $b(i,j)$ & $b(i,\bar{j})$ Erroneous
No Error	Error	Error	No Error	Double Error, $b(i,j)$ & $b(\bar{i},\bar{j})$ Erroneous
No Error	Error	Error	Error	Double Error, $b(i,j)$ Erroneous

$$t = (p+i-j \; mod \; p) \; mod \; p$$

Augmented Product Code

The proposed code is constructed from the conventional orthogonal product code by adding a set of diagonal parity bits. The new code is called *augmented product code* (APC), since in addition to the regular $p+q+1$ parity bits in the product code, it employs an additional set of q parity bits to correct double errors in the entire code word.

Definition 3: For all $0 \le u \le m_1-1$ and $0 \le k \le s-1$, the string of binary bits $\{c^k_{u,0}, c^k_{u,1}, \cdots, c^k_{u,m_2-1}\}$ forms an augmented product codeword, if m_2 bits are organized into $p \times q$ rectangular array of information bits and a linear array of $2p+q+1$ parity check bits such that if $b(i,j) = c^k_{u,iq+j}$ and $l(t) = c^k_{u,pq+t}$ then for all $0 \le i \le p-1$, $l(i) = \oplus\sum_{s=0}^{q-1} b(i,s)$, for all $0 \le j \le q-1$ $\quad l(p+j) = \oplus\sum_{r=0}^{p-1} b(r,j)$, for all $0 \le i \le p-1, 0 \le j \le q-1$ $\quad l(p+q+t) = \bigcirc\sum_{s=0}^{q-1} b((t+s) \; mod \; p, \; s)$, where $t = (p+i-j \; mod \; p) \; mod \; p$. The Kronecker product of $C_1 \otimes C_2$, where for all i, $C_1 = \{b(i,0), b(i,1), \cdots, b(i,q-1), l(i)\}$ and for all j, $C_2 = \{b(0,j), b(1,j), \cdots, b(p-1,j), l(p+j)\}$ along with $l(p+q+t)$ for all i and j, defines an augmented product code, APC($pq, 2p+q+1$), where $pq = m$ and $m_2 - m = 2p+q+1$.

The entire m_2 bits in a DRAM word line are organized into a $p \times q$ array of information or data bits, p bits of horizontal parity, p bits of diagonal parity, q bits of vertical parity, and one bit of overall parity, such that $m_2 = pq+2p+q+1$. The overall parity bit π is computed over pq information bits as $\pi = \oplus\sum_0^{p-1}\sum_0^{q-1} b(i,j)$. The diagonal parity bit corresponding to the data bit

$b(i,j)$ is computed by reading out the data bits in a cyclic path given by

$$b(i,j) \rightarrow b(i+1,j+1), \quad \text{if } j<p-1$$
$$b(i,p-1) \rightarrow b(i-p,0), \quad \text{otherwise.}$$

For an n-bit DRAM organized into s square subarrays, the proposed coding will need about $3(sn^3)^{1/4}+(sn)^{1/2}$ redundant bits and will have approximately $3(s/n)^{1/4}$ coding efficiency. Thus a 16M-bit DRAM organized into 16 subarrays will need about 9% extra redundant bits.

In order to read the cell $C^k_{u,iq+j}$ (i.e, $b(i,j)$), the values $c^k_{u,pq+i}$, $c^k_{u,pq+p+j}$, $c^k_{u,pq+p+q+t}$ (where $t = (p+i-j \bmod p) \bmod p$) and $c^k_{u,pq+2p+q}$ are compared with $l(i)$, $l(p+j)$, $l(p+q+t)$ and π, respectively. Let $X(i)$, $Y(j)$, $D(t)$ and Π, respectively be the results of comparisons. In order to write a data on the cell $C^k_{u,iq+j}$ (i.e, $b(i,j)$), the value of $c^k_{u,iq+j}$ is compared with the data. If they are different, then the values $c^k_{u,pq+i}$, $c^k_{u,pq+p+j}$, $c^k_{u,pq+p+q+t}$ (where $t = (p+i-j \bmod p) \bmod p$) and $c^k_{u,pq+2p+q}$ corresponding to the parity bits $X(i)$, $Y(j)$, $D(t)$ and Π, respectively are complemented. Since corresponding to each data bit, there are four parity check bits, the APC(pq, $2p+q+1$) is a distant five code, and can correct two-bits errors in the codeword. The different error syndromes have been represented by Table 2, and it can be seen that double errors can be corrected from the error patterns.

The augmented product code can be easily implemented within the DRAM chip. A typical implementation of a DRAM with an m-bit bit lines will contain $3\sqrt{m}+1$ redundant bits per bit line. A DRAM where each word line has 9 data bits (organized into 3x3 matrix) and 10 parity bits is shown in Fig. 5. The distribution of horizontal, vertical and diagonal parities on the word line is also shown. The vertical parities are easily computed by the parity trees (shown by PARITY in the diagram). The horizontal and diagonal parities are computed by the external Ex-Or gates shown in the Fig. 5. The suitable bit is selected by the MUXes. If there is no error in the selected bit line, all the outputs $X(i)$, $Y(j)$ and $D(t)$ are 0's.

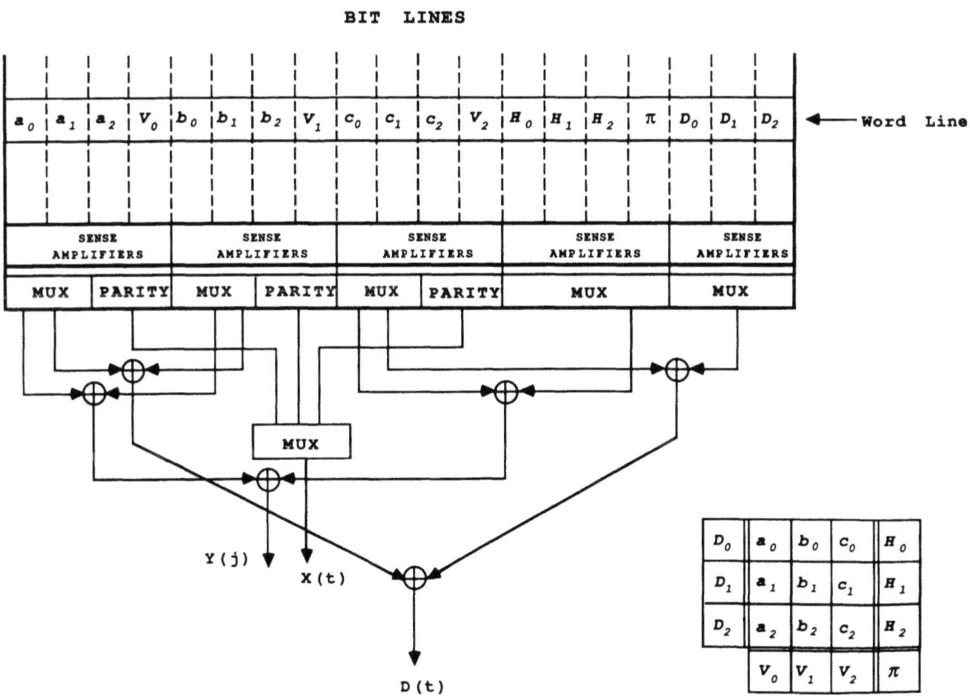

Fig. 5. Augmented Product Code APC(9,10) Organization in a DRAM

Table 3. Comparison of Four Coding Schemes

Criterion	Parity Code	Product Code	Cubic Code	Aug. Prod. Code
Cell Redundancy	$1/\sqrt{n}$	$2/n^{1/4}$	$3/n^{1/3}$	$3/n^{1/4}$
Code Distance	4	4	4	5
Error Correction	SEC	SEC	SEC	DEC
Number of Faults	$\sqrt{n}$ /chip	$\sqrt{n}$ /chip	$\sqrt{n}$ /chip	$2\sqrt{n}$ /chip
Address Computation	Simple	Medium	Complex	Medium
Layout Complexity	Low	Medium	Very High	High
Decoding Time	Small	Medium	Large	Medium

CONCLUSIONS

This paper has discussed the problems of error correction in multimega-bit dynamic random-access memory (DRAM) chip. A large number of alpha-particle-induced soft errors manifest themselves into double-bit errors when the alpha particle strikes on the intervening space between two vertically-mounted trench-type capacitors. The conventional SEC/DED codes, such as Hamming code, cannot correct these double-bit/word-line faults. A new augmented product code, which employs diagonal parities in addition to the horizontal and vertical parities, is proposed in this paper to correct the double-bit errors. It can be easily implemented within the 3λ pitch width limits of DRAM chip. The Projective Geometry Code (PGC) can also be utilized to correct two-bit errors [8]. But unlike the proposed code, the PGC cannot be easily implemented in a multimega-bit memory. The PGC requires special decoders that compute over Galois Fields, and also its double-error correcting logic is very complex. The proposed code can tolerate up to $2\sqrt{sn}$ soft errors, and thereby it can correct sense amplifier fault.

In addition to double-bit ECC, three schemes for single-bit ECC are also discussed. These ECC circuits can be easily integrated into the memory chips. The mean time between failures (MTBF) of the chip improves by a factor of $1.25n^{1/4}$ if one of these single-error correcting schemes is used, and the MTBF improves by a factor of $1.62n^{1/3}$ if the double-bit ECC is used. Thus, for a 16M-bit DRAM, using trench-type capacitors where the double-bit errors occur frequently, the reliability can improve by 32.5% if the double-bit ECC is used as opposed to the single-bit ECC's described here. The above four coding schemes are compared in Table 3.

In addition to detecting the on-line error, the error-coding circuit can be utilized to improve the fabricational yield. In defective memory chip the faulty cells can be automatically bypassed by the error-correction circuit and the resulting memory can be used as nonredundant and nonfault-tolerant memory. But, in practice, it will be a better idea to employ a few extra rows and columns to bypass the fabricational defects, and to utilize the error-correcting circuit exclusively for correcting the soft errors.

REFERENCES

1. T. C. May and M. H. Woods, "Alpha-Particle-Induced Soft Errors in Dynamic Memories," *IEEE Transactions on Electron Devices* **ED-26** pp. 2-9 (January 1979).

2. R. J. McPartland, "Circuit Simulations of Alpha-Particle-Induced Soft Errors in MOS Dynamic RAM's," *IEEE Journal of Solid-State Circuits* **SC-16**(1) pp. 31-34 (February 1981).

3. G. A. Sai-Halasz, M. R. Wordeman, and R. H. Dennard, "Alpha-Particle-Induced Soft-Error Rate in VLSI Circuits," *IEEE Journal of Solid-State Circuits* **SC-17**(2) pp. 355-362 (April 1982).

4. P. Mazumder and J. H. Patel, "A Novel Fault-Tolerant Design of Testable Dynamic Random-Access Memory," *Proceedings of International Conference of Computer Design,* (October 1987).

5. W. W. Peterson and E. J. Weldon, *Error Correcting Codes,* MIT Press, Cambridge, MA (1972).

6. Shu Lin, *An Introduction to Error-Correcting Codes,* Prentice-Hall, Inc., Englewood Cliffs, New Jersey (1980).

7. R. M. Tanner, "A Recursive Approach to Low-Complexity Codes," *IEEE Transactions on Information Theory* **IT-27**(4) pp. 533-547 (September 1981).

8. R. M. Tanner, "Fault-Tolerant 256K Memory Designs," *IEEE Transactions on Computers* **C-33**(4) pp. 314-322 (April 1984).

5 DEFECT MONITORING AND YIELD PROJECTION

MEASUREMENT AND DISTRIBUTION OF FAULTS ON DEFECT TEST SITE CHIPS

Reinhard Glang

IBM Corporation
Federal Systems
Manassas, VA 22110

INTRODUCTION

In the development of Very Large Scale Integration (VLSI) processes, yield losses due to chip design errors and inadequate control of device parameters are usually corrected early in the development cycle. Thereafter, the yield of product chips is determined by the number of pattern defects picked up in the lengthy fabrication sequence. Monitoring defect levels associated with various process steps and reducing them systematically to achieve competitive costs remain necessary activities for as long as the product is manufactured.

Following terminology introduced by Stapper[1] a defect on a chip that causes an electrical failure is called a fault. The number of faults per chip can be determined by electrical testing. For large numbers of like chips, the number of faults per chip must have some distribution. Earlier studies showed that fault distributions on RAM chips may be described by compounding Poisson distributions [1,2,3]. This leads to a two-parameter distribution involving the Gamma function:

$$P(k) = \frac{\Gamma(\alpha + k)}{k!\bullet\Gamma(\alpha)} \bullet \frac{[\lambda/\alpha]^k}{[1 + \lambda/\alpha]^{\alpha+k}} \qquad (1)$$

where P(k) is the probability of finding k faults on a chip; λ is the average number of faults per chip, and α is the so-called cluster factor. The cluster factor reflects the fact that fault densities vary locally within wafers as well as from wafer to wafer. In other words, faults are not simply randomly distributed but tend to cluster in some regions.

Given such a distribution, the yield of good chips follows from Equation 1 as the probability of chips with k=0 faults:

$$Y = P(0) = (1 + \lambda/\alpha)^{-\alpha} \qquad (2)$$

Equation 2 is referred to as the negative binomial yield model. The complete distribution as given by Equation 1 may take very different forms depending on the values of λ and α.[4] Earlier investigations found α to vary from lot to lot but to be nearly constant when averaged over multiple lots[4,5]. The purpose of this investigation is to examine the behavior of α and to test the ability of the negative binomial yield model to project actual product chip yields.

THE DEFECT TEST SITE CHIP

The presence of faults on a VLSI chip is easily established by electrical testing. However, to monitor defect levels the faults must be located so that they may be analyzed, attributed to causes, and counted. These tedious and labor-intensive procedures may be used on memory chips, but pinpointing faults on non-array chips is not practical. To overcome this limitation, special defect-monitoring test patterns have been developed. The most common patterns consist of parallel lines and spaces as narrow as the product design rules permit[6]. These patterns are independently testable for every layer of conductor and insulator so that electrically detected faults can be attributed to specific process steps without microscopic inspection and search.

To relate the number of faults measured on test patterns to the defect limited yield (DLY) of product chips, one has to know the relative susceptibilities of test patterns and product chips to defects. This susceptibility is called the critical area. For simple patterns consisting mostly of parallel lines, critical areas may be determined by analytical methods[7]. More universally applicable are computer simulation methods in which large numbers of defects of various sizes are randomly placed on the chip layout and then compared against pre-established rules to decide if they would cause a fault[4]. The latter approach has been used in this work to project DLYs for product chips on the basis of faults found on test patterns.

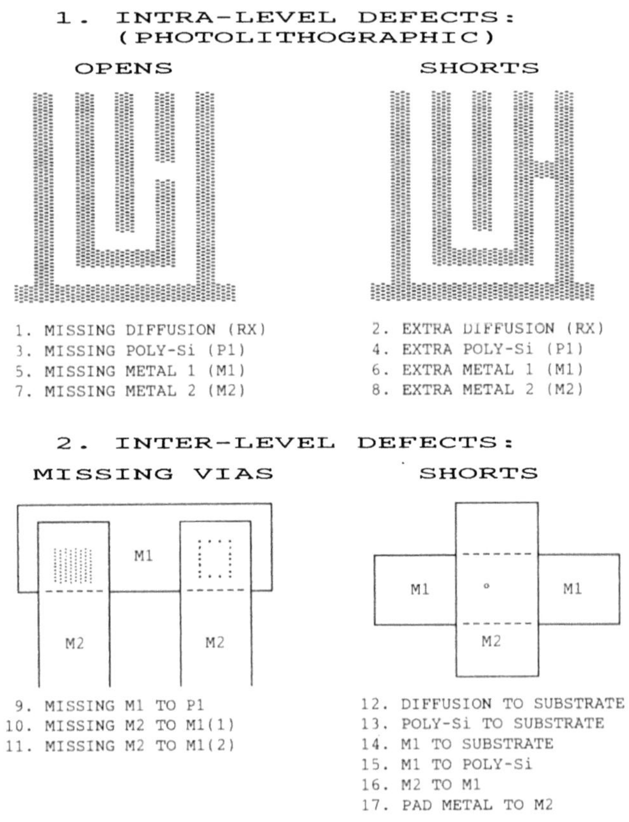

Figure 1. Defects (Faults) in VLSI Patterns

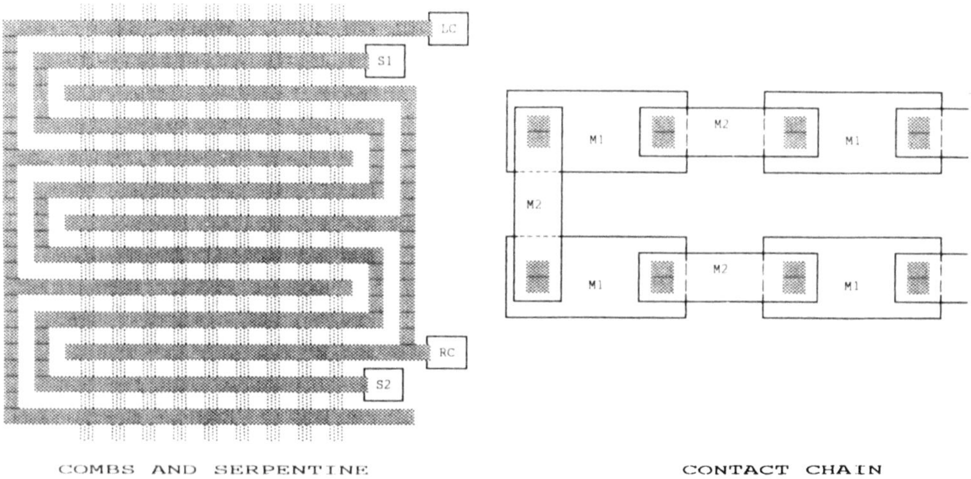

Figure 2. Defect Test Patterns

The technology for which the following data was gathered is a single-polysilicon Complementary Metal Oxide Semiconductor (CMOS) process with two levels of metal wiring, an additional level of metal pads for external connections, and 1µm minimum feature size. Seventeen types of defects have been routinely monitored; they are shown schematically in Figure 1. The first group is thought to occur as a result of particles causing image contrast reversal during the lithographic or subtractive etch processes. Missing via hole defects arise if a large particle blocks an opening during exposure or etching. Interlevel shorts may occur as a result of local contamination or damage between any two conductive layers. As defined here, all these defects would cause electrical fails and should properly be called faults. However, most authors--this one included--do not make this distinction consistently.

The test patterns are either of the combs and serpentine type or chains of via holes between two levels of conductors as shown schematically in Figure 2. The meandering line patterns are stacked on top of each other with 90° rotation from one layer to the next; they are used to determine all the intralevel defects, numbers 1 through 8 in Figure 1. The overlap areas between adjacent layers permit the measurement of interlevel shorts as well, numbers 12 through 17 in Figure 1. The contact chains consist of strings of about 10,000 via holes of minimum diameter. There are separate chains for every level of insulator film as indicated by defect numbers 9 through 11 in Figure 1. Some chains can also be tested against lower levels of conductors to yield information about interlevel shorts.

The test patterns were designed to primarily indicate such faults as would be caused by particles introduced accidentally during the process. Different structures are required to capture specific process-induced defects such as lack of step coverage, misalignment, or variability of critical dimensions[8]. The absence of such test structures limits the information that can be obtained from the defect monitor chip described here.

Individual combs and serpentine or contact chain structures are designed to fit into $0.4 \times 1.6mm^2$ layout areas called macros. Figure 3 shows how these macros, interspersed with contact pad areas, are combined into a defect test chip. Most structures occur twice, over N-well and over P-epi areas. Some patterns consist of more than one macro; others differ in critical dimensions or in topographical features. The numbers within the macros in Figure 3. relate to Figure 1 and indicate defect types for which the structures can be tested. There are 39 independently testable combs and serpentine and 14 contact chain macros. After each metallization step, an automatic test program executes about 600 tests per chip. Data reduction produces about 220 macro yield data that relate to the 17 defect types shown in Figure 1.

For routine evaluation, all data pertaining to one defect type are combined. For example, there are 18 macros with different critical areas that measure faults of the missing metal 1 type. A test chip is good if none of the 18 macros has a fault, or it is defective if one or more has a fault. Thus, only one defect limited yield, DLY_i, is derived for each of the 17 defect types. Equation 2 is used to convert DLY_i into faults per chip, $(\lambda_i)_{TS}$. The defect density of the i-th defect type is then derived by dividing the faults per chip through the sum of the critical areas, $(A_i)_{TS}$, of all the macros included in the yield number for this defect type:

$$DD_i = (\lambda_i)_{TS}/(A_i)_{TS} \tag{3}$$

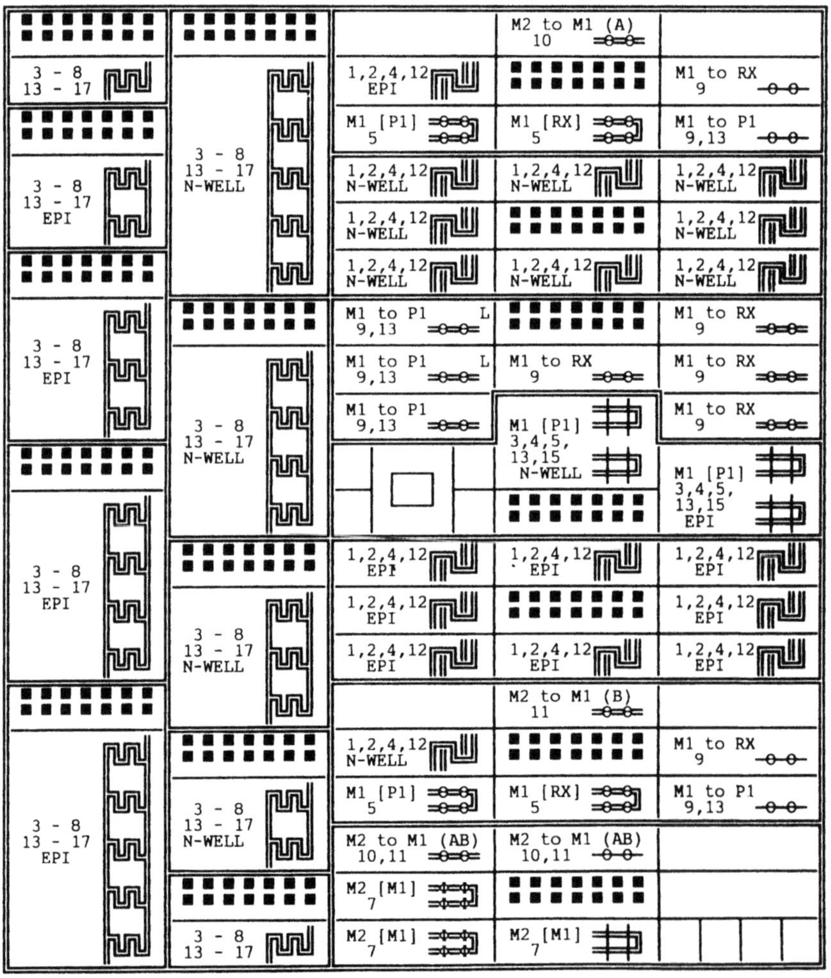

Figure 3. Schematic of Defect Test Chip Consisting of Macros with Indiviudal Test Patterns and Macros with Contact Pads

Typically, six of these test sites are included and processed along with product chips on each wafer. Since they experience the same process steps and conditions, it is reasonable to assume that test and product chips will acquire the same defect densities, DD_i, at each level. Hence, the expected average number of faults of the i-th type on product chips with a critical area of $(A_i)_{PC}$ can be predicted:

$$(\lambda_i)_{PC} = DD_i \cdot (A_i)_{PC} \tag{4}$$

The DLY of product chips, based on the number of faults found on the defect test chips, may then be projected as

$$DLY = \prod_i [1 + (\lambda_i)_{PC}/\alpha]^{-\alpha} \tag{5}$$

for i from the first to the 17th defect[4]. A universal value of $\alpha = 2$ has been used in Equation 5 since lot-specific cluster factors are generally not known.

OBSERVED FAULT DISTRIBUTIONS

To study the distribution of faults among wafers, a more detailed analysis has been performed. In this method, multiple faults found on different macros of the same defect type have been recorded in a grid as shown in Figure 4. The horizontal axis identifies the defect types, the vertical axis lists the wafers and their 6 chips from a particular lot. Each line shows how many faults and of what type a particular test chip had. Twenty lots comprising 104 wafers have been so analyzed.

Figure 4 shows two examples of this analysis. Obviously, not all faults are distributed randomly. The top lot in Figure 4 includes wafers with an unusually high incidence of polysilicon-to-substrate (P1/SX) or gate oxide shorts. A closer analysis of individual macros reveals further that structures in epi areas are more prone to be shorted than N-well macros. The relevant process difference is that the two types of structures receive different source/drain ion implantations at different places in the process sequence. Furthermore, it has repeatedly been observed that wafers with large numbers of gate shorts on test sites have product chip yields that do not reflect this problem. We may conclude that these faults are due to occasional process failures that are pattern-dependent and not caused by a random defect mechanism.

The bottom lot in Figure 4 illustrates another nonrandom problem, namely multiple metal defects within the same chip. Microscopic inspection identified incomplete pattern resolution at the first metal level as the primary defect. In addition to causing intralevel shorts at the M1 level, the residual metal penetrates the dielectric layers above and causes faults in the upper metal layers. The coupling of defects is very strong; it is estimated that more than 50% of all upper level metal defects are due to first metal problems. Again, this is an occasional process failure and not due to random incidence of particles. The occurrence varies from wafer to wafer and even within a wafer, thus exemplifying one mechanism of defect clustering.

The last column in Figure 4 gives the total number of faults per test chip. Histograms of four "good" lots are shown in Figure 5; these lots had 34 to 54% defect-free chips, and their average fault levels were between 1.0 and 2.3 faults per chip. Using these parameters, cluster factors $\alpha < 1$ have been computed from Equation 2. The dashed lines in the four histograms represent the compounded Poisson distributions as defined in Equation 1.

WAFER No.	CHIP No.	OPEN LINES: RX	P1	M1	M2	SHORTED LINES: RX	P1	M1	M2	MISSG. CONT. LEVEL 1	2	3	INTER-LEVEL SHORTS: RX SX	P1 SX	M1 P1	M2 M1	M3 M2	TOTAL FTS.
		NUMBER OF FAULTS PER CHIP, BY DEFECT TYPE:																
51	1													4				4
	2													5				5
	3													5				5
	4													3				3
	5			1	1			1				1		5		1		10
	6	1												5				6
52	1																	0
	2																	0
	3																	0
	4													2				2
	5													2				2
	6													5				5
57	1													1				1
	2																	0
	3													2		1		3
	4													3				3
	5									1				2				3
	6													1				1
61	1			1													1	2
	2																	0
	3													1				1
	4	1																1
	5							1										1
	6										1							1
56	1																	0
	2																	0
	3																	0
	4			1			1									1	1	4
	5																1	1
	6													1				1
64	1	1						1										2
	2													3				3
	3													1				1
	4																	0
	5							1		1				2	2			6
	6				2		1		1					2				6
59	1																	0
	2				1	1				1								3
	3													1		1		2
	4													1				1
	5																	0
	6	2		1		1	1											5
70	1																	0
	2													1				1
	3									1				1				2
	4													2				2
	5													1				1
	6																	0
TOTALS:		5	0	4	4	2	3	4	1	4	1	1	0	62	2	4	3	100

WAFER No.	CHIP No.	OPEN LINES: RX	P1	M1	M2	SHORTED LINES: RX	P1	M1	M2	MISSG. CONT. LEVEL 1	2	3	INTER-LEVEL SHORTS: RX SX	P1 SX	M1 P1	M2 M1	M3 M2	TOTAL FTS.
		NUMBER OF FAULTS PER CHIP, BY DEFECT TYPE:																
63	1		2															2
	2																	0
	3	3					1			1					1	1		7
	4	1					2									1		4
	5	1							1	1						6		9
	6						1									4		5
62	1																	0
	2						2											2
	3					1		1								1		3
	4				1													1
	5						1											1
	6				1											1		2
64	1																	0
	2															1		1
	3																	0
	4													1				1
	5														1			1
	6							1										1
59	1			6	2			5	3	1						5	5	27
	2			4	5			5	9						1	9	8	41
	3			10	5			3	8							8	8	42
	4			2	2			5							1	3	4	17
	5							1	1							1		3
	6			2	2				2							2		8
52	1																	0
	2							1	3									4
	3			1														1
	4																1	1
	5				1													1
	6																	0
TOTALS:		5	2	25	19	1	7	22	27	3	0	0	0	1	4	43	26	185

Figure 4. Occurrence of Faults on Defect Test Chips.
Top: 8 wafers from lot #5 in Figure 9
Bottom: 5 wafers from lot #10 in Figure 9

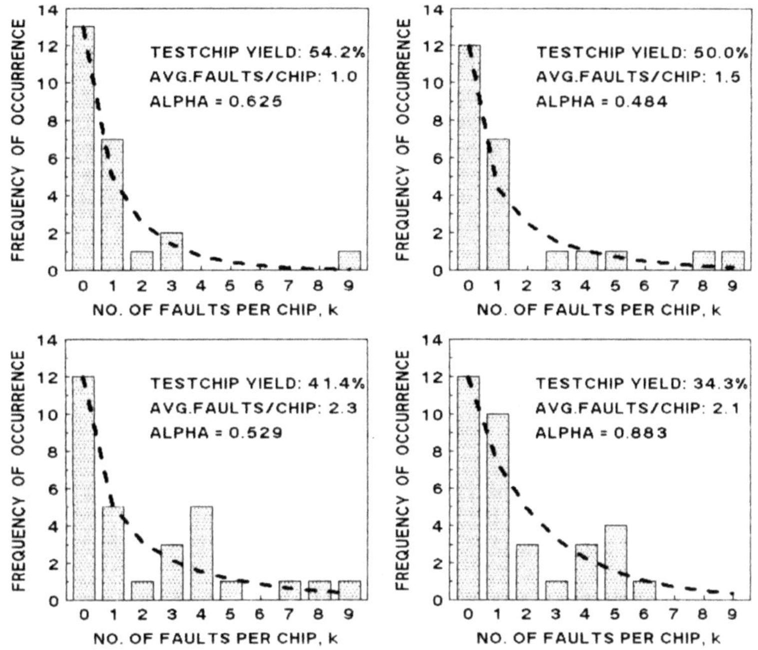

Figure 5. Fault Distributions Among Test Site Chips: Four "Good" Lots. Lot numbers: upper row: #6 and #11; lower row: #7 and #8.+

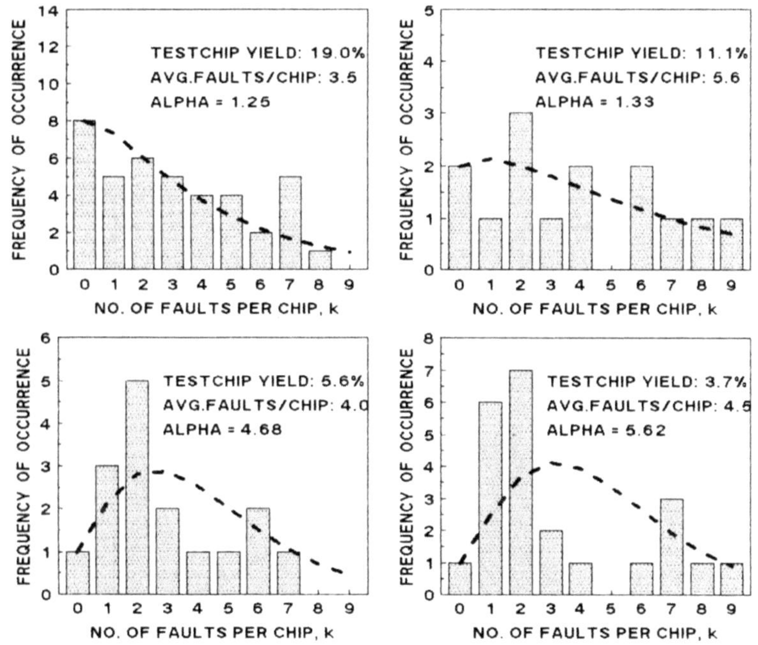

Figure 6. Fault Distributions Among Test Site Chips: Four "Fair" Lots. Lot numbers: upper row: #4 and #1; lower row: #3 and #13.+

+The lot numbers refer to the processing sequence in Figure 9

Figure 6 shows histograms of four lots that are only "fair" in regard to defect levels. The test chip yields range from 4 to 19%, and the average fault levels vary from 3.5 to nearly 6 per chip. The cluster factors are all > 1, and for the lowest yielding lots they are around 5. The compounded Poisson function that described very well the fault distributions in the "good" lots does not fit the histograms of the "fair" lots equally well. This may be due in part to the relatively small sample size, but one lot has a somewhat bimodal distribution that cannot be modelled by Equation 1.

In reviewing the data for all the lots analyzed, it became apparent that large values of α are associated with low-yield lots whereas "good" lots tended to have values of $\alpha < 2$. This is shown in Figure 7. The relationship may be understood by rewriting Equation 2:

$$\alpha = \frac{-\ln Y}{\ln(1 + \lambda/\alpha)} \tag{6}$$

The coefficient λ/α varies only from about 1 to 4 while the yield ranges from about 4 to 60%. A similar relationship between the cluster factor α and the yield of memory chips has also been reported by Stapper[9].

These data show that the cluster coefficient α is not a constant factor which, once determined, can be applied to individual lots to make highly accurate yield predictions. However, as has also been reported by Stapper[4], α is less variable if fault distributions of multiple lots are evaluated. This result is confirmed by the histograms in Figure 8. Here the cluster factors have been derived by a least square fit of the experimental data to the fault distribution Equation 1. The fitted curves are shown as dashed lines in Figure 8. The fit is very satisfactory, and the cluster factors for the three time periods vary only within a narrow range. Thus, it is possible to project defect limited yields for multiple lots using a predetermined average value of α.

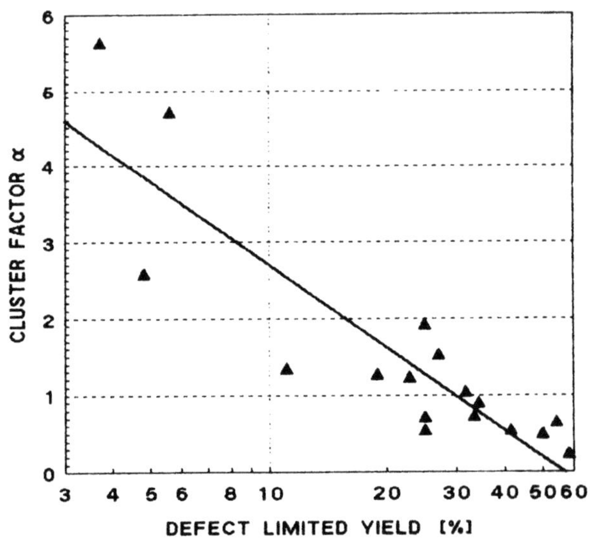

Figure 7. Relationship Between Cluster Factor α and Defect Test Chip Yield.

The cluster factors α as shown in Figure 8 have been determined by taking the entire fault distributions into account. A much simpler procedure is to derive α by inserting the average number of faults per chip and the chip yields into Equation 2. Table 1 shows a comparison of cluster factors derived by the two methods. The differences are negligible.

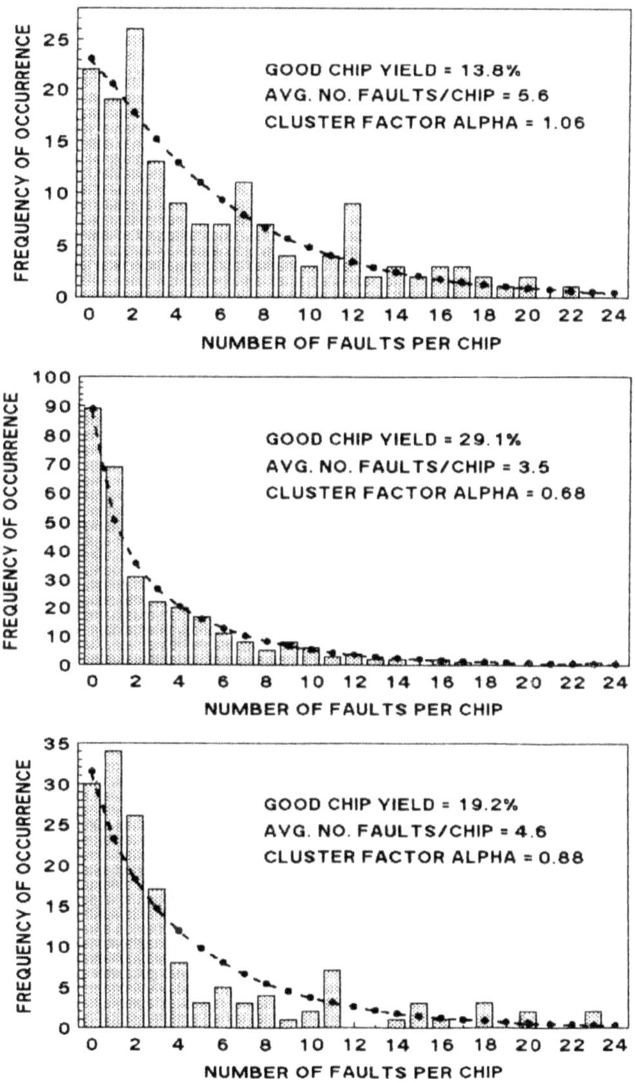

Figure 8. Fault Distributions on Test Site Chips from Multiple Lots.
Top:4 lots, 27 wafers, 160 chips; time period: 1stQ.
Middle: 10 lots, 51 wafers, 306 chips; time period 2ndQ.
Bottom: 6 lots, 26 wafers, 156 chips; time period 3rdQ.

Table 1. Cluster Factor α Calculated by Different Methods

Data from time period:	1stQ	2ndQ	3rdQ
Total no. of faults	888	1,081	710
Avg. faults per chip, λ	5.55	3.53	4.55
Test chip yield [%]	13.8	29.1	19.2
α calculated from:			
Equation 1 using least square fit	1.06	0.68	0.88
Equation 2 using avg. λ and yield	1.11	0.67	0.93

YIELD PROJECTIONS FOR PRODUCT CHIPS

As described earlier, the individual defect densities DD_i derived from the test chips may be used to project DLYs for product chips with known critical areas A_i by applying Equation 5. If there are no design errors on the chip and all device parameters are within their specifications, the projected DLY should approximate the actual test yield for electrically functional chips.

Defect limited yields have been projected for individual wafers of those lots for which defect grids as shown in Figure 4 had been made. The results are disappointing because reasonable agreement between DLY and functional test yields was obtained for only about 25% of all wafers.

Another attempt at projecting DLY was made by using the average defect densities of all wafers from one lot and combining the various faults according to Equation 5. This comparison included 48 lots representing 10 different chip designs. The results are shown in Figure 9. The agreement between projected and actual yields varies with time: up to lot #24, the projected DLY is generally higher than the functional test yield. From lots #25 through 36, the projected DLY is mostly lower than the functional test yield. For the last 12 lots, DLY and test yields track nearly always within ±5%.

In analyzing the reasons for projecting incorrect yields, we can dismiss chip design errors because each one of the designs involved produced high-yielding lots at one time or another. Incorrect device parameters may also be disregarded as major yield detractors because they have been monitored and found to be within acceptable limits. This leaves pattern defects as the principal yield-determining factor.

In describing the test site chip, it was pointed out that the defect test patterns were designed primarily to capture particle-induced faults. Compatible with this concept is a yield model that assumes a random distribution of faults. It is known, however, that pattern defects may also arise from marginal processes. Evidence for defects caused by localized process failures has been routinely collected by microscopic inspection of test structures that failed the electrical tests. In a sample of 21 lots with about 1200 electrical faults on metal 1 and metal 2 test patterns, 50 to 55% were attributable to localized process failures. Only 20 to 25% showed evidence of particle-induced faults, and 9% showed mechanical damage such as scratches.

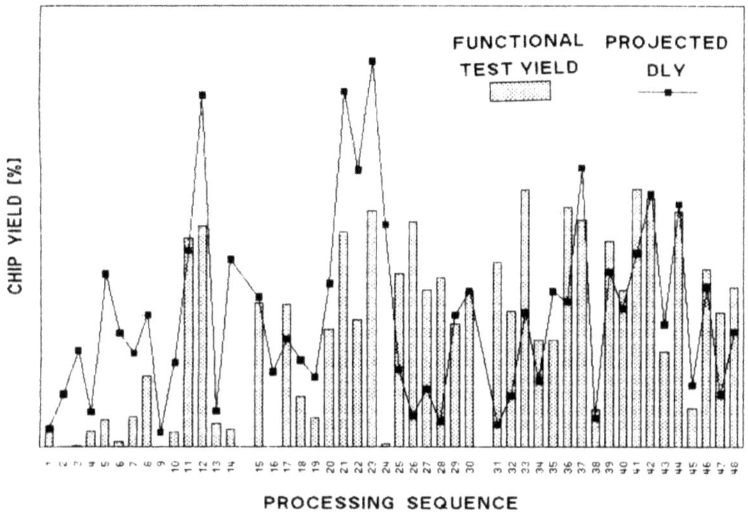

Figure 9. Comparison of Projected Defect Limited and Functional Test Yields for Product Chips. Lots 1 through 14 are from 1stQ. lots 15 through 30 from 2ndQ, and the rest from 3rdQ.

The majority of the observed process failures affected the metal wiring and insulation patterns. Overall, two thirds of all faults were associated with these layers. The most frequent types of faults were unresolved spaces between metal lines, breaks in metal films where narrow lines step down into a contact hole, and narrow metal lines without adjacent lines that tended to fade out and disappear. These defects were usually not evenly distributed but tended to cluster on certain wafers. Within a wafer they occur only on certain structures. Fading metal lines, for example, were rarely seen on defect test chips because these have closely spaced parallel lines but no singletons. Breaks in metal lines at via hole edges were far more likely to cause an open line on product chips than on test site chips due to a small difference in the contact design. On the other hand, the defect test chips have many more parallel metal lines with minimum spacings than the product chips. Therefore, the test chips are more prone to faults from unresolved line spaces.

Another type of process-induced faults originates from overetched via holes if these are large enough to straddle two adjacent lines underneath. While some product chip designs are sensitive to this problem, the test chip does not have a suitable pattern. A last example are gate shorts discussed earlier in context with Figure 4. These have periodically affected test chip structures without causing proportionate yield losses on product chips.

The yield data in Figure 9 reflect the gradual elimination of these and other process failures. The functional yield of the first 24 lots was severely diminished by metal opens at sharp steps and fading single lines--problems which the test chip was not designed to detect. Once these problems were overcome, the dominant defect type became unresolved spaces between metal lines. In this category, the test chip is more sensitive than the product chip. Therefore, it showed a disproportionately large number of faults, and the DLY projections for lots #25 through 36 in Figure 9 are too low. Further changes in the metal patterning processes reduced these failures to levels where the yields of product and test chips are determined by detractors that are not design sensitive.

SUMMARY AND CONCLUSIONS

Test chips containing a variety of defect-sensitive structures have been used to detect faults on CMOS VLSI wafers. The distributions of faults over many chips can be described by compounded Poisson functions, and the negative binomial yield model is applicable. The cluster factor α, which is the second adjustable parameter in these distributions, varies significantly from lot to lot. For groups of lots, however, α was found to vary only over a narrow range from 0.7 to 1.1.

Clustering of defects is often due to marginal processes that fail to produce certain structures reliably. These failures are sensitive to details of pattern design and layout. Therefore, the occurrences of such faults on product and test chips do not correlate very well. Once a process is fully optimized and firmly controlled the particle-induced faults become the yield-limiting factor. If this is the case, it becomes possible to project product chip yields based on defect data from special test structures. The last 12 lots in Figure 9 prove this point. Another example of successful yield projections based on test site data has been reported by Newhart[10] who predicted DLYs within ±3% of actual test yields.

ACKNOWLEDGEMENTS

The defect test chip has been designed by A. S. Rosen and P. A. Hanni. The electrical test and data reduction programs have been written by P. A. Hanni. The critical area determination program was written by J. C. Carter and A. S. Rosen. Microscopic inspection data have been gathered by P. M. Harris. The author gratefully acknowledges these contributions and the continued support from his IBM colleagues.

REFERENCES

1. C. H. Stapper, A. N. McLaren, and M. Dreckmann, Yield Model for Productivity Optimization of VLSI Memory Chips, IBM J. Res. Develop., 24:398 (1980).
2. C. H. Stapper, On Yield, Fault Distributions, and Clustering of Particles, IBM J. Res. Develop, 30:326 (1986).
3. C. H. Stapper, Correlation Analysis of Particle Clusters on Integrated Circuit Wafers, IBM J. Res. Develop. 31:641 (1987).
4. D. M. H. Walker, "Yield Simulation for Integrated Circuits," Kluwer Academic Publishers, Boston (1987).
5. S. Gandemer, B. C. Tremintin, and J.J. Charlot, Critical Area and Critical Levels Calculation in I.C. Yield Modeling, IEEE Trans. on Electron Devices, 35:158 (1988).
6. W. Lukaszek, W. Yarbrough, T. Walker, and J. Meindl, CMOS Test Chip Design for Process Problem Debugging and Yield Prediction Experiments, Solid State Technology, March 1986:87.
7. C. H. Stapper, Modeling of Defects in Integrated Circuit Photolithographic Patterns, IBM J. Res. Develop., 28:461 (1984).
8. A variety of defect test structures have been described recently in Proc. IEEE 1988 Int. Conf. on Microelectronic Test Structures, Long Beach, California.
9. C. H. Stapper, Modeling Redundancy in 64k to 16Mb DRAMs, Proc. 1983 IEEE Int. Solid-State Circuits Conf. p. 86.
10. R. E. Newhart and E. J. Sprogis, Defect Diagnostic Matrix - a Defect Learning Vehicle for Submicron Technologies. Proc. IEEE 1988 Int. Conf. on Microelectronic Test Structures, Long Beach, California, p. 108.

PROCESS DEVELOPMENT AND CIRCUIT DESIGN INTERACTIONS IN VLSI YIELD IMPROVEMENT

James J. Hammond, Brian Boerman, and **Fred W. Voltmer**

Intel Corporation
3065 Bowers Avenue
Santa Clara, CA 95051

ABSTRACT

Success in VLSI manufacturing requires accelerated yield learning and higher steady-state yields than ever before on state-of-the-art monolithic VLSI systems. These requirements, coupled with increasing system complexity and accelerated product introduction, are extremely difficult to satisfy simultaneously. Historically, emphasis in design has been placed on system complexity and new product introduction. Recently, growing attention has been paid to yield and manufacturability issues during the design phase. There is a perceived need to incorporate yield and manufacturability concerns into design optimization.

We present here our history in rapid yield improvement on new technologies for VLSI logic systems. The nature and behavior of primary yield limiters for three stages of process maturity, the priorities for process improvements, and the design options that may impact yield are highlighted. The currently available yield modeling tools and their application during each stage of process development are evaluated.

INTRODUCTION

The trends toward increased number of logic gates per chip and toward accelerating new product introduction are not new, nor are the general needs for accelerated yield learning and higher "ultimate" or steady-state yield. However, the cost pressures that had previously driven smaller, high-volume commodity IC's are now driving large monolithic logic systems ("micro-computers on a chip"). In addition, the well-known relationships among die size, defect density, and percent good die per wafer show that as die get very large ($\geq$ 1 cm^2) both high yield and stability become extremely difficult to achieve.

To address the yield problem on large die, our work focuses on yield engineering during the development, ramp, and maturity phases of the technology life-cycle. The decision to focus on new technologies proceeds from our experience, indicating that for state-of-the-art VLSI chip designs, new technologies are necessary to attain increased functionality in silicon.

This paper describes the large-die yield problem from the process engineering perspective. The intention is two-fold. First, we describe our experience in yield engineering on large die built on advanced VLSI processes. We illustrate a typical yield learning history, and describe the general yield, quality and manufacturability issues that arise as a function of process maturity. The yield limiting mechanisms change dramatically throughout the process life hence the appropriate yield improvement methodologies must also change. Second, we describe specific actions that VLSI design and product engineers can take to accelerate yield learning and raise the ultimate yield potential of large-die VLSI system products.

Throughout the paper, we will present the various yield modeling and analytical tools available, and comment on their efficacy both in accelerating yield learning and raising the final "steady state" yield of VLSI processes.

CHOICE OF TECHNOLOGY FOR A NEW MONOLITHIC VLSI SYSTEM PRODUCT

Table 1 shows the trade-offs involved in choosing a technology for a new monolithic VLSI system. It is important to note that this choice is not always available to the designer. Functional complexity of the product, clock rate, die size, power consumption, and reliability (as a function of power density) may dictate aggressive design rules which are only possible using process enhancements such as multi-layer interconnect, CMOS, and silicided poly gates ie. a new technology.

When more than one technology is feasible, the newer technology will be preferable in the long run, provided that the market presence is not delayed too long. Percentage of good die per wafer is a secondary consideration to net good die per wafer and overall die cost. Higher performance in general provides higher revenue. However, the pressures towards early market presence and rapid decrease of product cost will be elevated for a newer process.

A HISTORY OF YIELD IMPROVEMENT ON NEW PRODUCTS AND DEVELOPMENT PROCESSES

The process lifetime can be segmented into three phases: development, ramp, and volume production. Figure 1 depicts the learning curves for small memory and large logic die, and serves to illustrate the three phases of process history.

TABLE 1

OLD vs NEW TECHNOLOGY TRADE-OFFS
WHEN TWO TECHNOLOGY GENERATIONS ARE AVAILABLE

OLD Technology	NEW Technology
higher initial % good die	lower initial % good die
lower ultimate net good die	higher ultimate net good die
lower performance	higher performance
earlier market presence	better performance and cost

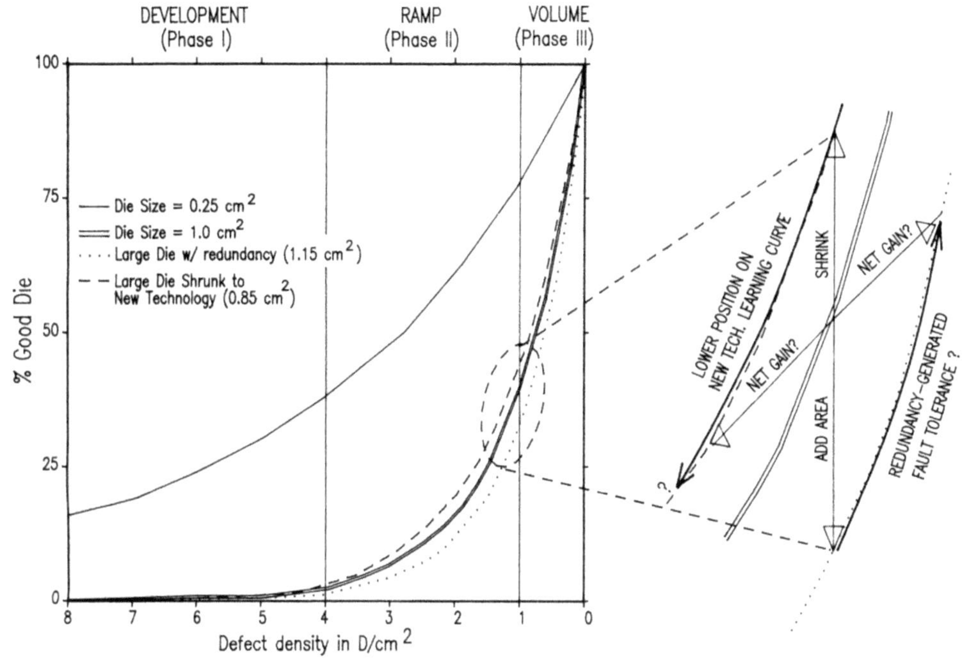

FIGURE 1. Net die vs. defect density for a learning vehicle and a large logic die using a simple Poisson model.

The development phase (Phase I) begins with technology definition and startup and it ends with the release of the first product for revenue shipments. In general the overall yield should be at or below about 4 defect/cm^2 by the end of this phase. If the first product is a very large die, it will yield near 0 at this defect level.

The ramp phase (Phase II) begins when a product may be shipped for revenue. At the end of the ramp the yield approaches an asymptote, the value of which will depend on the inherent robustness of the process and the fundamental cleanliness and quality of the factory and equipment. This slowing of the learning should occur at about 1 d/cm^2.

By the volume phase (Phase III), the process has reached maturity. Yield improvements are still sought and implemented but each change, no matter how small, must be evaluated in exhaustive detail using strict statistical process control (SPC).

The following sections describe in detail each of the three phases.

Phase I: DEVELOPMENT

1) Process Development Priorities. Initial design rules are defined with iterations as process and equipment specifications are tested against those rules. Major equipment choices are made and the targets and controls

are characterized for each process module. As soon as design rules are committed, extensive characterization of circuit elements follows. Device models for the circuit simulators are generated from these data. Achievement of first level (of many in a continuous improvement effort) reliability verification occurs. New product designs are verified in silicon and prototypes are generated for customers.

2) Product Design Priorities. During design rule evolution, chip plans are being formed and modified. Once design rules are frozen and circuit simulator models are available, standard cell drawing begins. Significant portions of the large-die product are drawn based on design rule and device feasibility with virtually no data on process yield. For these early design projects, design-for-manufacturability cannot be based on fault mechanisms or defect populations.

3) Yield Improvement Priorities. The yield engineer responds to one key mission during Phase I: which yield problems require redefinition or redirection of the process development effort? Open contacts for example can be caused by poor metal step coverage, which is a fundamental problem, or by particulates which may not be. Fundamental yield limiters must be understood and designed out of rather than into the process. Even though equipment modules and individual process steps are not fully developed, some understanding of future yield capability must be projected. An example of a future yield limitation is the use of composite sputtering targets like TiW which by their construction present a particle problem for the entire process lifespan.

FIGURE 2. Development vehicle with 9 test pattern sites and 3 EPROMs (tight, intermediate and loose cell designs).

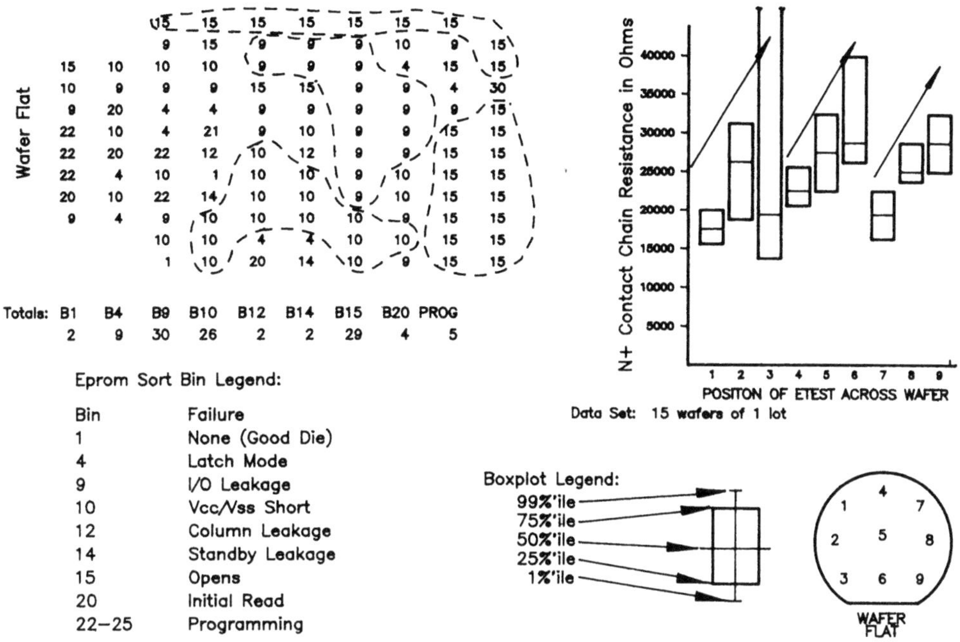

FIGURE 3. GFA defects (3a) and a Parametric defect (3b: N+ contact chain resistance variation across wafer).

4) Tools and Methodologies. During Phase I, wafer volumes are limited and yield vehicles must be small and easily analyzed for failures because the yields are low. Simply-arrayed logic and memory circuits best meet these requirements. Large, complex parametric test structures provide an additional tool. Some in-line process monitors are in use, and others are being defined and developed for use in the ramp phase. Figure 2 shows a non-volatile memory yield vehicle containing test patterns and three EPROMS of varying cell density.

Methodologies are required to distinguish fundamental from incidental yield limiters in the technology. We find that simplistic fault classification schemes are most appropriate. Defects can be classified as random or systematic. (We make no distinction here between defects and faults since their causes must be eradicated in any case.) Random defects are mostly due to particles in deposited films before, during, and after etching. Systematic defects can be subdivided into GFAs (Gross Functional Area defects) and parametric defects. In a GFA, yield loss occurs in a portion of a wafer in which the area shape is a direct indicator of the loss mechanism (see Fig. 3a). A Parametric defect is continuous, affecting the whole wafer although the effect may not be severe enough to cause yield loss throughout (see Fig. 3b). These classifications allow rapid identification of defect mechanisms.

Since wafer volumes are very low, and because the process may change from batch to batch (by design), statistical and numerical tools are of little value. Decisions on the failure mechanisms, whether Parametric, Random, or GFA must be made based on small data volumes and finely detailed failure analysis. The analyst must be expert in the interaction among circuits, devices and processes on the circuits being built. Exhaustive failure analysis is done on multiple wafers of almost every batch processed. Again, the priority is not necessarily to find the largest yield issue(s), but rather the most fundamental ones that impact technology direction decisions.

Phase II: RAMP

1) Process Development Priorities. Yields, volumes, number of products in production, and manufacturing costs change rapidly during the ramp. A new factory may start the process to provide additional capacity in parallel with the manufacturing arm of the development site (at Intel, processes are developed in factories, not in laboratories).

Fundamental changes in the process are difficult to justify and implement. Even minor changes that improve yield must be carefully evaluated since customer re-qualification might be required. As yield learning progresses, improvements become harder to find and their net effect diminishes. Ramp risk models are employed to minimize the risk of missing an improvement[1].

2) Product Design Priorities. Product proliferation also occurs early in this phase. Verified product designs are intensively characterized and samples are sent out as engineering prototypes. Some units may be shipped for revenue early in Phase II. Later in Phase II, the race begins between declining costs in the factory and declining sales prices in the field. This is the major motivator for aggressive yield ramp programs.

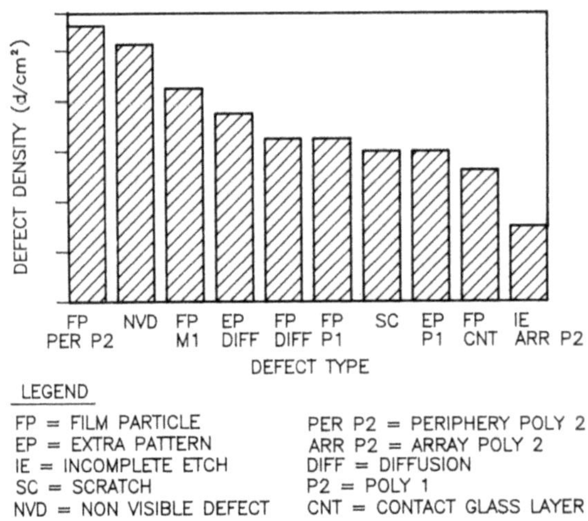

FIGURE 4. Violation of the Pareto Principle for an EPROM technology: the top ten defects observed for 15 consecutive weeks of production during phase II.

3) Yield Improvement Priorities. Concurrency in yield improvement programs is absolutely required. The often quoted Pareto Principle which states that most of the problem is concentrated in only one or two mechanisms fails during the ramp (see Fig. 4). Pareto charts tend to categorize defect types by process layer. Experience teaches that each defect category usually contains several defect sources, each requiring its own solution. The relative height of each category is not indicative of the relative contribution of each defect source. Eliminating a few big categories does not comprise the overwhelming majority of yield improvements during phase II. Further, we cannot afford to make yield improvements serially, nor can we simply solve only the largest problems. Lesser more subtle problems require more time and energy to understand. Since time is the primary enemy, the longer term yield programs must progress in parallel with the search for "low-hanging fruit."

New variables in the yield equation are being added, namely additional equipment and factories, that offer new opportunities for head-to-head comparisons. Careful designing of yield experiments and capitalizing on the growing volume of available yield data are significant challenges for the yield engineer.

4) Tools and Methodologies. The tools and methodologies must provide rapid learning iterations with very high concurrency. Hierarchically designed yield tool sets refine the whole process successively into individual yield-limiting steps (see Fig.5). Such a tool set allows for fast iteration on small portions of the process while elucidating the relationships between individual step defect densities and full process yields.

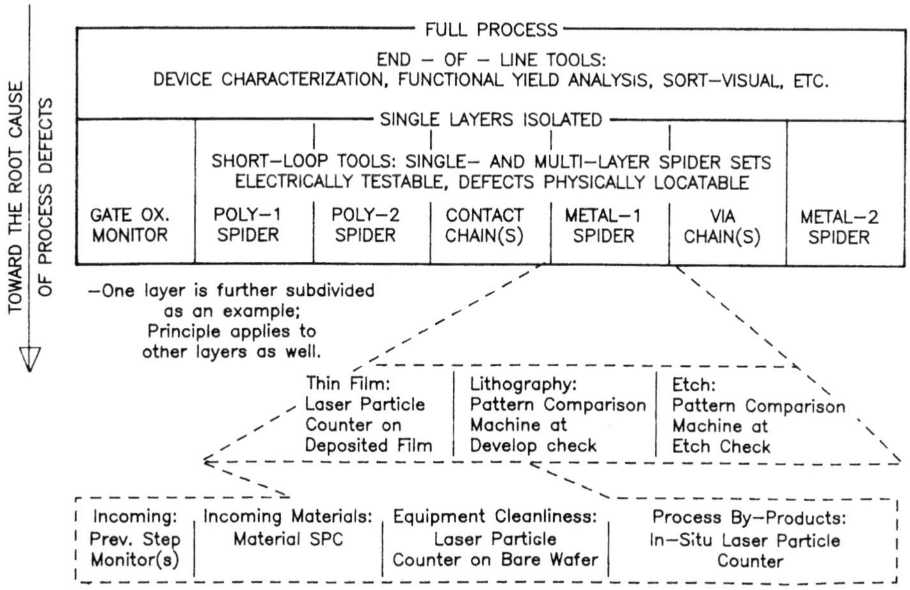

FIGURE 5. Hierarchical yield improvement methodology.[2]

The primary element of the tool set is the "spider mask" (composed of interdigitated serpentines). As a single or few-layer defect structure it requires only a subset of the process to fabricate. Spiders are capable of very high resolution defect density measurements on (or between) conducting layers. Quantitative correlation to full-process defects is accurate as the spider conforms to the layout sensitivities of the product. These structures are exhaustively fault-analyzed electrically, visually, and by other techniques. They also are used to calibrate the defect-monitoring equipment that are not electrically testable (eg. particle counters and pattern comparison machines).

As these monitors are implemented, and as manufacturing ramps quickly, we also must provide for data collection and databasing tools and methodologies which will be essential for statistical understanding of the process at higher volumes. The expense of collecting and archiving data must be balanced against that of reproducing them later by experimentation.

A learning iteration on a spider or step monitor takes a few hours or days, whereas full process iterations require several weeks. Large test pattern vehicles like those in Fig. 2 are almost useless in responding quickly to a rapidly changing process environment. Clearly, product redesign or process modeling efforts to capitalize on yield learning are ill-advised when the yield picture is changing this rapidly. A redesign cycle for a large die may span all of Phase II.

Phase III: MATURITY

1) Process Development Priorities. In Phase III, avoiding yield variations is at least as important as raising yield. Commitments to customers have been made so predictability is a high priority. For a large die, very small movements in net defect density will result in huge swings in net die yield per wafer. In the case of a 1 cm^2 die, yielding at about 0.8 d/cm^2, an excursion of 0.5 d/cm^2 will be equivalent to eliminating 40% of the factory's capacity.

The defect reduction rate slows as the overall defect density reaches a low level ($<1d/cm^2$). Yield improvements are possible but more difficult to obtain. Each change must present absolutely minimal risk necessitating a very different risk model than that used during the ramp[1].

2) Product Design Priorities. Products are shipping in volume and profit margins have diminished. Optimal process targets for speed, power, and yield are well understood. By this time defect populations and size distributions are well known and fairly stable. A redesign of the chip for higher yield is possible but the redesign must now be weighed against the alternative of shrinking the chip onto a newer technology and against the implications for customer requalifications and new characterization.

3) Yield Improvement Priorities, Tools, and Methodologies. The yield priorities follow the process priorities: maximum process control, absolute avoidance of process excursions, and immediate response to any violations of control algorithms. In this yield regime we begin to understand the limitations of the factory itself including the level of automation, the quality of the microcontamination control, and the discipline of the manufacturing sector. Since extant yield gains are small, numerical and statistical techniques are mandatory in identifying and quantifying them.

Factory capacity is optimized for revenue shipments to customers. Consumption of silicon for experimental (non-revenue) purposes requires greater justification. Therefore we must capitalize on the data collection from the manufacturing line. Our knowledge of the process and its further potential may be primarily limited by the database tools and methods.

GENERALIZATIONS

From the above process history we discuss the limitations of the large die yield problem and suggest a few improved approaches.

1) Fundamental Yield Limitations of Large-Die VLSI Systems

Fundamental yield limitations and the yield itself of large die change with time. During Phase I, yield knowledge is restricted because systematic yield issues must be solved first. Random defects are only important if they are a result of fundamental process or equipment limitations that impact the ultimate yield potential.

During Phase II, the yield is rate-limited. We expect to move up faster on the yield learning curve with each new generation and in fact we have done so (see Fig. 6). The slopes of these curves have increased during the same time ranges even with added process complexity: more mask layers and new process modules. Remaining defects are increasingly random rather than systematic in nature. Under such fast yield ramp conditions, specific defect information used for yield modeling becomes obsolete before products can be redesigned.

During Phase III, the total yield, defined as units shipped per wafer start, is limited as much by instabilities as by incremental yield improvement. Therefore, all proposed changes must present a near zero risk of unforeseen downside; this will limit changes dramatically. Process control improvement has priority over yield enhancement.

2) The Transition from Small Die to Large Die

To get the technology from near zero to some moderate yield level requires a small die vehicle. Large die have no chance of yielding until we advance up the learning curve to some intermediate point. In some cases, the yield sensitivities between the large logic die and the small learning vehicle (a memory) are quite different. Memory chips tend to be

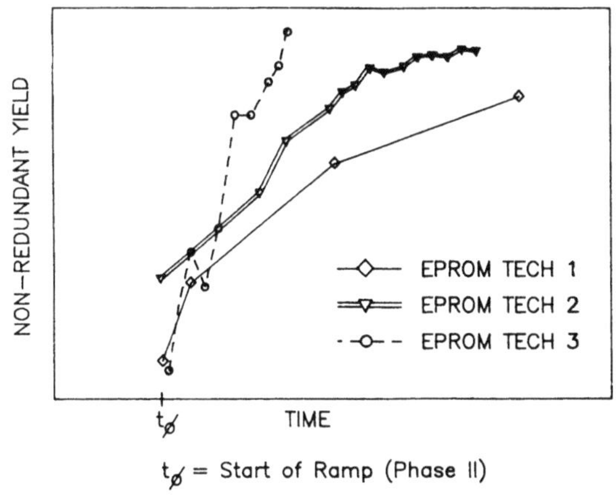

FIGURE 6. Successive EPROM technology learning curves.

packing density limited by poly-to-poly space, poly-to-gate space (for EPROMs), by well-to-diffusion spaces (CMOS SRAMs), and so on. Meanwhile, the large VLSI logic devices tend to have different layout limitations; frequently they are interconnect intensive.

The disadvantage is that some yield learning on the small die was not transferrable to the large die. Nonetheless, there are distinct advantages to having differential yield sensitivities among the vehicles. EPROMs can detect a much lower level of gate oxide defects and SRAMs a lower metal interconnect defect density compared to large random logic systems.

To capitalize on these differences we have begun to use critical area calculations[3-5]. The specific layout of the large die is not known during the design of the primitive yield vehicles. However, extrapolations from layout data of large die on previous generations can be used adjusting for predictions about the new product from the system designers. The result is that in some cases, the learning vehicle layout can be modified to make it better correlate with the yield sensitivities of the large dice to follow.

The critical area technique also can be applied rigorously to product sets in production to verify whether yield differences are due to critical areas alone or to product-specific parametric sensitivities. DRC (Design Rule Check) flows can be adjusted to calculate the precise area averaged over the entire product for any given set of layer manipulations. N--channel gate oxide area, Metal 1 bridging sensitivity as a function of interconnect space, and contact to gate total area are examples of "critical areas" that can be determined for each chip. Since these calculations are performed on hierarchical CAD layout data, these areas can be calculated for each major functional circuit block and then summed at the end. This gives the yield analyst some quantitative estimate of functional block sensitivity to various random defects found in-line with particle detection methods. As mentioned above, the use of these analytical techniques is only for diagnostic purposes since altering design rules is not possible by the time the necessary data are available to prove any sensitivities.

A great advantage in yield learning would be gained if fault diagnosis of the large-die product were sufficiently powerful to create its own defect sensitivity composite. Only a minimal expense in added Silicon area could be justified to achieve this. One example could be the Built-in Current measuring block[6] which can be inserted into logic circuits to monitor excessive current flow in CMOS circuitry. The corresponding memory circuit functions are the leaky individual and multiple column test modes. These test circuits can measure process marginalities in contact integrity, isolation, punch-through, and contact-to-gate that might be missed by logic functional tests. These failures could represent potential reliability problems. If we could approach on large die, the defect analytical capability of the many test modes used in small memories, a great advantage in yield improvement would be gained.

Our default solution is to continue to use rasterable dice (and reverse engineering) together with large-area test patterns. These structures reside on engineering plate sets that also contain a large die product. This type of plate set allows direct yield comparison studies and analyses of process changes and variations. However, it is an expensive alternative to those suggested above and cannot be used to respond rapidly to changing process conditions. Since failure analysis is extremely difficult for the large-die product, we must rely on smaller dice for end-of-line defect data.

3) Redundancy and design-for yield require some defect characterization

For memory products, redundancy implementation is cost-effective at lower yields. However, the redundancy algorithm must take advantage of the actual defect types and distributions in the technology for maximum benefit[7]. Unfortunately, from the process history above, such knowledge is unavailable when the memory design decisions are made. The same dilemma exists in some cases for VLSI logic designers. The only choices are: to guess, to wait and redesign, or to do both. Waiting for the knowledge means missing the most benefit from redundancy. During the ramp, hopefully the need for redundancy becomes obsolete because the limiting defects are minimized or eliminated.

Given the inexact knowledge of the yield limitations, design-for-yield may follow two paths. In the first path the inherent yield limitations are predicted from the process modules selected during Phase I. Some process and equipment choices are made with the knowledge that defect problems exist but viable alternatives are unavailable eg. reactive sputter processes, resist-on-resist processes, and atmospheric glass reactors. An example of a process flow having built-in defects is one allowing multiple transistor device types; this type of flow can place extreme demands on gate oxide quality. These process architecture and module choices are known early in development, possibly allowing a standard cell designer to modify layouts to avoid these more sensitive process features.

The second path is to predict, wherever possible, the ultimate limitations of yield based on the specific factories where the process will be committed. These are largely environmental issues such as water and airborne particle filtration systems, equipment layout, etc. Some equipment from a previous generation will be used on the new process, and its defect behavior is known (eg. poly Silicon reactors). Thus, the limitations of a factory demonstrated over multiple generations of technologies, may be leveraged by the cell designer.

4) At Phase III, we know enough, but would we redesign?

By maturity there is sufficient knowledge of fundamental limitations due to the product, process, and the factory. However, the decision to redesign on the existing technology must be weighed against the cost of redesign, and against the alternative to shrink to a newer technology. A redesign for yield insensitivity may be feasible based on well characterized defect populations, but will only be profitable for a short while compared to the longer term benefits of shrinking the die to a newer technology. The dotted lines on Figure 1's learning curve for a large die represent the +/- 15% change in die area resulting from redundancy and conversion to a new technology respectively. The resulting yield increases are balanced by concomitant negative effects: redundancy by larger defect capture area and fewer die placements per wafer and the new technology by the shift towards the lower part of the learning curve.

CONCLUSIONS

We have described our strategies for rapid yield improvement of large VLSI die within the framework of our three phases of process history: development, ramp, and maturity. Four main generalizations result from this analysis. In the first, we find that the yield improvement priorities depend on the phase: in development we must ensure that the design of the technology will not limit the future yield potential, in the ramp we must be working on yield problems in parallel, and by maturity we must ensure process stability.

The second is that small (memory) die are necessary vehicles to bootstrap process learning up the yield curve so that large die can yield. Failure analysis on large die is extremely difficult. Rasterable memories coupled with reverse engineering are required to determine defect sensitivities. Critical area analysis can help quantitatively relate these small die yield sensitivities to large die. On-chip circuits that measure excessive leakage currents represent one step toward on-chip defect predictors.

Thirdly, redundancy and design-for-yield schemes can only utilize previous generations' process defect data as zero order input. New process modules necessary to the new technology can completely invalidate defect data from previous technologies. Shrink strategies are based on performance and cost (ie. die size) rather than previous generations' defects. During development when many large die designs are committed, no defect data are available. When data are available during the ramp they are too rapidly varying and caused by too many defect mechanisms acting in parallel to use explicitly in the design.

Finally, by the time that defect populations are both stable and well characterized it is usually too late to be cost competitive to redesign on the current technology.

REFERENCES

1. Diamond, W.J., "Practical Experiment Designs for engineers and scientists", VNR, 1981.

2. Rajkanan, K. and Boerman, B., "Application of Various Yield Improvement Tools and Techniques as a Function of Yield Maturity", ITM (Intel Technical Memo).

3. Gandemer, S., Tremintin, B.C., and Charlot, J., "Critical area and Critical levels calculation in I.C. Yield Modeling", IEEE Trans ED-35 (2) 158 - 166 (1988).

4. Maly, W. "Modeling of Lithography Related Yield Losses for CAD of VLSI circuits", IEEE CAD-4 (3) 166 - 177 (1985).

5. See references in D. Walkers book "Yield Simulation for Integrated Circuits" Kluwer Acad. Pub. 1987.

6. Maly, W. and Nigh, P., "Built-in Current Testing - Feasibility Study", submitted to the International Conference of CAD.

7. Engelbrecht, L. "Redundancy analysis on a 1u 256K Eprom", ITM.

YIELD PROJECTION BASED ON ELECTRICAL FAULT DISTRIBUTION AND CRITICAL STRUCTURE ANALYSIS

P. Schvan, D.Y. Montuno*, and **R. Hadaway**

Northern Telecom Electronics
***Bell Northern Research**
Ottawa, Ontario K1Y 4H7, Canada

INTRODUCTION

With the growing complexity of semiconductor processing technologies, the ability to predict the manufacturing yield of an integrated circuit at the design stage has become increasingly important. This allows the designer to evaluate the feasibility of a particular design from an economical point of view, and to optimize the partitioning of a system into individual ICs. The yield projection methodology described here has the added capability of distinguishing the contribution of each failure mode to the total yield loss. This is a valuable feature for fault-tolerant circuit design and the development of efficient high fault coverage test programs.

Earlier papers on yield modeling concentrated on determining the defect distribution and its effect on yield[1,2]. More recently increasing attention is paid to the sensitivity of the circuit to defects, through the analysis of critical structures[3-5].

In this paper we describe a yield projection methodology that is based on process dependent electrical fault distributions and circuit dependent critical structure statistics.

DEFECTS AND ELECTRICAL FAULT

The primary cause of yield loss are defects generated during the fabrication process. Defects can be defined as any kind of disturbance of, or deviation from the ideal device structures. Most of these defects are small compared to the die size and affect only a localized area of the circuit. They are usually referred to as random defects since their probability of occurrence increases with the die size.

When a larger area of the wafer shows faulty behavior, it is usually caused by excessive variations in the process parameters or other gross errors in wafer handling. The frequency of this type of failure does not depend on die size and is expected to contribute to yield loss to a much lesser extent. Therefore, this type of failure will not be considered here.

Most yield models concentrate on describing the random defect limited yield. When a defect distribution is used as input data, it is necessary

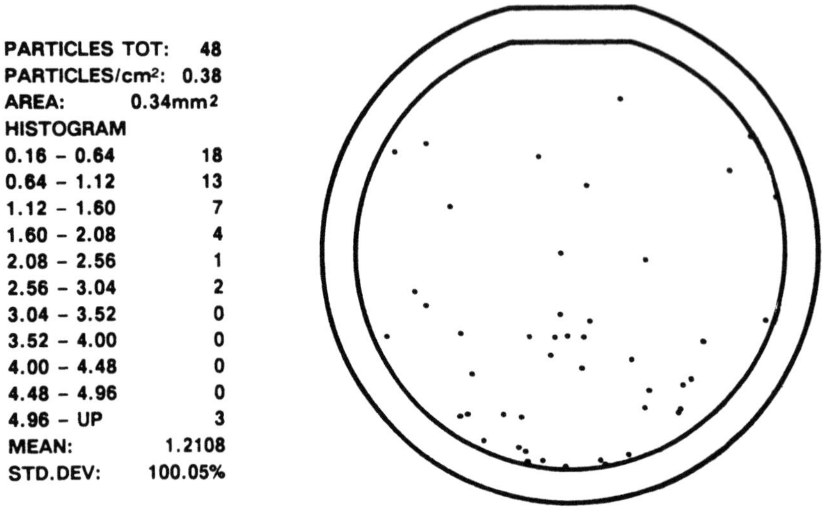

Figure 1. Computer enhanced image of defect distribution based on optical defect monitoring.

to observe the defects directly[6]. The most widely practiced technique for direct monitoring of defects within the manufacturing process is by the use of product representative unpatterned wafers processed through identical equipment and procedures. Pattered wafer optical defect monitoring is also available but usually the sensitivity is significantly less than required for current technologies. Defects manifest themselves optically by scattering light from an otherwise specularly polished surface. Detection of this scattered light leads to the spatial distribution of defects on the

Table 1 Failure Modes and Associated Critical Structures

FAILURE MODES	CRITICAL STRUCTURES
Intralayer short - metal2 - metal1 - polysilicon	closely spaced lines, bus system.
Interlayer short - metal1 to metal2 - polysilicon to metal1 - gate to substrate - diffusion to substrate - contact to substrate	overlapping layers at different potential MOS gate, diffusion contacts, crossing buses, diffusion contact, etc.
Discontinuity - metal1 - metal2 - polysilicon - diffusion	narrow conductive lines bus etc.
Missing contacts - via - diffusion contact - polysilicon contact	contact chains

wafer, as well as to defect size categorization. A typical defectivity map generated by this technique is shown in Figure 1.

Defects introduced during the manufacturing process may lead to electrical faults either directly or through interfering with subsequent process steps. Whether an electrical failure occurs depends on the physical and chemical properties of the defect and its exact location in the circuit. This uncertainty and potential complexity of failure generating mechanisms make it difficult to determine the actual yield loss based on optically characterized defects alone.

To circumvent this problem the yield model presented here uses electrical fault distribution data. Electrical faults can be monitored using a carefully designed set of test structures which include all the critical circuit structures where faults can occur.

A key issue in developing an accurate yield model is the determination of important failure modes and the design of the test structures. A series of fault monitoring experiments with a wide range of test structures were done to select a minimum set of structures that represents all the important structures where faults occur for our CMOS technology. Several such structures have been discussed in various publications[7,8]. A list of the failure modes and associated critical structures are shown in Table 1.

YIELD MODEL

We adopted the yield formula

$$Y(f) = (1 + S(f)*p(f)/\alpha(f))^{-\alpha(f)} \qquad (1)$$

derived from the negative binomial distribution to model both the observed test structure yield and the partial yield of an arbitrary circuit[9]. Here f describes the failure mode for which the yield loss is calculated. S(f) is the ratio between the size of the critical structures in the arbitrary circuit and in the smallest test circuit, which can be expressed in area, length, or number of polygons, according to the type of the structure. The two other parameters are related to the distribution of the faults, p(f) is the probability that a fault occurs in the associated test structure and $\alpha(f)$ is the clustering parameter. The product S(f)*p(f) is the average number of faults. The fault probability, p(f), in general, is a function of both the defect density that can affect the given structure and the probability that a defect will cause an electrical failure. The total yield, Y_{Tot}, is

$$Y_{Tot} = \prod_{\text{for all faults}} Y(f)$$

Although there are many theoretical arguments supporting this model[9], its validity was tested against actually occurring fault distributions. Here we show results for three important failure modes only; these are intralayer and interlayer shorts involving the first and second metal layers. The test structure used was the well known interdigitated serpentine and finger structure, shown in Figure 2. In this experiment the entire wafer was covered with this test structure to derive accurate information on the distribution of faults. The test structure was partitioned so that faults could be located within a small area. First the yield versus area curves were generated for all the failure modes monitored, then the fault probability and clustering parameters were determined by curve fitting the yield function (1). The measured yield curve for the three failure modes combined, was also determined. The average measured yield curves for 50

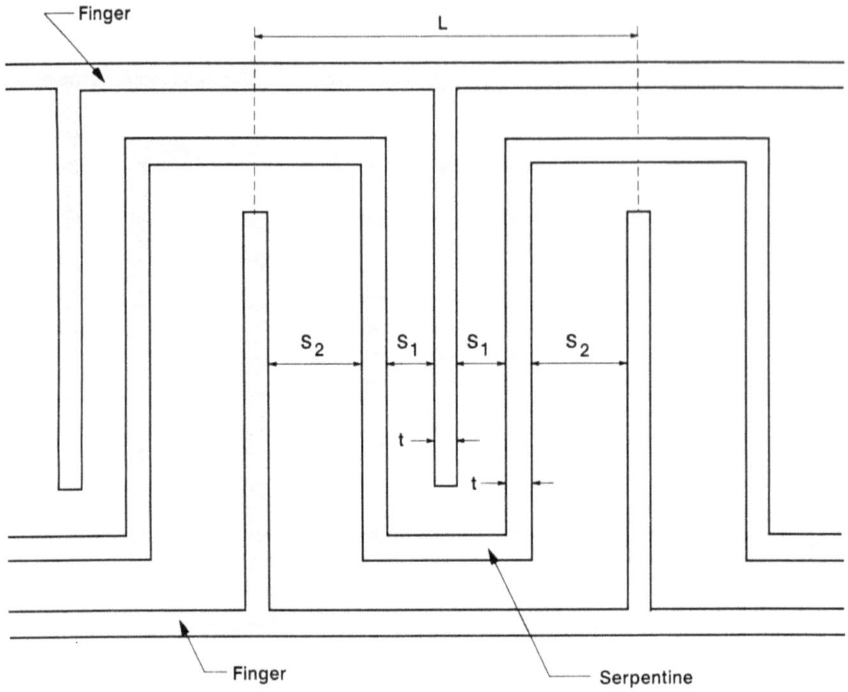

Figure 2. Interdigitated serpentine and finger test structure for the monitoring of intrelayer shorts.

wafers and the theoretical yield curves obtained by curve fitting including the total yield are shown in Figure 3.

There are two important observations that can be made based on these results. Firstly, the actual fault distribution can be adequately modeled by the yield formula (1). Secondly, the combined yield can be obtained by multiplying the individual yields for the various failure modes. The error was less than 4%, in our investigation, indicating only a small degree of correlation between the different types of faults. This type of correlation can be expected when a certain type of defect can result in two different faults. In this case a closer investigation of the fault distributions showed that a defect in the intermetal insulator can cause shorts both between two metal layers and within one layer.

Similar investigations have been carried out using other test circuits to determine the fault probabilities and clustering parameters for all the important failure modes.

CRITICAL STRUCTURE ANALYSIS

In order to estimate the fabrication yield of an arbitrary circuit the probability that a fault occurs in this circuit must be expressed with the probabilities of fault occurring in the test circuits containing the critical structures. For a given failure mode the partial yield loss depends on S(f) which is the frequency or size of the associated critical structure in the circuit.

A critical structure extraction program was developed to determine the size of critical structures in a circuit layout. For each failure mode first a sequence of Boolean operations are performed between the primary layers (such as metal, polysilicon, device well, contact etc.) to generate a new layer that contains all the critical structures associated with the given failure mode. Once the critical layer is generated, the size of the critical structure is determined in terms of area, length or the number of polygons depending on the failure mode.

Interlayer Short Type Failure

For most of the critical structures this extraction process is relatively straightforward. For interlayer shorts between "layer_a" and "layer_b", the critical layer "crit_layer" is generated by a simple AND operation.

crit_layer = layer_a AND layer_b

It is assumed that interlayer shorts are caused by pinhole type defects which are small enough so that the probability of a fault is proportional to the overlap area between the two layers. To account for the possible dependence on the aspect ratio of these overlap polygons the test structure can be designed to represent typical overlap geometries found in circuits. It is also possible to calculate the perimeter of the polygons.

The critical structure related to shorts between metal1 and metal2, polysilicon and metal1, gate and substrate, and diffusion and substrate belong to this category. The size of the critical structure is determined by the area of all the polygons on layer "crit_layer". The size of the critical structure

S(interlayer_short) = Area(crit_layer)/Area(crit_test)

where Area() represent the area function for the given layer, and "crit_test" is the critical layer on the appropriate test structure.

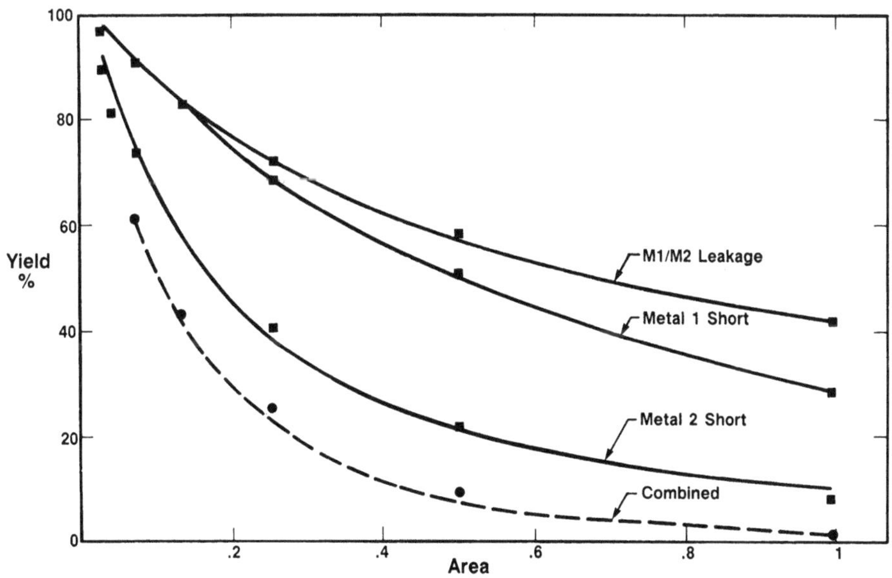

Figure 3. Experimental (squares and dots) and theoretical (solid and dashed lines) yield curves for selected failure modes.

Contacts Related Failures

For contacts related faults the extraction algorithm depends on the failure mode. For missing contact type fault the probability is assumed to depend on the number of contacts. However contacts that connect the same pair of layers should not be included more than once. This can be achieved by generating the following layer.

independent_cont = contact OVERSIZED BY d/2

Where d is the minimum separation between contacts, and "contact" denote the layer of contacts either to the diffusion or to the polysilicon, or between metal1 and metal2. The size of the critical structure

S(independent_cont) = Num(independent_cont)/Num(crit_test)

Here the function, Num(), gives the number of polygons on the layer.

Short can also occur through a diffusion contact to the well. Since that will only lead to electrical fault if the diffusion is biased, contacts to biased diffusions must be separated using the following algorithm

```
ptype       = contact  INSIDE  pdope
ntype       = contact  INSIDE  ndope
biased_cont = ntype    AND     ptype
```

Here "pdope" and "ndope" are the p and n type doping masks respectively. The size of the critical structure related to short across a contact is determined by the number of contacts on layer "biased_cont".

Interlayer Short Type Failure

The algorithm to determine the size of the critical structure related to intralayer shorts is somewhat more complex. The critical area for intralayer shorts is defined as the area of the chip where, if the center of a defect falls, it can short two conductive lines. This area depends on the separation between the conductive lines and on the effective size of the defect. For a simple repetitive layout the critical area or the probability of short can be expressed in a closed form[4] using the assumption that the effective area of the defect is a circle. For our test structure shown in Figure 4, we can derive the following expression for the probability, P, that a defect of diameter D, will cause a fatal fault

```
P(s1,L,t,D) = 0            for          D < s1
            = 2*(D-s1)/L   for   s1       < D < 2*s1+t
            = (D+t)/L      for   2*s1+t   < D < L-t
            = 1            for   L-t      < D
```

where L=2*s1 + 2*s2 + 4*t, using the notation in Figure 2.

This probability function, which characterizes the sensitivity of the structure to defects, is different from the fault probability p(f), which is also a function of the defect density.

In order to compare the sensitivity of an arbitrary circuit layout to the sensitivity of the test structure, a similar probability function must be calculated for the arbitrary layout. The following iterative approximation algorithm was used to determine the critical area related to intralayer shorts for an arbitrary layout.

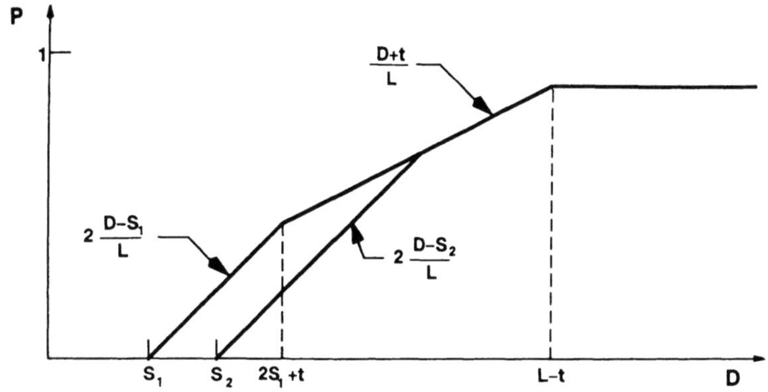

Figure 4. Probability function of fatal defect for the intralayer short monitoring test structure.

First a channel layer, "channel", is extracted where a defect of diameter, D, can cause short in the layer "conductor".

```
channel(0) = conductor    OVERSIZED  BY D/2
channel(0) = channel(0)   UNDERSIZED BY D/2
channel(0) = channel(0)   AND NOT    conductor
```

This channel is shown in Figure 5(a). Since the critical area depends on the width of the channel it must be generated for each "section" of the channel separately. Let w(i) i=1,2 .. n denote a series of spacings, where w(i)<w(j) for i<j, starting from w(1)=minimum spacing design rule for this conductive layer to w(n) < D defect size, and w(0)=w(1).

```
     crit_layer = EMPTY

FOR i=1,n

     channel(i)  = channel(i-1) UNDERSIZED BY w(i)/2
     channel(i)  = channel(i)   OVERSIZED  BY w(i)/2
     section     = channel(i-1) AND NOT channel(i)

     margin = (D - w(i)-w(i-1))/2

        IF margin > 0 THEN
        crit_section = section OVERSIZED BY margin
        ELSE
        crit_section = section UNDERSIZED BY ABS(margin)
        END IF

     crit_layer = crit_layer OR crit_section

END FOR
```

For each w(i) a "section" of the channel is extracted which is narrower than w(i) but wider then w(i-1), shown in Figure 5(b). Following that the "crit_section" is generated, shown in Figure 5(c), assuming that the exact channel width is (w(i)+w(i-1))/2. The layer "crit_layer" contains all the critical structures for the given defect size. Since the exact incremental increases of the channel width is not known an error is introduced into the critical area calculation. This error is a function of the w(i)-w(i-1) steps. By increasing n, thus reducing the step size, the error can be reduced below any given value.

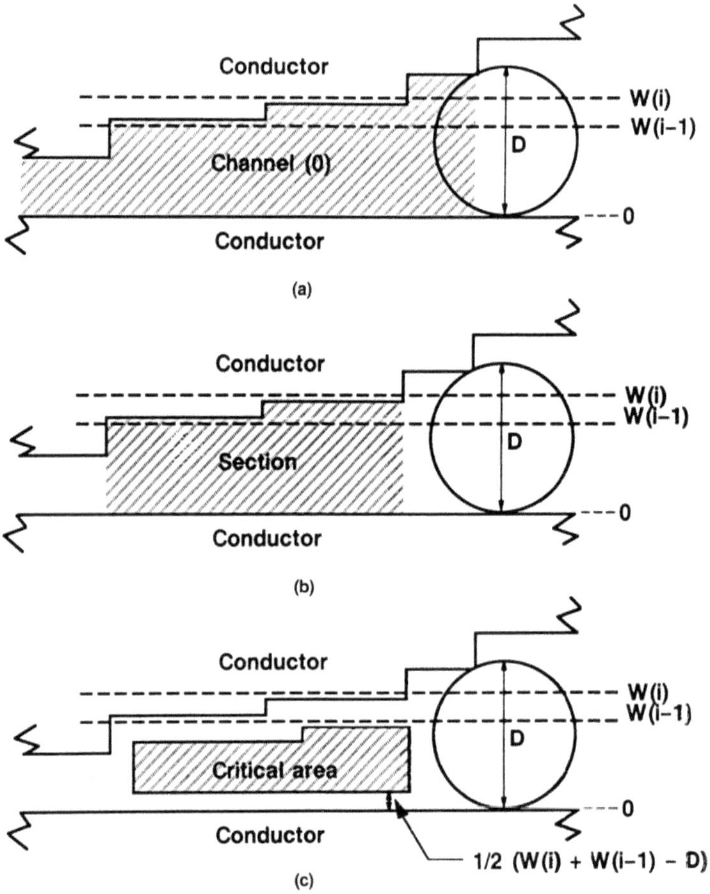

Figure 5. Schematic representation of the critical area extraction algorithm for intralayer shorts type failure.

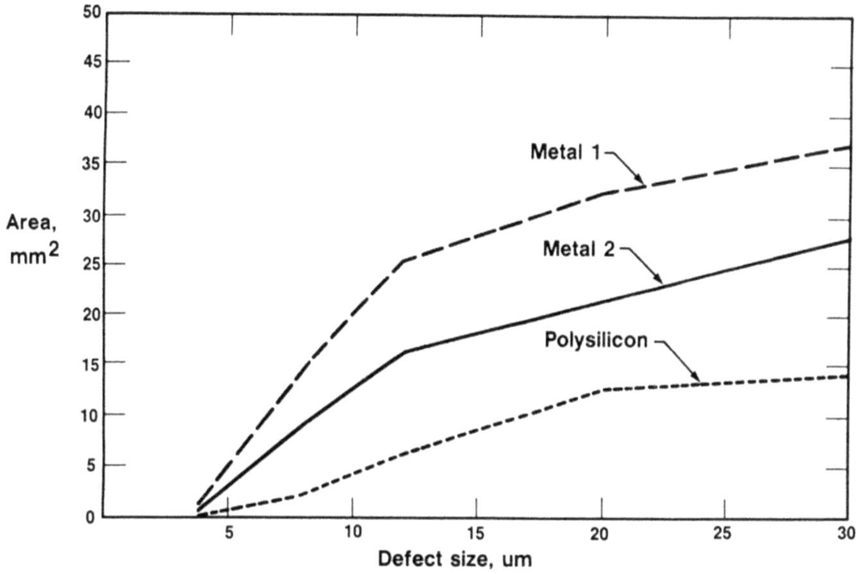

Figure 6. Calculated critical area as a function of defect size of the metal1, metal2 and polysilicon layers of a mixed type integrated circuit.

The critical area versus defect size curve that represents the sensitivity of a given layout to defects is different for the various classes or circuits, such as random logic, memory, PLA etc. Figure 6 shows the critical area curves of metal2, metal1 and polysilicon layers for a circuit which contains both random logic and memories.

The probability that a defect causes fault is related to the critical area, Area(crit_layer), which is a function of the defect size. It is possible to approximate this probability function with a linear combination

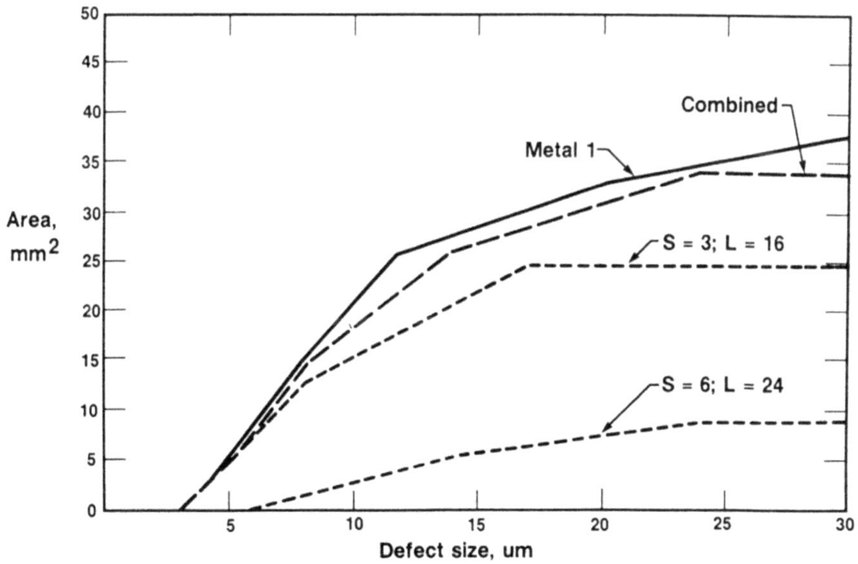

Figure 7. Approximation of the critical area function with the linear combination of test structure short probability functions.

of probability curves, shown in Figure 4, for m test structures with appropriately selected spacings. An example is shown in Figure 7, which uses two test structures to approximate the critical area curve of the metall layer shown in Figure 6. Consequently the probability of fault for an arbitrary circuit can be expressed in the following form;

$$P(D) = \text{Area(crit\_layer,D)/Die\_area} = \sum_{k=1}^{m} C_k{*}P\ (s_k, L_k, D)$$

Where $P(s_k, L_k, D)$ denotes to the probability function for the k-th test structure, with geometries s_k, and L_k, and c_k is the appropriate scaling factor. Die_area is the area of the arbitrary circuit. This implies that from the interlayer short related yield loss point of view, the circuit layout can be substituted by a mixture of regular test structures each with an area of c_k*Die_area. This provides the S_k(intralayer_short)=C_k*Die_area size parameters for the yield formula. Since the yield Y_k for these regular structures can be calculated based on experimentally determined fault distribution parameters p_k, and α_k, the partial yield of the circuit for intralayer short type failures can be calculated in the following way:

$$Y(\text{intralayer\_short}) = \prod_{k=1}^{m} Y_k(\text{intralayer\_short}, S_k)$$

It is important to note that only the actually measured test structure yield data is used for the calculation of the partial yield without any assumption regarding the defect size distribution.

Similar methods can be used for calculating the partial yield for discontinuity types of failure, where the critical area is also a function of the defect size distribution. However, since this type of failures were found to be less important than intralayer shorts this algorithm will not be discussed.

CONCLUSION

We have shown how fault distribution data obtained from electrical measurements on simple test structures can be used to calculate the yield of an arbitrary circuit. An important result of this work is the critical structure extraction algorithm which allows the quantitative characterization of the sensitivity of an arbitrary layout to defects for the various failure modes. The accuracy of this method depends greatly on the selection of the failure modes which are to be included in the model and the design of the test structures. After the technology dependent fault distribution parameters are determined for a mature technology, the projected yield has been found to show good agreement with the measured yield.

The yield model has been used to study the ratio of different type of faults, which is important for the development of more efficient test algorithms. The ability to predict the yield of various sub-circuits can be exploited for fault-tolerant circuit design. It has been shown[10] how the yield of a fault-tolerant circuit can be predicted from the yield of the various parts of the circuit.

Acknowledgment

The authors wish to thank J. Powell for the valuable discussions on optical defect monitoring techniques.

References

1. Seeds, R.B., "Yield, Economic and Logistical Models for Complex Digital Arrays", IEEE Int. Conv. Rec. 60 (1967).

2. Stapper, C.M., "LSI Yield Modelling and Process Monitoring", IBM J. Res. Development 228 (1976).

3. Stapper, C.M., "Modelling of Integrated Circuit Defect Sensitivities", IBM J. Res. Development 461 (1984).

4. Ferris-Prabhu, A. V., "Role of Defect Size Distribution in Yield Modeling", IEEE Trans. Electron Dev. 32, 1727 (1985).

5. Grandemer, S., et al., "Critical Area and Critical Levels Calculation in I.C. Yield Modeling", IEEE Trans. Electron Devices 35, 158 (1988).

6. Koizumi, M., et al., "Automated Visual Inspection Technique for the Detection of Small Contaminants in LSI Manufacturing Process", Hitachi Review 35, 355 (1986).

7. Weber, C., "Standard Defect Monitor" IEEE Proc. on Microelectronic Test Structures 1, 115 (1988).

8. Newhart, R.E., et al., "A Defect Learning Vehicle for Submicron Technologies" Proc. on Microelectronic Test Structures 1, 147 (1988).

9. Strapper, C.H., et al., "Integrated Circuit Yield Statistics" Proc. of IEEE, 71, 453 (1983).

10. Schvan, P., et al., "Defectivity and Yield Analysis for VLSI and WSI" Proc. on IEEE International Conf. on Computer Design 89 (1986).

YIELD MODEL FOR YIELD PROJECTION FROM TEST SITE

Steven Magdo

IBM General Technology Division
East Fishkill, NY 12533

INTRODUCTION

Defect identification, monitoring and yield play an important role in VLSI circuit manufacturing.

It is very important to project both limited yields and final test yield for the product in order to rank the failure modes and assure that no failure mode is undetected. For this purpose, test sites provide a large automatically generated data base which can be related to the product via product to test site critical area (A_{cr}) ratios. The data has to be generated right after inline testing and fed back to the processing and product design areas in order to accelerate defect learning.

TEST SITE DESCRIPTION

It is important that the test structures should minimize unwanted failure mode interferences and maintain product attributes with respect to the defects to be monitored.

Fig. 1 shows a product cell edited for high density transistor chains. The transistor from the original product cell is taken and "brick walled" into a dense array. The emitters and bases are wired into parallel chains. The collector contacts are not wired, instead a substrate contact is used for contacting collectors. This eliminates metal short interferences with the emitter-collector leakage test. Since all the identical contacts in the transistor chain are lined up, the wiring-metal design rules can be relaxed. This minimizes metal short and open interferences with emitter-base leakage test. Schottky diode (SBD) chains and resistor chains (not shown) are formed in a similar way. These densified chains have about four times higher critical area than chains formed with the original cells.

It is important that everything which influences the failure mechanism to be monitored, such as emitter shape, emitter to isolation separation, etc., remain the same during editing.

Fig. 2 shows the metal maze test structure for monitoring metal opens and metal shorts. The mazes are formed from the first metal word lines and second metal bit lines of the product by linking them at both ends into continuous serpentines. In

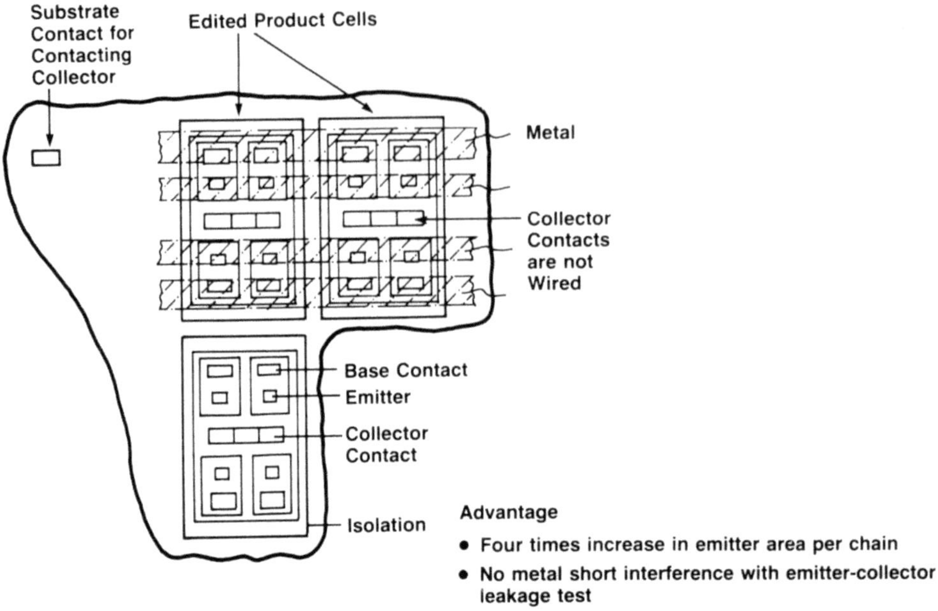

Fig. 1 Transistor Chain Test Structure.

order to assure that adjacent metal lines on the same level belong to a separate serpentine, for short monitoring, underpasses and overpasses are used at the ends. The resulting product mazes have exact product dimensions and topology except device contacts are not opened.

The test site methodology is shown in Fig. 3.

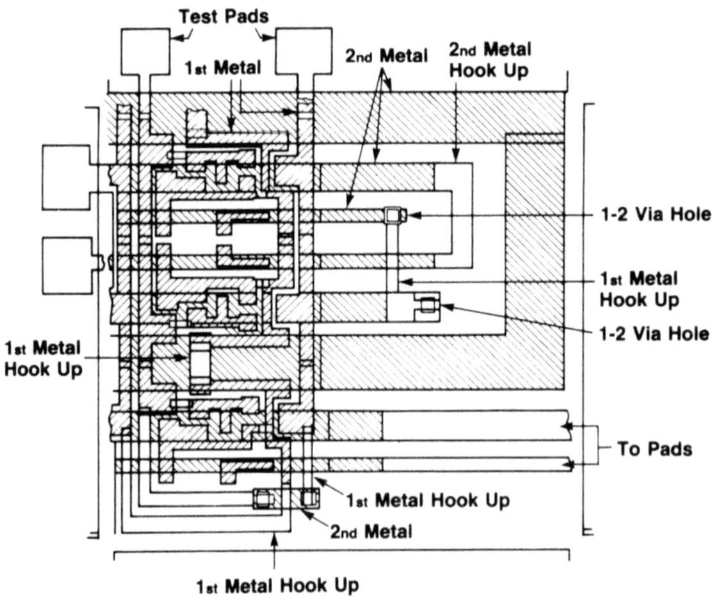

Fig. 2 Metal Maze Test Structure

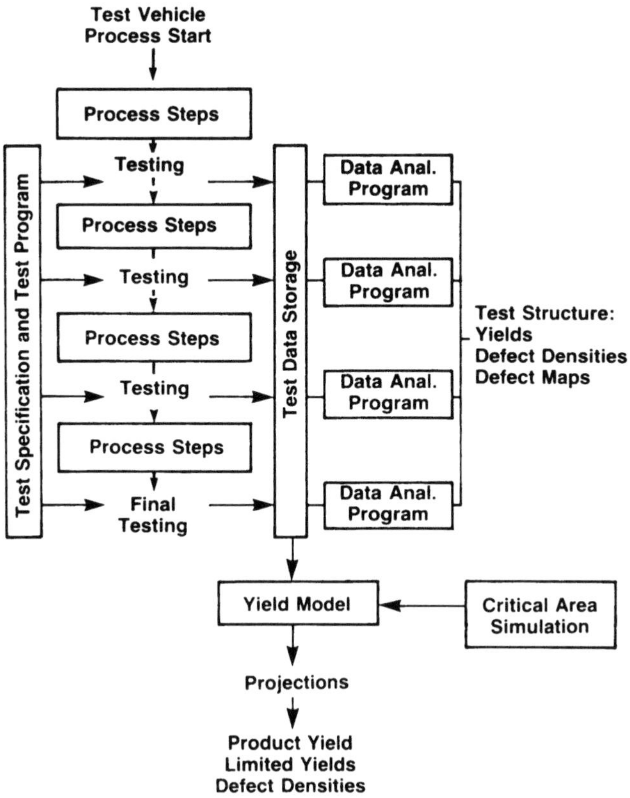

Fig. 3 Test Vehicle Methodology

COMPARISON OF YIELD MODELS

All the yield models have a common shortcoming, i.e. defect double counting. Defect double counting occurs when a physical defect causes two different electrical failures. For instance, metal penetration can cause both emitter-base and emitter-collector shorts. Double counting results in pessimistic yield projections. The double counting can be eliminated by grouping together yield detractors which exhibit tracking among themselves and computing the aggregate or combined yield. For example, cell yield is an aggregate yield. Following is a comparison of several yield models.

POISSON YIELD MODEL

$$Y_p = Y_o\, e^{-\lambda A_{crp}} \tag{1}$$

where: Y_o is the non random, parametric yield for product
A_{crp} is product critical area
Y_p is product yield
λ is defect density

Yield projection using this model is pessimistic because defect clustering is neglected[1]. It requires a linear regression curve for each defect type.

NEGATIVE BINOMIAL YIELD MODEL

$$\beta = \frac{ln\left[\frac{ln\ Y_d}{ln\ Y_s}\right]}{ln2} \quad (2)$$

where: α is the cluster parameter

Yield projection, using this model is accurate because defect clustering is included[1]. It is difficult, though, to use the model since a non-linear regression technique is required to force fit the model to experimental data for each defect type.

DEPRESSED EXPONENT POISSON YIELD MODEL

This model takes into account defect clustering by depressing the exponent (critical area ratio) of the POISSON model when the yield is extrapolated for the product.

$$Y_p = Y_o\ Y_t^{\left[\frac{A_{crp}}{A_{crt}}\right]^{\beta}} \quad (3)$$

where: A_{crp} is product critical area
A_{crt} is test site critical area
Y_t is test site yield
Y_p is product yield
β is exponent depressant

The exponent depressant, β, can be computed from Eq. 4:

$$Y_p = Y_o(1 + A_{crp}\frac{\lambda}{\alpha})^{-\alpha} \quad (4)$$

where: Y_s is limited yield of single test structure
Y_d is limited yield of double test structure
$\ln 2 = 0.693$

The model gives very good approximation for the NEGATIVE BINOMIAL MODEL. No regression technique is needed, thus this model is also much simpler to use.

The yield projection methodology is summarized in Table 1. The defect types which show tracking are grouped together and the aggregate or combined yield is computed for the test structures. There are three basic aggregate yields: cell yield, metal yield and contact yield. Other defects, such as trench integrity, vias and resistor, do not show any tracking. All the above test structure yields, aggregate and non aggregate, are projected for the product. Multiplying these limited yields together gives the random defect portion of the product yield. The nonrandom portion of the product yield, the Y_o in equations 1-3, is obtained from discrete device yield.

Table 2 summarizes the limited yield projections which gives the yield loss breakdown and failure mode ranking for the product. The yield for each defect type is computed separately which may include double counting. Double counting for limited yield has to be allowed since without physical analysis it is not known which one of the simultaneously occurring but different electrical fails should be screened out.

Table 1: Final Test Yield Projection Methodology

$$Y_p = Y_o Y_t \left(\frac{Ap}{At}\right)^{\beta}$$

Project Aggregate Limited Yields For Product	← Compute Aggregate Limited Yields For Test Site	Defect Types
		Chains
Y_p Cell	Y_t Cell ←	Em. - Coll. Lkg. Em. - Base Lkg. Base - Isol. Lkg. Anode - Isol. Lkt. Pem. - Isol. Lkg. SBD Lkg. Pem. - Coll. Lkg. Poly B - Poly B Short
Y_p Per	Y_t Per ←	Peripherial TX's Peripherial SBD's
Y_p TR	Y_t TR ←	Trench Integrity
Y_p R	Y_t R ←	Resistor
Y_p M	Y_t M ←	First Metal Open First Metal Short Second Metal Open Second Metal Short 1st - 2nd M. Short
Y_p S	Y_t S ←	1 - 2 Via
Y_p N	Y_t N ←	2 - 3 Via
Y_p C ↓	Y_t C ←	Emitter Contacts Collector Contacts Base Contacts P Emitter Contacts
Random Def. Yield $Y_p = \prod_{i=1}^{n} Y_i$		**Discrete Devices**
	Y_o R	Resistor Value
	Y_o T	TX Forward Drop
Nonrandom Def. Yield $Y_o = (Y_o R)(Y_o T)(Y_o S)(Y_o G)(Y_o C)$	T_o S	SBD Forward Drop
	Y_o G	TX Current Gain
	Y_o C	Contact Res.
$Y_{FT} = Y_o Y_p$		

Table 2: Limited Yields Projection Methodology

$$Y_p = Y_o Y_t \left(\frac{Ap}{At}\right)^{\beta}$$

Defect Types	**Compute Limited Yields For Test Site** →	**Project Limited Yields For Product**
Chains		
Em. Coll. Lkg.	Y_t CE	Y_p CE
Em. - Base Lkg.	Y_t EB	Y_p EB
Base - Isol. Lkg.		
Anode - Isol. Lkg.	Y_t Overlay	Y_p Over
Pem. - Isol. Lkg.		
SBD Lkg.	Y_t SBD	Y_p SBD
Pem. - Coll. Lkg.	Y_t PNP	Y_p PNP
Poly B - Poly B Short	Y_t P-P	Y_p P-P
Peripherial TX's	Y_t LTX	Y_p LTX
Peripherial SBD's	Y_t LSBD	Y_p LSBD
Trench Integrity	Y_t TR	Y_p TR
Resistor	Y_t R	Y_p R
First Metal Open	Y_t 1OP	Y_p 1OP
First Metal Short	Y_t 1SH	Y_p 1SH
Second Metal Open	Y_t 2OP	Y_p 2OP
Second Metal Short	Y_t 2SH	Y_p 2SH
1st - 2nd Metal Short	Y_t 1-2	Y_p 1-2
1 - 2 Via	Y_t S	Y_p S
2 - 3 Via	Y_t N	Y_p N
Emitter Contacts	Y_t E	Y_p E
Collector Contacts	Y_t C	Y_p C
Base Contacts	Y_t B	Y_p B
P Emitter Contacts	Y_t PC	Y_p PC
Discrete Devices		
Resistor Value	Y_o R	Y_o R
TX Forward Drop	Y_o T	Y_o T
SBD Forward Drop	Y_o S	Y_o S
TX Current Gain	Y_o G	Y_o G
Contact Res.	Y_o C	Y_o C

DISCUSSION OF EXPERIMENTAL RESULTS

Fig. 4 compares the DEPRESSED EXPONENT POISSON MODEL, the NEGATIVE BINOMIAL MODEL and experimental data. It can be seen that both models give a very good fit to experimental data. Fig. 5 compares cell yield projections with both the DEPRESSED EXPONENT POISSON MODEL and the POISSON MODEL. The error caused by neglecting clustering in the POISSON MODEL can be as high as 100%. The error due to neglecting defect tracking can also be as high as 100%.

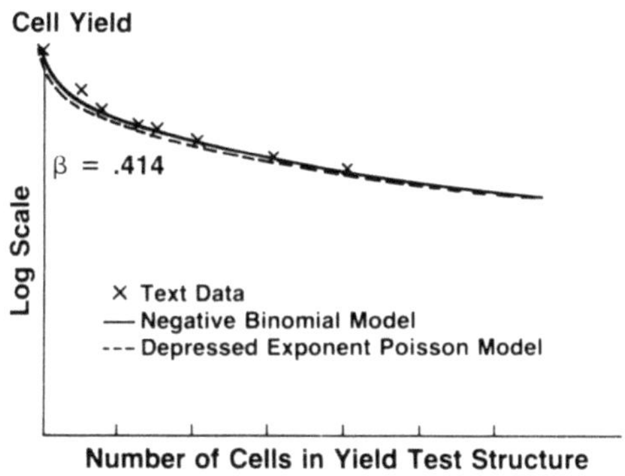

Fig. 4 Comparison of Negative Binomial and Depressed Exponent Poisson Distributions

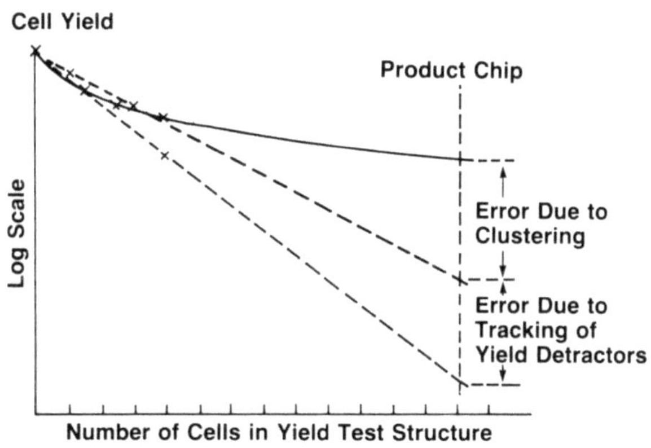

Cell Yield Detractors:

Em.-Coll. Lkg.
Em.-Base Lkg.
Base-Isol. Lkg. ⎤
Anode-Isol. Lkg. ⎬ Overlay
PNP Em.-Isol. Lkg. ⎦
SBD Lkg.
PNP Em.-Coll. Lkg

--- Poisson Model
— Depressed Exponent Poisson Model

Fig. 5 Yield Projection Comparison

Fig. 6 shows the projected and actual final test yields of an array product. As can be seen, the agreement between projected and actual final test yield values is quite good. The projected limited yields for JOBS 1 and 3 in Fig. 6 are shown in Figs. 7 and 8, respectively. Both projected and actual final test yields are also shown. These yield values include redundancy. A comparison of Fig. 7 and Fig. 8 indicates that the lower final test yield of JOB 1 compared to JOB 3 is mostly caused by the lower nonrandom defect yield.

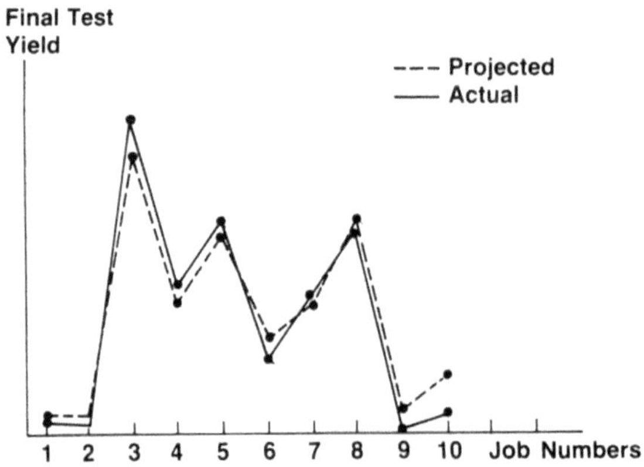

Fig. 6 Projected and Actual Final Test Yield Comparison for a Product

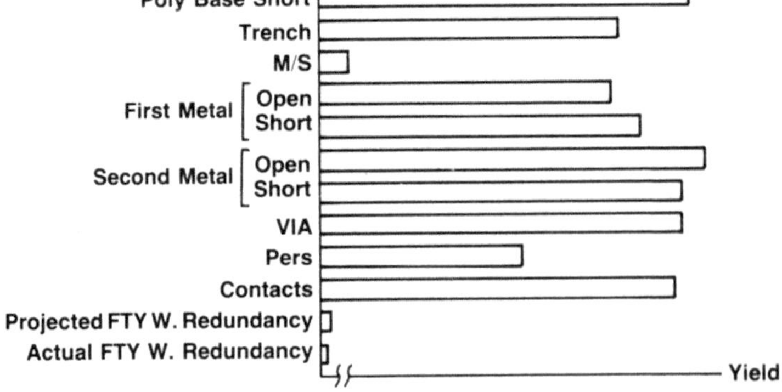

Fig. 7 Projected Limited Yields for Job 1

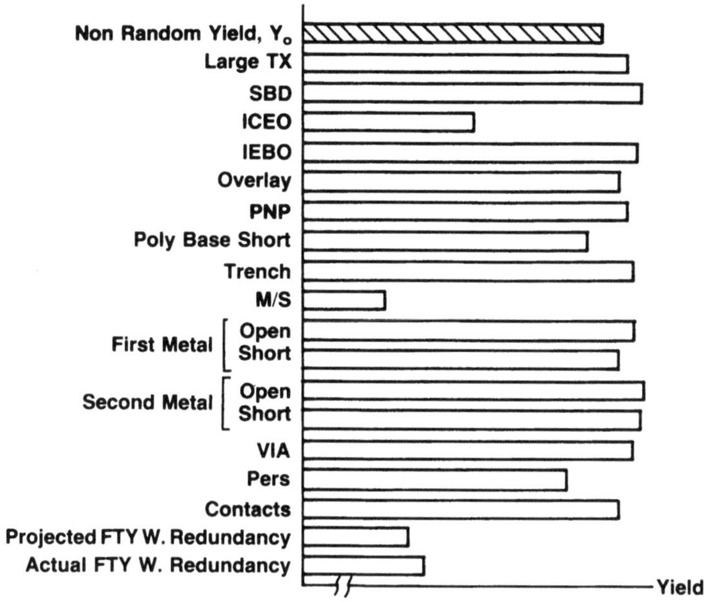

Fig. 8 Projected Limited Yields for Job 3

SUMMARY

The DEPRESSED EXPONENTIAL POISSON MODEL provides:

- Accurate final test yield projections.
- Failure mode separation for both random and non random defects.
- Greater simplicity that the NEGATIVE BINOMIAL MODEL but equal accuracy.

The test site design using edited products cells minimizes unwanted failure mode interference and thus improves yield projection accuracy.

REFERENCES

1. Stapper, C.H., "Yield Statistics for Large Area ICs", *ISSCC Digest of Technical Papers*, pp. 168-169, February 1986.

6 TESTING AND TESTABLE DESIGNS

TEST METHODS FOR WAFER-SCALE INTEGRATION

Koichi Yamashita, Shinpei Hijiya, Gensuke Goto, and **Nobutake Matsumura***

Fujitsu Laboratories Ltd. Atsugi
10-1, Morinosato-Wakamiya, Atsugi 243-01, Japan

***Fujitsu Limited**
1015, Kamikodanaka, Nakahara-ku, Kawasaki 211, Japan

INTRODUCTION

Wafer-scale integration (WSI) is expected to yield higher speed and better reliability and to drastically improve circuit density. The essential problem is that WSI cannot be fabricated without defects because the number of defects increases with circuit area. WSI systems require redundant silicon areas to enhance yield, and must be reconfigured by detecting faulty parts and substituting them with spares. It is important to obtain adequate yield by decreasing the redundant area as much as possible.

WSI testing is also a severe problem because of redundancy and incresing complexity of circuits. WSI testing presents three problems: (1) Test costs are high because a very large number of subsystems must be tested, and then the WSI system formed by connecting the good subsystems must itself be tested. (2) Inner nodes are difficult to observe because the pad-to-gate ratio is very low and a WSI system does not have enough external test pads for large circuits to be tested. (3) Conventional VLSI testers cannot be used, thus special test equipment is required. The optimum solution to these problems would be to build self-diagnostic facilities into the WSI system itself.

This paper discusses WSI test methods. First we present a survey of WSI test methods which have been studied so far, and discuss the implementation of a built-in self-test (BIST) technique to the wafer-scale fast Fourier transform (FFT) processor, and then discuss WSI fabrication.

WSI TEST METHODS

Conventional probe tests should be used for final WSI system test because it is necessary to indicate that the power can be supplied to the system from external I/O pads and the system can be correctly operated before assembly. Since there are not enough external I/O pads to test the inner circuits due to test cost, area overhead, and test equipment, this method should be used in combination with the test methods described below.

A test method using laser cutting and linking has been reported.[1] The structure for a laser link is a sandwich of amorphous silicon between two metal layers. Laser links are made as follows: The second metal, an amorphous silicon, and part of the first metal melted by the laser spot create a silicon-aluminium alloy. Metal is also cut by laser. The connection paths for testing subsystem can be made and tested. After routing the entire WSI system, the connection paths for testing can be created and are cut after testing. Laser cutting and linking can be done at the end of manufacturing. The area overhead of linking and cutting points is required and this method cannot be performed repeatedly at the same point.

Test methods using MNOS (metal-nitride-oxide semiconductor) transistors have also been studied.[2] The MNOS transistors used as electrically programmed nonvolatile switches have a structure such that the usual gate oxide layer in a MOS (metal-oxide semiconductor) transistor is replaced by a silicon nitride layer over silicon dioxide. The high voltage applied to the gate electrode causes the thin silicon dioxide layer to become conductive. The signal paths for testing subsystem can be thus formed. The MNOS transistors can be repeatedly programmed without significant damages. The area overhead of transister switches, MNOS transistors and interconnects to external pads to control MNOS transistors is required.

Test methods using floating gate FETs programmed by electron beam (EB) have been reported.[3-4] A floating gate FET has a completely oxide-encapsulted polysilicon gate. Electrons are injected by the EB into the oxide overlying the gate and the charge is trapped on the floating gates. The floating gate FET turns off if an EB applied to the gate. The charge can be erased with an ultraviolet light exposure. This mechanism is reliable and is commonly used in erasable programmable read only memories (EPROMs). The signal paths for testing subsystems can be formed. The floating gate FETs can be programmed repeatedly. The area overhead of floating gate FETs is required.

Test methods using logical circuits have been studied.[5-7] Electrical logical switches are used to test subsystems and to configure the WSI system composed of these working subsystems. The connection and test of configuration logic are first performed, and each subsystem is tested. Thus the WSI system is formed by serially connectiong subsystems with external software control. This method provides repeated and dynamic testing. The area overhead of configuration logic is required. The problem with this method is that the system cannot be formed if the configuration logic suffers a power-short failure.

Testable design utilizing the regularity of WSI systems, for example the systolic array, has been reported.[8] Test vectors enter from one side of the array and identical results from some signal paths are compared taking into account delay time at the other side of the array. This method strongly depends on the system architecture.

The BIST technique for WSI has also been proposed.[9-10] This method generates the test vectors and compares the result. The on-chip hardware to perform the BIST is provided in the circuit itself, and circuits are partitioned into a number of testable units, for example combinational logic units. Although overhead costs for BIST circuitry is required and testing time is longer, the BIST has the advantages described below. The inner subsystems can be tested by a small number of external I/O pads. The hierarchical test, for example PE test and entire WSI system test, can be performed. No new process technology and special test equipment are required. It is not necessary to generate test vectors. Thus the BIST technique itself is an effective test method and is efficient if it is combined with test methods as mentioned above.

BIST FOR FFT PROCESSOR

We used the BIST technique for an FFT processor because it can test both the PEs and the entire WSI system using a few test pads without new process technology and special test equipment. Figure 1 shows a block diagram of the WSI system and the processing element (PE). The WSI system is a perfect-shuffle, complex 8-point, radix-two FFT system composed of 48 identical multiplier-accumulators. The 8-point FFT processor has a three-stage pipeline. We selected a multiplier-accumulator as a PE, taking into account the size and yield of the PE. The PE has a multiplier-accumulator and BIST circuitry. The BIST circuitry is included in each PE to insure regularity in the WSI system. The WSI system has no other test circuitry to reduce overhead of BIST circuitry. The BIST circuitry is designed to test the WSI system.

The parallel 16-bit, 2800-gate multiplier-accumulator that processes 16-bit two's complement binary numbers, has a 2200-gate multiplier, a 260-gate subtractor, a 240-gate adder, and a 100-gate data register.

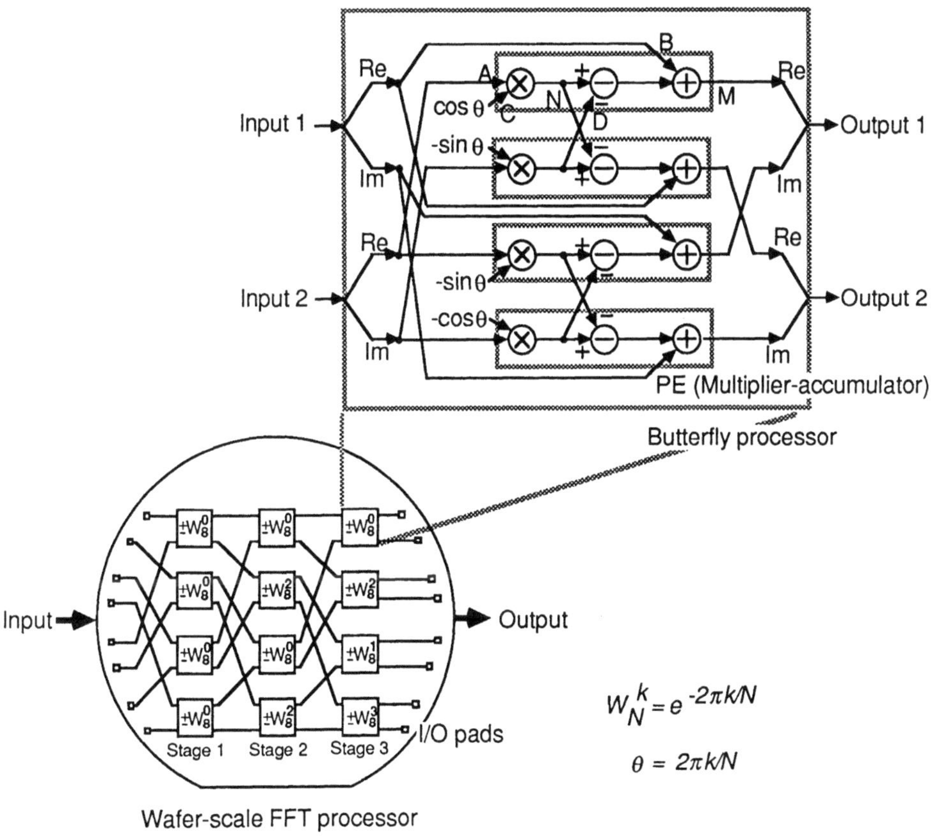

Fig. 1. Block diagram of a perfect-shuffle 8-point complex FFT system composed of 48 multiplier-accumulators.

Sixteen-bit fixed-point data is presented in a format that has a sign bit, an integer bit, and fourteen fractional bits. The binary point signifying the integer and the fractional field division is located between the second and third most significant bit. The multiplier-accumulator has four inputs (A, B, C, and D) and two outputs (M and N) as denoted in Figure 1. Inputs A and B are a 15-bit signal from input pads or the multiplier-accumulator in the previous stage. Data signals from one stage to the next are a 15-bit signal because the data input to the multiplier-accumulator is divided in two to prevent fixed-point data overflows in FFT. Input C is a 16-bit phase factor signal. Input D is a 16-bit signal from output N of the neighboring multiplier-accumulator. Output M is a 15-bit signal to output pads or inputs A or B of the next stage multiplier-accumulator.

Our approach is based on random pattern generation from linear feedback shift registers (LFSRs) and signature analysis. A test method utilizing the regularity of the PE array can be used in our configuration if input C is not fixed mask-programmed data. Test vectors from LFSRs are fed to all PEs in the left column of the PE array and test signatures can be obtained from all PEs in the right column of the array. All PEs in the same stage must yield the same test response if they are not faulty. Since input C of each PE is identical, we used simulated signature patterns.

Figure 2 shows a block diagram of the PE. The 700-gate BIST circuitry has a random pattern generator, a pattern compressor, a multiplexer, and a test controller.

A random pattern generator is 29-bit LFSRs. Since inputs A and B are 15-bit data, and inputs C and D are 16-bit data, a 62-bit test vector is needed to test the multiplier-accumulator. We selected 29-bit LFSRs to test the multiplier-accumulator because the gate density on the PE is minimized. Fourteen-bit data of 29-bit data is for inputs A and C and 15-bit data is for inputs B and D. When a 14-bit random pattern from

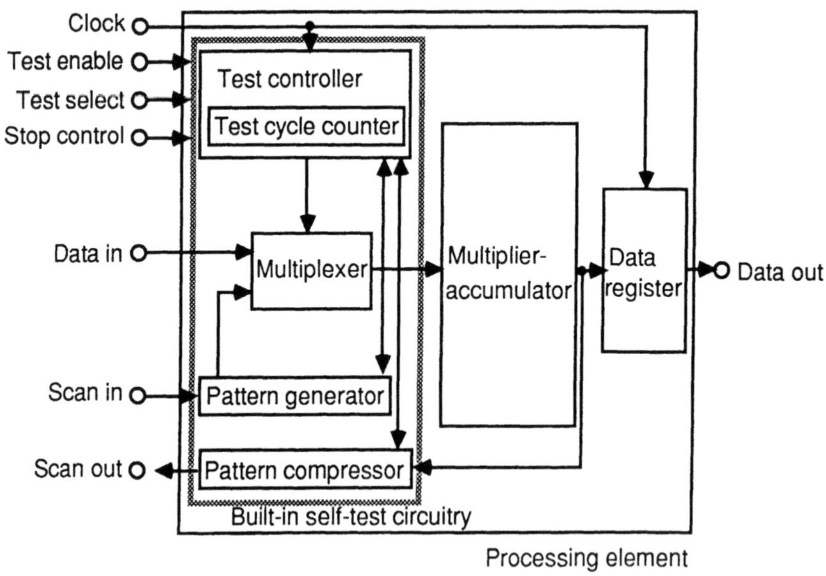

Fig. 2. Block diagram of processing element.

LFSRs is considered to be fixed-point data, this data value is less than 2 and greater than -2. All fixed-point inputs except input B must be less than 1 and greater than -1. By right-shifting this data down one bit, 15-bit data which is less than 1 and greater than -1 is fed to inputs A and C. Sixteen-bit data is supplied to input B and D as mentioned above.

The feedback connection in 29-bit LFSRs comes from the primitive polynomial $X^{29} + X^{28} + 1$ to generate up to $2^{29}-1$ different patterns before repeating the cycle. Figure 3 shows a block diagram of 4-bit LFSRs to explain the pattern generator. Four-bit LFSRs have the feedback connection from the primitive polynomial $X^4 + X + 1$. A multiplexer selects a signal from the scan-in when S = 0 and from the feedback signal when S = 1. LFSRs perform as linear shift registers in mode S = 0, and operate as a pseudo-random pattern generator in mode S = 1. LFSRs can be set in the initial state by the reset signal. The initial state only defines the first pattern of cyclic data stream. LFSRs can also be set in the other states by inputting data stream serially to registers using operating as shift registers.

The output data from multiplier-accumulator is compressed to the signature by the pattern compressor. The pattern compressor is 15-bit multiple input shift registers (MISRs). Signature analysis reduces the amount of test response data. It is obvious that the fault in the multiplier-accumulator is visible at input M. Fifteen-bit MISRs in the BIST circuitry have the characteristic polynomial $X^{15} + X^{14} + 1$. Figure 4

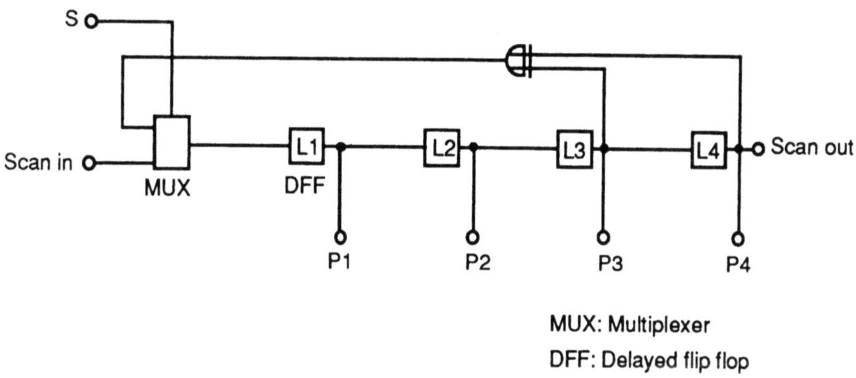

Fig. 3. 4-bit pseudo-random pattern generator.

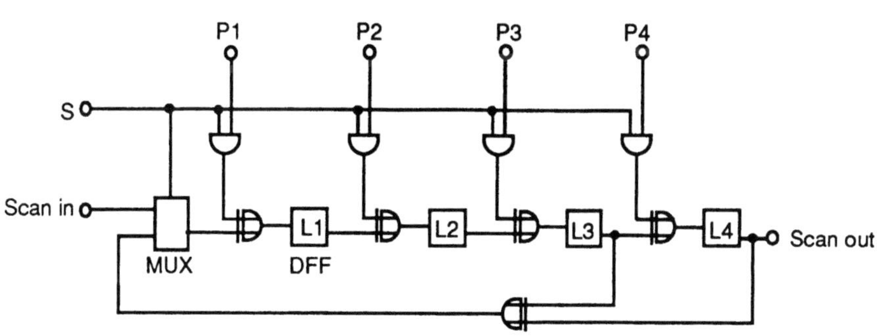

Fig. 4. 4-bit pattern compressor.

is a block diagram of 4-bit MISRs. MISRs perform as linear shift registers that can serially output signature pattern after testing in mode S = 0, perform parallel signature analysis in mode S = 1, and all the registers can be reset.

The test controller generates control signals, for example test clock, test enable, test select, input select, and test stop, and has a binary counter to stop pattern generation when sufficient test patterns are fed to the circuit.

The test scheme is as follows. (1) First in mode S = 0, all registers in the PE are scan-tested using a serial signal pattern from I/O pads in each PE. (2) The LFSRs and the MISRs are set in the initial state, and each PE is tested by random pattern generation and signature analysis in mode S = 1. The signature is output to I/O pads serially and compared with predicted signature patterns in mode S1 = 0. (3) In mode S = 0, all registers in the entire WSI system are scan-tested. (4) The LFSRs and the MISRs are set in the initial state. The WSI system is tested by random pattern generation and signature analysis in mode S = 1, selecting the test vector input in stage 1 and the data signal from previous stage for other stages. Then the signature is output to I/O pads serially in mode S1 = 0.

Figure 5 shows a simulated signature pattern in a self-test with 4096 input patterns; below this is the actual signature pattern from an active PE in the self-test. Both signature patterns are identical, indicating that the multiplier-accumulator under test has operated correctly. It takes only 1.2 ms to test the PE for a 300-ns system clock. After good PEs are connected, the WSI system is tested using the same BIST circuitry. Figure 6 is a block diagram of the WSI testing system. Simulated signature patterns with 4096 input patterns of the final stage (in hexadecimal) are also shown. Pseudo-random patterns generated from the LFSRs of the PE in first stage are fed serially through multiplier-accumulators in three stages. When the signature pattern from MISRs in the final stage is as expected, all PEs and interconnctions in

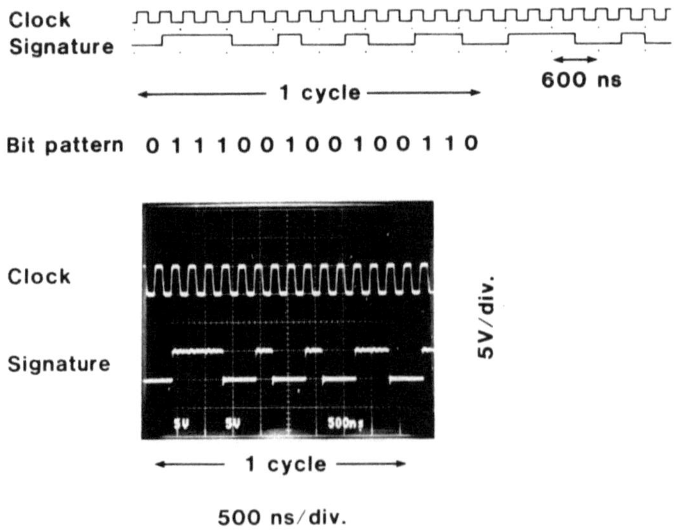

Fig. 5. Simulated signature and output signature waveform.

the 8-point FFT system are fault-free. Troubleshooting of these interconnects can also be done by comparing signatures from all the PEs with the anticipated ones.

This approach significantly reduces the test time and costs needed to test all PEs and the WSI system. The area overhead of BIST circuitry is only 8 percent of the FFT processer circuit area.

WSI FABRICATION

The prototype FFT processor was made for a WSI feasibility study.[11] In the WSI, a parallel logic system is advantageous to random circuits because a system with a regular structure needs a smaller redundant silicon area than random circuits. Discretionary wiring is used in WSI system reconfiguration because of high reliability and low contact

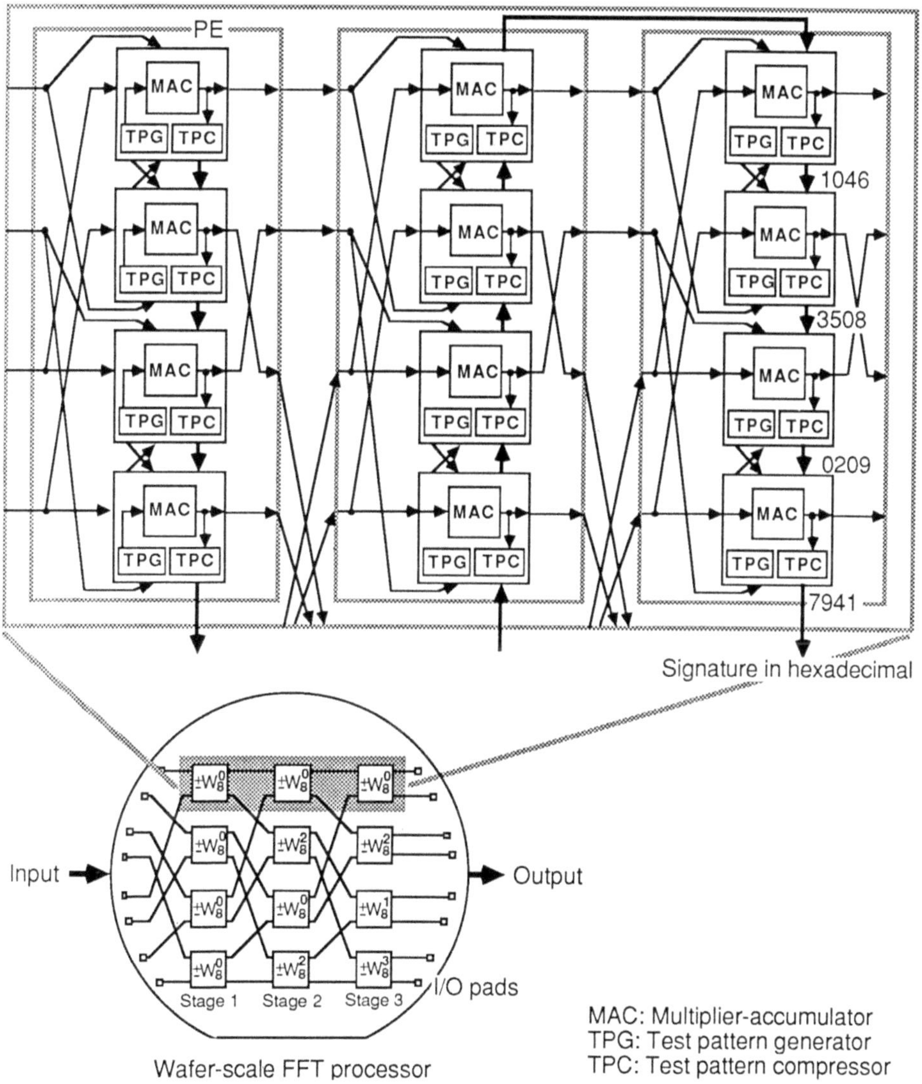

Fig. 6. WSI system test.

resistance and does not require special process technology. A probe test during wafer processing is required to test each PE. No other reconfiguration methods are employed. Although mechanical probing is employed, wafer processing to clean metal suface is done after probing.

Since customization cost for discretionary wiring should be reduced, only one mask is used to program the individual wafer. The WSI system is physically designed by using a single building block, containing a PE and interconnects. Coventional wafer process technology can be used by treating a building block as a VLSI chip. The PE is fabricated with two metal layers and is tested by probing. Good PEs are connected to form the entire WSI system by a third-layer metal.

Figure 7 shows how working PEs are connected by programmed contact holes. The squares in bold lines represent faulty PEs and squares marked with circles represent good PEs. Many regular patterns containing three metal interconnect layers are embedded around each PE. These interconnect patterns are designed to connect arbitrary terminals by placing contact holes between second- and third-metal layers at the program points. Thin lines represent interconnect wires. Interconnects between good PEs are represented using thick lines. Dots represent programmed contact holes. Many programmed contact holes based on BIST results are generated by a newly developed CAD system.

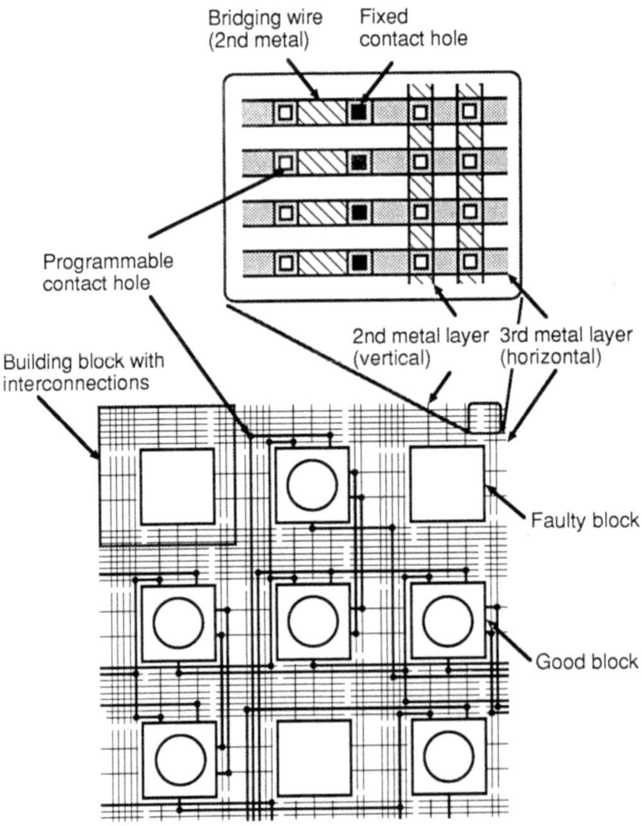

Fig. 7. Connection of selected good PEs.

Figure 8 shows a photomicrogragh of a wafer-scale 170k-gate FFT processor made on 4-inch wafer using triple-metal 2.3 μm CMOS technology. The system performs parallel 16-bit, 8-point 3.3 MHz complex FFT. It consists of 48 good PEs selected out of 88 PEs. A building block containing a PE and interconnects is about 8 x 8 mm^2. The building block region occupies about 70 percent of the entire wafer. PEs occupy about 80 percent of building block region and interconnect regions about 20 percent. The 725 I/O pads are located outside the building block region and used for final testing and assembly. The wafer is mounted facedown on the 11.8 x 11.8 cm^2 substrate. Solder bumps are used for bonding because a large number of contacts must be formed over a large area.

CONCLUSION

We studied the BIST technique for a wafer-scale FFT processor. Our approach is based on random pattern generation and signature analysis.

The BIST circuitry is implemented to PE itself and is designed to test both the PE and the WSI system. Interconnect troubleshooting can also be done. WSI testing can be performed without expensive special equipment such as a VLSI tester.

The BIST technique can reduce the test time for generating test vectors because of random pattern generation. The important thing with the BIST technique is to balance the test time with fault coverage in the limited silicon area.

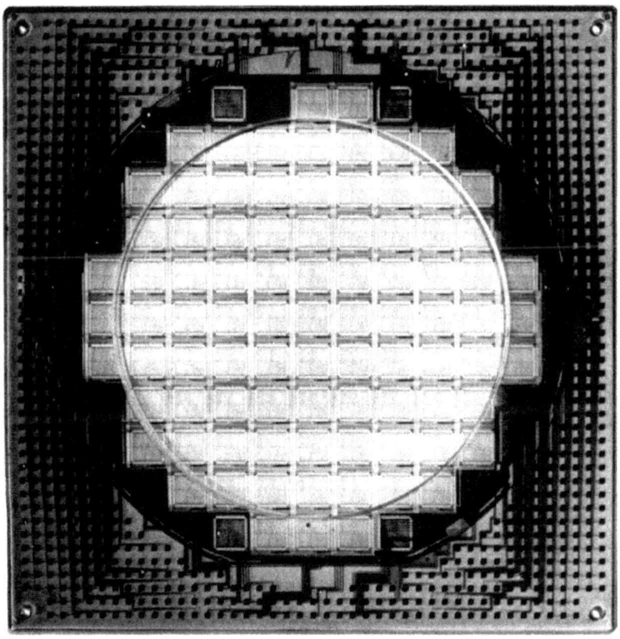

11.8 cm

Fig. 8. Photomicrograph of wafer system.

The area overhead of BIST circuitry was only 8 percent of the circuit area. The circuit density of the BIST increases with its LFSRs bit width, but can be assumed to be small because the gate density of the PE usually increases with its PE bit width.

Although the BIST technique is used for the array processor that is composed of combinational logic, it can also be used to test more complex systems, such as an array processor composed of microprocessors or digital signal processors partitioning some testable units that has combinational logic and memory between registers.

REFERENCES

1. A. H. Anderson, Computer aided design and testing for RVLSI, in: "Wafer scale integration," C. Jesshope and W. Moore ed., Adam Hilger, Bristol and Boston (1986).
2. Y. Hsia, G. C. C. Chang and F. D. Erwin, Adaptive wafer scale integration, Proc. 11th Conf. Solid-State Devices, pp. 193-202 (1980).
3. D. C. Shaver, Electron-beam techniques for integrated circuit testing and customization, Proc. of CICC, pp. 606-609 (1984).
4. P. Girard, F. M. Roche and B. Pistoulet, Electron beam effects on VLSI MOS conditions for testing and reconfiguration, in: "Wafer scale integration," G. Saucier and J. Trilhe ed., Elsevier Science Publishers B.V., Amsterdam (1986).
5. R. C. Aubusson and I. Catt, Wafer-scale integration-A fault tolerant procedure, IEEE J. of Solid-State Circuits, vol. SC-13, no. 3, pp. 339-344 (1978).
6. N. H. MacDonald and G. B. Neish, An algorithmic approach to the testing of a wafer scale integrated (WSI) circuit, Proc. of Test Conf., pp. 590-600 (1982).
7. M. J. Shute and P. E. Osman, COBWEB-a reduction architecture, in: "Wafer scale integration," C. Jesshope and W. Moore ed., Adam Hilger, Bristol and Boston, (1986).
8. W. R. Moore, A. P. H. McCabe and V. Bawa, Fault tolerance in a bit-level systolic array, in: "Wafer scale integration," C. Jesshope and W. Moore ed., Adam Hilger, Bristol and Boston (1986).
9. G. E. Dixon, J. F. Dickson, J. Fox, A. K. J. Stewart, and G. Sumerling, A test methodology for a parametrised cell design approach to fault-tolerant VLSI and wafer scale integration, in: "Wafer scale integration," C. Jesshope and W. Moore ed., Adam Hilger, Bristol and Boston (1986).
10. C. Joy, J. P. Eynard and G. Saucier, Test facilities and test plans is WSI 2D-array, in: "Wafer scale integration," G. Saucier and J. Trilhe ed., Elsevier Science Publishers B.V., Amsterdam (1986).
11. K. Yamashita, A. Kanasugi, S. Hijiya, G. Goto, N. Matsumura, and T. Shirato, A wafer-scale 170000-gate FFT processor with built-in test circuits, IEEE J. of Solid-State Circuits, vol. SC-23, no. 2, pp. 336-342 (1988).

FAULT DIAGNOSIS OF LINEAR PROCESSOR ARRAYS

Donald Fussell, Sampath Rangarajan, and **Miroslaw Malek***

The University of Texas at Austin
Department of Computer Sciences
*Department of Electrical and Computer Engineering
Austin, TX 78712

Abstract

We present a comparison-based algorithm for identifying faulty and fault-free elements in a wafer-scale linear array of processors (or other logic elements). Only nearest neighbor communication is assumed to be possible between the processors in the array. Because the algorithm is simple and requires no storage of test vectors or test outcomes, it is ideally suited for implementation on the wafer to provide the capability for built-in production (or post production) testing. We show that surprisingly this algorithm achieves high accuracy of diagnosis over a wide range of yields even though the diagnosis may be based on a high proportion of results produced by faulty processors. Quantitative and qualitative reasoning validate the efficiency of the algorithm.

INTRODUCTION

A number of methods have been proposed for diagnosing failures in homogeneous processor arrays [1,2,3,4,5,6,7,8]. The homogeneity of processors (or other elements) lends itself to comparison-based schemes, where the outputs of the processors for test inputs are compared and decisions made on their fault status. Methods based on a comparison approach have been proposed both for on-line and off-line diagnosis of systolic arrays based on the expected result for a particular computation. These methods assume a single failure in the array[1] or have other restrictions[2] which make them unsuitable for production testing of wafer-scale arrays. The method proposed in [3] for a mesh of processors assumes that there are only t failures in a neighborhood. The method presented in [4] achieves high accuracy without any restrictions on the number of failures, but requires the use of reconfigurable networks for comparison of outputs between processors.

*This research was supported in part by the Office of Naval Research under Contracts N00014-86-K-0597 and N00014-86-K-0763.

In this paper, we present a comparison-based approach which requires communication only between adjacent processors in a linear array and which achieves higher diagnostic accuracy than previous approaches. This method is very well suited to production testing of IC wafers where a linear processor connectivity is pre-fabricated on the wafer during the production process. Even if failures occur in these connections, an algorithm such as the one proposed in [9] could be used to detect them and distinct small linear arrays could be recovered using the fault-free connections. The testing algorithm proposed here could then be run on this system of linear arrays. Due to its simplicity, this technique could also be used for off-line testing of post-production failures in linear arrays even though such a use does not require the full power of our scheme since faults are expected to occur only rarely. We show that given certain reasonable assumptions, the scheme proposed here will diagnose all fault-free and faulty processors correctly. If these assumptions do not hold, we show that the method has a vanishingly small probability of mis-diagnosis. Diagnostic errors have a much higher probability of identifying fault-free processors as faulty than of identifying faulty processors as fault-free. Since the incorporation of faulty processors into a processor array can be fatal, this property is important in a diagnostic algorithm intended for use in a wafer-scale system. We compare our scheme with the scheme proposed in [4] in order to evaluate our efficiency gains.

In the next section, we describe a functional fault model which will be used in the analysis of the proposed algorithm. In the subsequent three sections, the algorithm is presented and analyzed. Conclusions are presented in the last section.

A FUNCTIONAL FAULT MODEL

In a VLSI or WSI environment, we assume that we are not in a position to repair faulty circuits and therefore that our tests need only tell us whether or not a circuit functions correctly and not how or where it may have failed. This justifies the development of a functional fault model to be used in such a situation.

A *target circuit* C_t is a function mapping a domain I of input values to a range O of output values. C_t is intended to represent an ideal or correct circuit, in other words, the function that is computed by a fault-free physical realization of the circuit. Let C_a donate an *instance* or *realization* of a circuit C. C_a is thus also a function from I to O, but not necessarily equal to C_t. When C_a is not equal to C_t, we say that C_a is *faulty*. In testing these circuits, our goal is to check whether or not a given C_a is faulty i.e. to check whether or not $C_a = C_t$. Of course, the obvious way that this can be done is to exhaustively generate $C_t(e)$ and $C_a(e)$ for every element e in I and then check whether $C_a(e) = C_t(e)$, but frequently the size of I makes this *a priori* impractical.

A *failure model* for a target circuit C_t is a family of functions $\{C_t, F_1, F_2, ..., F_n\}$, where the subset $\{F_1, F_2, ..., F_n\}$ is the set of possible functionally distinct faulty realizations of C_t. For $0 < i \leq n$, the ith *fault state* F_i is a unique function on domain I of C and is therefore uniquely characterized by the subset $I_i = \{e | F_i(e) \neq C_t(e)\}$ of I, i.e. the set of input values for which faulty outputs are produced. Each member of I_i is called a *characteristic input* for faulty circuit F_i. For any particular input value $e \in I$ there is a *fault coverage set* $s = \{i | e \in I_i\}$ of faulty realizations of C_t for which e is a characteristic input. Then $\frac{|s|}{n}$, the fraction of the total number of fault states that input e characterizes, is called the *coverage* of e. A fault coverage set s can be partitioned into a set of *failure modes* where each failure mode is the set of all faulty circuit realizations in s which produce the identical faulty output on input e.

A *test* set for a circuit C is a sequence $T = \{t_1, t_2, ..., t_r\}$ of *tests*, each of which is a sequence of elements from the input set I of C_t. Each such test t_i has an associated fault coverage set s_i. We can let $S = \cup_{i=1}^r s_i$, the union of these fault coverage sets, be the fault coverage set of T. The *coverage of a test set* is

$$\frac{|\cup_{i=1}^r s_i|}{n}$$

which is the cardinality of its fault coverage set divided by the number of fault states. The coverage of an individual test t_i is denoted by c_i and is equal to $\frac{|s_i|}{n}$. We say that a test covers a circuit if that circuit generates a faulty output when this test is applied to the circuit. Coverage of a test or of a test set is *complete* if its value is one.

If the test set being used to determine the correctness of a circuit is all of I, then its coverage is obviously complete, but as noted above the size of this set generally makes such exhaustive testing impossible in practice. Research on the generation of test sets is concentrated on methods that select small test sets with the highest possible coverage, but how these sets are generated is not our concern here. In fact we will find that not only the coverage of the test set, but also the coverage of individual tests is important for our schemes to be efficient.

TESTING A LINEAR ARRAY OF PROCESSORS

Let us now describe our comparison based algorithm which requires comparison between only the nearest neighbors in a linear array of processors. Consider a ternary evaluation predicate $V(O_L, O_R, O_M)$ on $\{O \times O \times O)\}$. Let a circuit be a processor denoted by P and the processors to the left and right of P be denoted by P_L and P_R respectively. The ith comparison carried out by processor P is $V(O_L(t_i), O_R(t_i), O_M(t_i))$. $O_M(t_i)$ is the output of P for test input t_i and $O_L(t_i)$ and $O_R(t_i)$ are the outputs of processors to the left and right of processor P respectively. In further discussions, we assume that processor P makes a status decision on processor P_R, although the algorithm could also be specified such that the decision is made on P_L. We assume that the above evaluation predicate evaluates to 1 for test t_i if $O_M(t_i) = O_L(t_i)$ *and* $O_M(t_i) \neq O_R(t_i)$. A *diagnosis* of a circuit realization C_a is a sequence $R = \{r_1, r_2, ..., r_r\}$, where $r_i = V(O_L(t_i), O_R(t_i), O_M(t_i))$, that is R is the sequence of r results of the tests performed by evaluating V on O_L, O_R and O_M for each of the tests. Processor P_R is deemed fault-free by processor P only if $r_i = 0, \forall i, 1 \leq i \leq r$, i.e., the diagnosis vector does not contain any ones.

The communication variables required for the implementation of the scheme are as follows, assuming that processors are numbered in ascending order from left to right.

MA_i- This is a single bit which is set to 0 if processor i matched its result with processor $i+1$ for a particular test and set to 1 otherwise. We use $\overline{MA_i}$ to denote the complement.
A_i- This is a single bit. This bit is set to 0 if processor i matched its output with processor $i+1$ on all test inputs.

$STAT_i$- This is also a single bit. This bit is set to 1 if for at least one test vector, processor i matched its result with the processor to its left, but did not match the result with the processor to its right. Otherwise, this bit stays at 0.

Let $MA_i(r)$, $A_i(r)$ and $STAT_i(r)$ specify the value of these variables after the application of the rth test.

The algorithm is as follows:

Step 1:

Set $MA_i(0) = 0$, $A_i(0) = 0$ and $STAT_i(0) = 0$ for each i, $1 \leq i \leq k$ where k is the number of processors in the array. Note that $i = 1$ specifies the boundary conditions for the leftmost element in the array.

Step 2:

Pipeline the test inputs through the array and for each test t_j and for each i:

If the result for test vector t_j between processors P_i and P_{i+1} did not match, then set $A_i(j)$ to 1, set $MA_i(j)$ to 1 and set $STAT_i(j) = \overline{MA_{i-1}}(j) \cup STAT_i(j-1)$. Else, set $MA_i(j)$ to 0 and leave all other variables unchanged.

At the end of the application of all r tests, if $STAT_i(r) = 1$, then P_{i+1} is faulty. Else P_{i+1} is fault-free.

□

Note that the diagnosis for the two processors P_1 and P_2 should be carried out from $P_k \ldots P_3$ or a circular connection should be established.

The intuitive reasoning behind the algorithm is as follows. A processor which correctly matches its result with the processor to its left has confidence in its diagnostic ability, and so, if it does not match its result with the processor to its right, it deems that processor faulty. If a processor *does not* correctly match its result with the processor to its left, it is not certain of its fault status, and so, for that test input, it does not make a decision on the processor to its right. Once a processor deems a processor to its right faulty on a test input, it does not change that decision on subsequent test inputs.

DETERMINISTIC BEHAVIOR OF THE METHOD

In this section we evaluate the accuracy of our method, given a set of reasonable assumptions on how the processors fail. If these assumptions hold, our algorithm will be *totally deterministic*. That is, all processors will be diagnosed correctly as faulty or fault-free. In the next section, we will provide a quantitative analysis of the probability of these assumptions being true. Also, from now on, we will denote a fault-free processor by ○ and a faulty processor by □. As shown in *Figure 1*, a sequence of three processors could be in any one of eight states.

A general assumption we make in this section and in the subsequent probabilisticanalysis is that each faulty processor will be detected by at least one test, which implies the test set is complete. This means a faulty processor generates a faulty output for at least one test in the test set. Further discussions in this paper assume that this always holds, since a faulty processor has to misbehave on at least one test input for it to be detected. The assumptions allowing deterministic behavior are as follows.

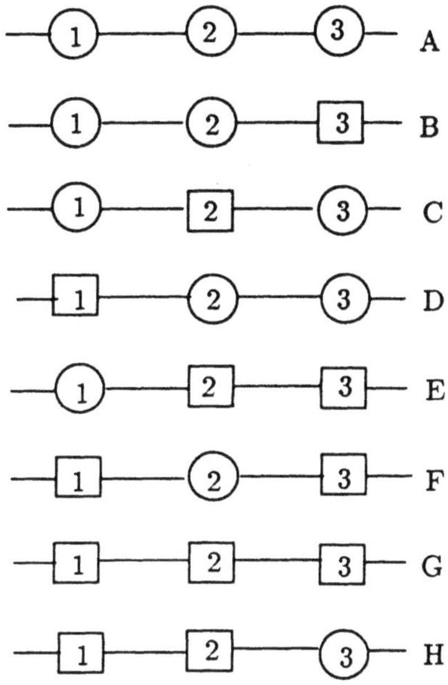

Figure 1: The eight states of a sequence of three processors.

Assumption 1: If there are three adjacent faulty processors then there is *at least one test* for which the first two processors give the same result and the third one gives a different result. Note that this may happen either due to the first two processors not being covered and the third processor being covered by the specific test or due to the first two processors generating the same faulty output due to both the processors having the same failure mode for that test input, whereas the third processor is either not covered or is covered but gives a different faulty output for that test input. For example, in *Figure 1* if all three processors are faulty (pattern G), then there is at least one test for which 1 and 2 give the same output while 3 gives a different output.

Assumption 2: If there are two faulty processors in a string of three processors, then there is *at least one test* which covers the first faulty processor but not the second faulty processor in the string. Thus, for patterns E and F in *Figure 1*, there is at least one test which covers processor 2 but not processor 3 in the first pattern and there is at least one test which covers processor 1 but not processor 3 in the second pattern.

Assumption 3: No two faulty processors which lie adjacent to each other in a string of three processors, with the rightmost processor being a fault-free processor, have the same failure mode for any test. Thus, for *any test* which covers both the processors, the processors generate different observable faulty outputs. For instance for pattern G in *Figure 1*, for any test which covers processors 1 and 2, the processors generate different faulty outputs.

With the above assumptions, we now proceed to analyze our scheme.

Lemma 1: In any string of three processors, if there are at least two processors which

are fault-free, the status of the rightmost processor is always correctly determined.
Proof: There are four different patterns of three processors with at least two processors being fault-free, as shown in ***Figure 1- A,B,C,D.***

Consider pattern A. It is obvious that there could not be any test for which the middle processor matched its result with the one on the left but not the one on the right. For pattern B, for some test where the rightmost processor generates the wrong result, the middle processor would correctly match its output with the processor on the left but not the one on the right. For pattern C, if for a test, the first two processors match their result, it has to be a test which does not cover the middle processor. In that case, the result has to match the result of the third processor also. For pattern D, if for a test, the first two processors match their result, it has to be a test which does not cover the first processor. In this case also, that result has to match the result of the third processor.

□

Thus, in each case the variable $STAT_i$ is set correctly and the status of the rightmost processor is determined correctly.

Theorem 1: The proposed algorithm identifies the status of all the processors correctly if Assumptions 1-3 hold.
Proof: According to Lemma 1, the scheme works correctly on strings of three processors where there are at least two fault-free processors. We are left with strings of three processors with at least two faulty processors. We have four such patterns, as shown in *Figure 1- E,F,G,H.*

For pattern G, according to Assumption 1, there is at least one test for which the first two processors match their results whereas the third one does not. For pattern H, from Assumption 3, there could be no test for which the first two processors generate the same faulty result. Then, the first two processors could match their results for a test only if both of them are not covered by that test. In that case, their results will match the output of the third processor also. So, there could be no test for which the first two processors match their results correctly whereas the third processor does not. For patterns E and F, from Assumption 2, we can conclude that there is at least one test for which the first two processors match their results correctly, whereas the third processor does not.

Thus we see that for a string of three processors with any fault pattern the middle processor correctly finds the status of the rightmost processor if Assumptions 1-3 hold, and thus each processor in the array identifies the status of its right neigbor correctly. So, the status of all the processors in the array is correctly determined.

□

PROBABILISTIC ANALYSIS

We now consider the possibility that our assumptions do not hold. We first define some performance parameters.

Locatability (L_s) is the probability that a fault-free processor is correctly located to be fault-free.

Optimistic Error Probability (O_s) is the probability that a faulty processor is mistakenly deemed to be fault-free by the algorithm.

In the analysis below, p denotes the probability of failure of a processor (if Y_e is the expected wafer yield, then $p = 1 - Y_e$), t denotes the number of tests applied during the test process, c denotes the coverage of individual tests and f denotes the total number of possible failure modes for any test. For ease of analysis, f and c are taken to be uniform over all test.

A fault-free processor is correctly deemed to be fault-free if

1. At least one of the other two processors to its left in the chain of three processors is fault-free, the probability of which is given by $(1 - p^2)$ or
2. Both of the processors are faulty (see pattern G) but they do not have the same failure mode for any test so that they do not give the same faulty output for any test (Assumption 3) the probability of which is $p^2(1 - \frac{c^2}{f})^t$.

So, locatability is given by

$$L_s = (1 - p^2) + p^2\left(1 - \frac{c^2}{f}\right)^t$$

In practice, selection of different mask sets for neighboring processors during production testing will lead to the probability of two faulty processors having the same failure mode for any test input becoming vanishingly small so that *Assumption 3* holds. In that case, $f \to \infty$ and so $L_s \to 1$. *Figure 2* shows the variation of L_s with the other parameters.

As a means of comparison of our method with other algorithms, we look at the algorithm presented in [4] for diagnosing failures in linear processor arrays. In the method proposed by Choi et al., a processor compares its output on a test with k other processors and deems itself fault-free if it matches its output with at least one of the k outputs. An upper bound on the locatability of this algorithm is $L_a = 1 - p^k$. We compare the locatability of our algorithm with the above method by equating the values and calculating the number of comparison processors k required by the above method to match the performance of our method.

$$1 - p^k = (1 - p^2) + p^2\left(1 - \frac{c^2}{f}\right)^t$$

For small values of $\frac{tc^2}{f}$, we get

$$1 - p^k = 1 - \frac{t(pc)^2}{f}$$

from which $k = \frac{\log(\frac{t}{f}) + 2\log(pc)}{\log(p)}$

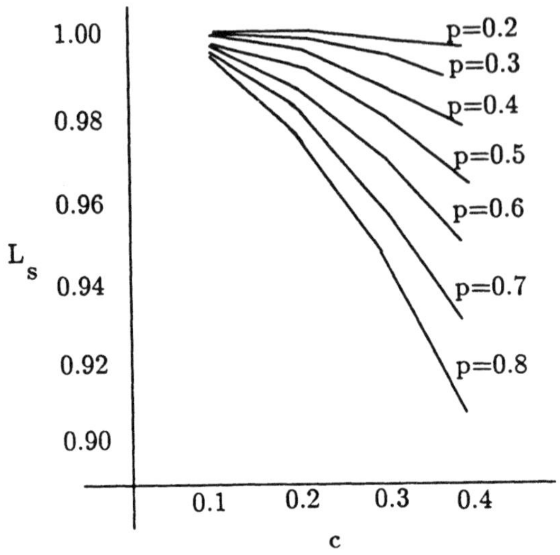

Figure 2: Variation of locatability with single test coverage (c) for different values of p with $f = t = 1000$.

Let us calculate k for $t = f = 1000$ for various values of p and c. $k = \frac{2\log(pc)}{\log(p)}$. *Table 1* shows these numbers. For example, for $p = 0.8$ and $c = 0.1$, more than 22 comparison processors are required, which means that long communication links to processors at a distance are required, whereas our method requires only nearest neighbor comparison. Furthermore, we will save time by only having nearest neighbor comparison.

Let us consider the optimistic error probability of our algorithm, We note that patterns E,F and G could lead to this type of error. Consider pattern G. For this pattern, an optimistic error could be made if assumption 1 does not hold. In this event there is *no test* for which

1. The first two processors are not covered and the third processor is covered the probability of which is $p_1 = (1-c)^2c$, or

2. The first two processors are covered and belong to the same failure mode so that they give the same faulty result, (and) whereas the third processor is either not covered *or* belongs to a different failure mode. The probability of this occurence is given by $p_2 = \frac{c^2}{f}(1 - \frac{c}{f})$.

So, the optimistic error probability due to pattern G is given by $(1 - (p_1 + p_2))^t$.

For pattern E, there will be an error made if there is no test input for which the second processor is not covered and the third processor is covered. For pattern F, an error will be made if there is no test for which the first processor is not covered and the third processor is covered. The probability of each of the above occurrences (Assumption 2 does not hold) is given by

$$(1 - c(1-c))^t$$

Table 1: Number of comparison processors (k) required to match our performance for different values of single test coverage c and failure probability p with $f = t = 1000$.

c \ p	0.5	0.6	0.7	0.8
0.1	8.64	11.0	14.9	22.6
0.2	6.64	8.30	11.0	16.4
0.3	5.47	6.71	8.75	12.8
0.4	4.64	5.58	7.14	10.2

which evaluates to

$$(1 - c + c^2)^t$$

Then the overall optimistic error probability will be given by the above probabilities weighted by the probability of occurence of such patterns as shown below.

$$O_s = p^2(1 - (p_1 + p_2))^t + 2p(1 - p)(1 - c + c^2)^t$$

It should be noted that the optimistic error probability derived above is maximum with respect to f when $f \rightarrow \infty$. In further discussions for upper bound calculations of optimistic error probability, we will let $f \rightarrow \infty$ in which case $p_2 \rightarrow 0$. *Figure 3* plots the value of the optimistic error probability with respect to the coverage parameter c for $p = 0.8$ and $t = 1000$ with $f \rightarrow \infty$. The error probability will decrease with increasing number of tests. It is seen that the error probability reaches a minimum for a c value around 0.3. For example, for a c value of 0.2, the error probability is around 10^{-59}. It was found that for values of p less than 0.8, the optimistic error probability was even smaller. In the above and subsequent examples, the value of t is chosen to be small in order to match the value of c, in the sense we consider *relatively large grain* tests (large c). Note that these two parameters will depend on what the *grain size* of a test input is.

Consider a test computer which runs tests on individual circuits and compares the generated results with precomputed test results. We compare our error probability with the probability that a test computer does not identify a faulty circuit. Assume that a test computer cannot identify a faulty circuit only if none of the tests in the test set cover the faulty circuit. In our fault model, the probability of this occurence will be given by $(1-c)^t$. We would like to factor our optimistic error probability equation into some equation of the form $(1 - c_s)^t$, so that we can calculate

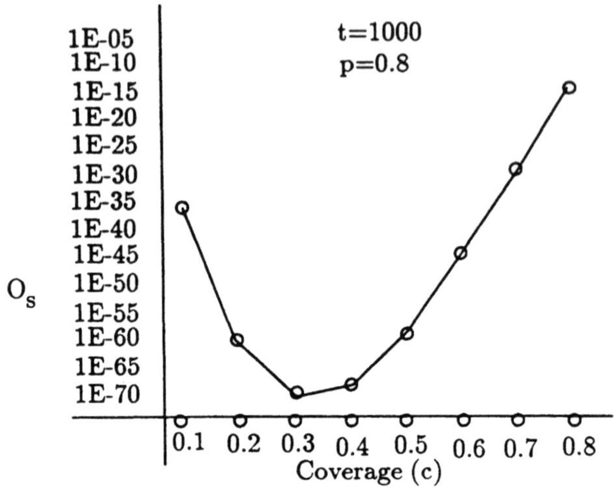

Figure 3: Variation of optimistic error probability (O_s) with single test coverage (c) for $p = 0.8$ and $t = 1000$.

the apparent coverage of a test due to our algorithm (which will be given by c_s). In terms of average performance, we know that the average number of tests that has to be applied by a test computer in order to identify a faulty circuit will be given by $\sum_{i=0}^{\infty} ic(1-c)^{i-1}$ which evalutes to $\frac{1}{c}$. In order to calculate the average number of tests required by our algorithm, let $\alpha = (1 - p_1) = (1 - (1 - c)^2 c)$ and let $\beta = (1 - c + c^2)$. Then, the average number of tests will be given by

$$p^2 \sum_{i=0}^{\infty} i\alpha^{i-1}(1-\alpha) + 2p(1-p) \sum_{i=0}^{\infty} i\beta^{i-1}(1-\beta)$$

$$+(1-p^2) \sum_{i=0}^{\infty} ic(1-c)^{i-1}$$

The last term $(1 - p^2)\sum_{i=0}^{\infty} ic(1-c)^{i-1}$ is duc to the fact that even if both the processors to the left are fault-free, the faulty processor has to be covered by a test before it could be identified as faulty. We did not have this term in our optimistic error probability because it was our general assumption that at least one test in the test set covered the faulty circuit. That is, a faulty circuit has to manifest its faulty behavior on at least one test input to be of concern. The above equation evaluates to $\frac{p^2}{(1-\alpha)} + \frac{2p(1-p)}{(1-\beta)} + \frac{(1-p^2)}{c}$. Substituting for α and β, this evaluates to

$$\frac{p^2 + 2p(1-p)(1-c) + (1-p^2)(1-c)^2}{c(1-c)^2}$$

It is clear from this that in terms of average performance, comparing the average number of tests (with coverage c) required, $(\frac{1}{c})$ in the test computer case with that of our algorithm, we can say that the algorithm performs equivalent to a test computer

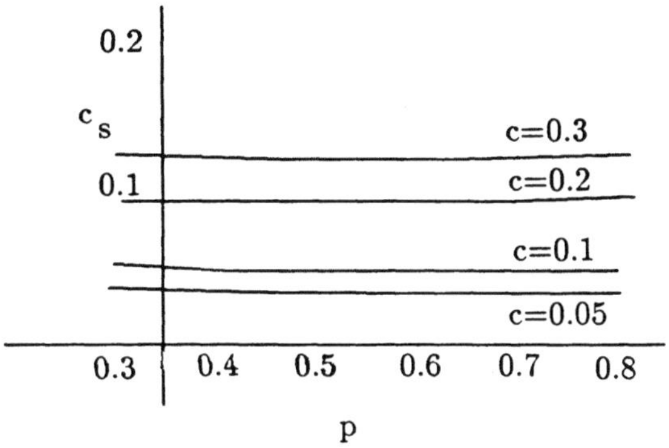

Figure 4: The apparent fault coverage of a test

where the *apparent* coverage of a test (c_s) is given by

$$\frac{c(1-c)^2}{p^2 + 2p(1-p)(1-c) + (1-p^2)(1-c)^2}$$

Figure 4 plots the apparent coverage with respect to c for various values of p. We note that c_s is close to c for small values of c and becomes much smaller than c when it increases. But, in the practical case, we expect c to be very small (≤ 0.1) and so, in practice we expect c_s to be very close to c. Of course, the graph for optimistic error probability shows that it is vanishingly small over the whole range of c.

CONCLUSIONS

We have proposed a comparison-based built-in diagnosis scheme for identifying faulty and fault-free processors in a linear array of such elements. The method is anticipated to involve very low implementation overhead due to the expected simple test circuitry requirements. While other methods have been proposed for testing linear processor arrays, this new method has been shown to be superior in accuracy as a result of more subtle use of comparison results. The technique proposed here is particularly suitable for built-in testing of wafer-scale arrays since its accuracy is robust across a wide range of processor yield variations and because it is particularly effective at reducing optimistic errors, which are by far the most serious form of misdiagnosis in such systems.

REFERENCES

1. Y-H. Choi, S. Han and M. Malek, "Fault Diagnosis of Reconfigurable Systolic Arrays," Proc. ICCD., 451:455 (1984).

2. H. Hellwagner, "Fault Detection in Partially Utilized Systolic Arrays," Proceedings of the First Intl. Conference on Comp. Technology, Systems and Applications, 716:719, Hamburg (1987).

3. A. Das, K. B. Lakshmanan, K. Thulasiraman and V.K. Agarwal, "Distributed Diagnosis For Regular Interconnected Systems," Second International Conference on Supercomputing, vol.3, 277:283 (1987). pp. 277–283.

4. Y-H. Choi, D. Fussell and M. Malek, "Token Triggered Systolic Diagnosis of Wafer Scale Arrays," Proceedings of the International Workshop on WSI, 246:258, Southampton, England (1985).

5. I. Koren, "A reconfigurable and fault-tolerant VLSI multiprocessor array," Eighth Annual Symposium on Computer Architecture, 425:440 (1981).

6. W.R. Moore, A.P.H. McCabe and V. Bawa, " Fault tolerance in a large bit-level systolic array," in: *Wafer-scale Integration*, Adam Hilger, Bristol, England (1986).

7. A. L. Rosenberg, "The Diogenes Approach to Testable Fault-Tolerant Array of Processors," *IEEE Trans. on Computers.*, vol.C-32, 902:910 (1983).

8. A. K. Somani and V. K. Agarwal, "System Level Diagnosis in Systolic Systems," Proc. ICCD., 445:450 (1984).

9. Y-H. Choi, D. Fussell and M. Malek, "Straight-Line Testing of Wafer-Scale Switch Arrays," Proceedings of the International Conference on Computer-Aided Design (1986).

FAULT DIAGNOSIS OF ARRAY PROCESSORS WITH UNIFORMLY DISTRIBUTED FAULTS

Michal Cutler, Minghsien Wang, and **Stephen Y.H. Su**

State University of New York at Binghamton
Department of Computer Science
Binghamton, NY 13901

INTRODUCTION

Array processors can be defined as computing systems consisting of a number of identical cooperating processing elements which are connected in a desired structure. The regular structure of array processors makes them suitable for implementing in very large scale integration (VLSI) or wafer scale integration (WSI) devices. Two classes of VLSI array processors, systolic arrays[1] and wavefront arrays[2], have been proposed for real-time digital signal and image processing. Recent developments and applications of systolic/wavefront arrays can be found in [3] [4].

Systolic arrays are highly parallel and pipelined processing systems with a large number of cells and regular connections, in which data are rhythmically computed, passed and synchronized by a global clock to achieve high throughput[1]. Wavefront arrays have similar properties except that they adopt the data-flow (or data driven) design by which computations of a cell is started as soon as the required operands have arrived. The requirement for the correct timing of data movements in systolic arrays is replaced by the correct sequencing of data in wavefront arrays. However, in wavefront processors, hardware and time overhead for handling the communication protocol between cells is required.

Due to the limited number of I/O pins and the high complexity of circuits in a VLSI/WSI device, it is very difficult to access test points inside the wafer. Therefore, the controllability and observability of VLSI/WSI array processors are extremely low. The problem of testing and diagnosing a VLSI/WSI array processor becomes a real challenge.

Since the hardware cost is decreasing, it is acceptable to dedicate some portions of circuits for testing and diagnostic purposes. The repetitive structure of array processors allows efficient implementation of self-testing and self-diagnostic capabilities. A solution for diagnosing VLSI/WSI array processors is to adopt the "built-in self-test" technique[5] [6] for the design of each cell and use the test results of cells to perform self-diagnosis at the array level[7] [8].

Preparata et al. proposed a graph-based diagnostic model[9], called the PMC model, for diagnosing multiprocessor systems in which processors (cells) are connected in an arbitrary way and each cell has the capability of testing its neighbors. A diagnostic unit collects the testing results from each cell to identify the faulty cells. Many proposed system-level diagnostic algorithms are based on this model[10] [11] [12] [13]. The PMC model assumes that any testing result from faulty cells is unreliable. Under this assumption, the *diagnosability* (i.e. the maximum number of faulty cells which can be diagnosed correctly by an algorithm) of an array is severely limited. In fact, it is limited by the degree of connectivity of cells in the array. For example, the diagnosability of a linear array is *two* since its degree of connectivity of cells in linear arrays is two. Therefore, this global diagnostic approach may not be appropriate for systolic/wavefront arrays which usually have simple interconnection structure, thus, have low diagnosability.

In this paper, for arrays with uniformly distributed faults, we develop diagnostic algorithms which do not trust the outcomes of faulty cells (similar to the PMC model) and still achieve high diagnosability. These diagnostic algorithms are distributed among the cells of arrays. The self-testing capability of each cell is obtained by modifying the general registers of each cell into pseudo-random test pattern generators or signature analyzers to produce a compressed test response called ***signature***. Each cell will compare its signature with those of its neighbors. The comparison results of each cell are the basis of our diagnostic algorithms.

OUR APPROACH TO BUILT-IN SELF-TEST AND SELF-DIAGNOSIS

The accessibility (i.e. controllability and observability) of internal cells of VLSI/WSI array processors is limited. Therefore, built-in self-test and diagnosis are believed to be better approaches than external testing and diagnosis. Built-in self-test allows all cells of an arrays to test themselves simultaneously, and thus reduces the test application time.

Built-in self-test can be achieved by replacing all input registers and output registers of each cell (see Fig.1) with BILBO-like registers [14] as shown in Fig. 2. In the test mode, each input register acts as a pseudo-random test pattern generator (TPG) to generate pseudo-random test patterns which are applied to the circuit of the cell. Each output register acts as a signature analyzer (SA) to compress test responses and generate a signature. A signature generated by the BIST circuit of a cell represents the status (fault-free or faulty) of the cell. The signatures of all cells are the basis for diagnosis. Fig. 3 shows a linear systolic array for convolution. Each cell of the array is modified as shown in Fig. 4 to have the self-test and self-diagnosis capabilities. Registers X, W and Y are replaced by BILBO-like registers. Multiplexers are utilized to select a signature in register X or W. A comparator is added for comparing the signature of the cell in Y with the selected signature. A diagnostic control unit is used for performing the distributed diagnostic procedure of each cell.

The length of pseudo-random test patterns required for achieving a desired fault coverage can be calculated by an analytical method[15] or by fault simulation. Both methods are not expensive for an array with fine-grain cells (i.e. cells with low circuit complexity). For random pattern resistant faults, Eichelberger[16] proposed a circuit-modifying approach for enhancing the testability. Compressing test responses will reduce the fault coverage (due to fault masking). This phenomena, called *aliasing*, has been studied by Smith[17].

Self-diagnosis is especially desirable for diagnosing an array processor in the field-operating stage, where no sophisticated test equipment is available, and the percentage of new faulty components in a test is very low.

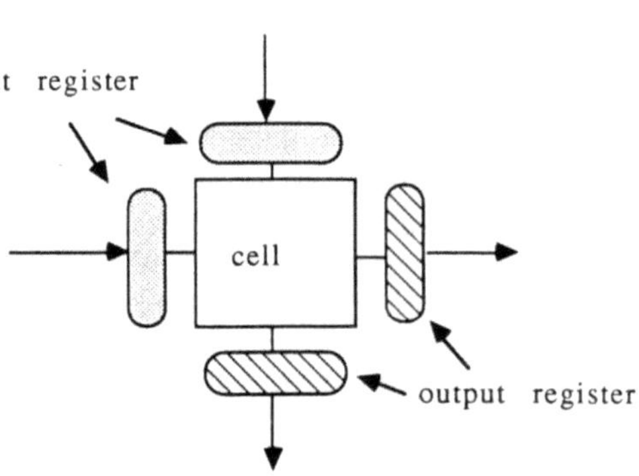

Fig. 1 A cell with input and output registers

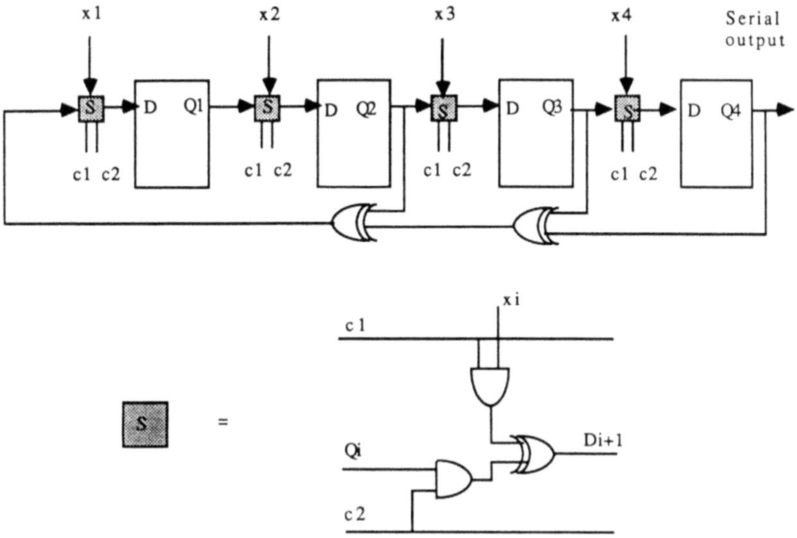

Fig. 2a A 4-bit BILBO-like register and the switch S

Mode 1(reset): c1=0, c2=0, All flip-flops are reset.

Mode 2 (latch): c1=1, c2=0

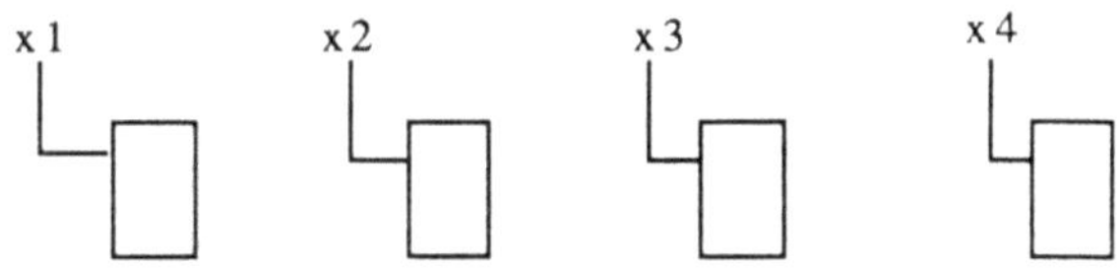

Mode 3 (multiple input signature analyzer): c1=1, c2=1

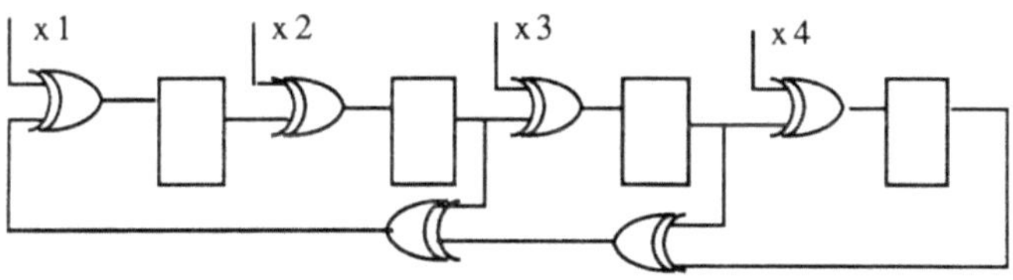

Mode 4 (pseudo-random pattern generator): c1=0, c2=1

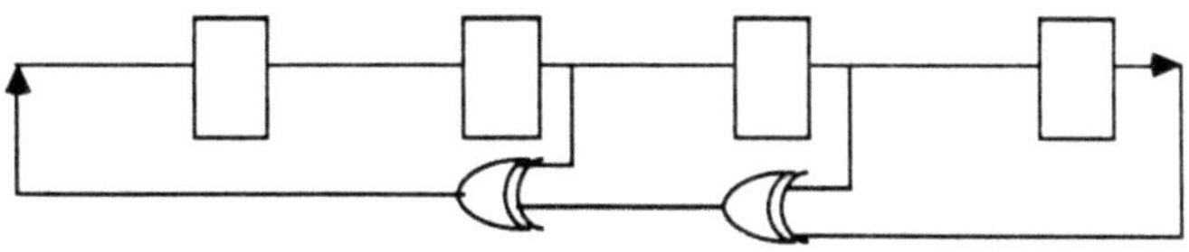

Fig. 2b Four modes of a 4-bit BILBO-like register

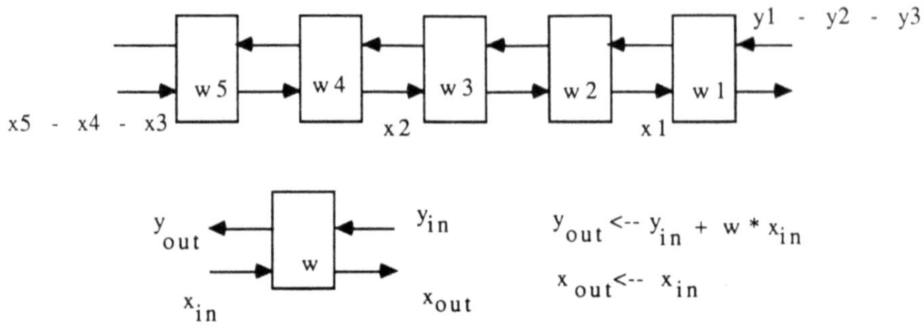

Fig. 3 A linear systolic array for convolution

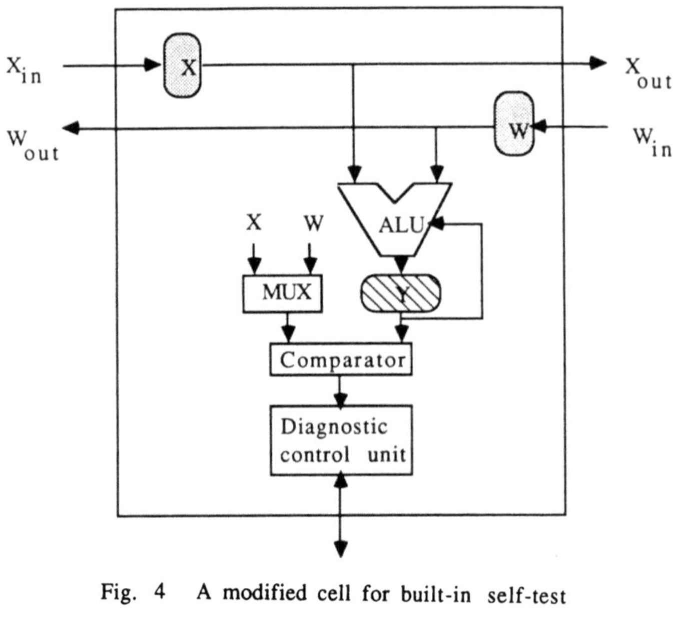

Fig. 4 A modified cell for built-in self-test

The distributed diagnostic algorithms presented here are mainly for wavefront arrays. To apply these algorithms to systolic arrays, they would have to be modified to include timing consideration. We believe that the same concept can be applied to other types of array processors, such as Massive Parallel Processor[18], with minor modifications.

An array processor is a *m-majority array* if and only if there are less than m/2 faulty cells in any group of m consecutive (adjacent) cells.

An array with small fault clusters can be modeled as a m-majority array. This fault model is suitable for uniformly distributed and sparse faults. In the field-operating stage, the faults are assumed to be distributed uniformly, and the percentage of faulty cells is low due to the periodical diagnosis and reconfiguration. In such an environment, an array can be assumed to be a m-majority array.

DIAGNOSIS OF M-MAJORITY ARRAYS

In the operational stage, faulty clusters in an array are usually small. Therefore, we can assume that arrays are m-majority arrays with small m (or with small clusters). This property suggests that each cell may reach a correct diagnosis result by simply exchanging its signatures and diagnosis information with its neighbors, without passing signatures and information through the whole array, provided that fault clusters are small enough. In this section, we shall develop distributed diagnosis algorithms for m-majority arrays by signature comparison.

The basic assumptions for our diagnosis algorithms are as follows:

1. A good cell always generates a good signature and performs its diagnostic algorithm correctly.
2. A bad cell always generates a faulty signature which is different from the good signature of the array. Two bad cells may have the same signature due to common mode failures or the aliasing effect.
3. A bad cell may perform the diagnostic algorithm incorrectly, but is not malicious. A bad cell is *malicious* if it incorrectly sets itself to good, or if it incorrectly sets its neighbor(s) to bad.
4. Communication lines among cells are fault-free.

According to the above assumptions, comparison results issued by a faulty cell is assumed to be unreliable. This is similar to the PMC diagnosis model. When a good cell compares its signature with that of another good cell, the comparison result is 0 (i.e., match). When a good cell compares its signature with that of a faulty cell, the comparison result is 1 (i.e., mismatch). When a faulty cell compares its signature with the signature of another cell, the comparison result is either 0 or 1 (i.e., unreliable). Therefore, when cell i compares its signature with cell j's signature, the result $E_i(j)$ is:

$E_i(j)=0$ when cells i and j are good, or
$E_i(j)=1$ when cell i is good and j bad, or
$E_i(j)=0/1$ (unreliable) when cell i is bad.

In the following, we shall present the diagnostic algorithms for 3-majority and 5-majority linear arrays with end-to-end connection (i.e. with ring structure).

Three-majority Arrays

The outline for the 3-majority diagnostic algorithm is as follows. Following the outline, the 3-majority algorithm will be given.

A. Initialize.

In step A.1 of the algorithm, each cell sets itself to unknown. Then each cell send its signature to its neighbors and receives a signature from both its neighbors in Step A.2. In Step A.3, each cell compares the received signatures with its own, and sends the comparison results back to the neighbors in Step A.4. In Step A.4, it also receives the comparison results sent by its neighbors.

B. Good cells are set to good.

In a 3-majority array, each bad cell has two good neighbors, and each good cell has at least one good neighbor. Thus a bad cell i receives $E_{i-1}(i)=1$ and $E_{i+1}(i)=1$, while a good cell receives $E_{i-1}(i)=0$ or $E_{i+1}(i)=0$ or both. Therefore, all good cells can set themselves to good in Step B.

C. The bad cell is set by its good neighbors to bad.

In a 3-majority array, each bad cell has two good neighbors that set its status to bad.

In regards to the 3-majority diagnostic algorithm, a bad cell is malicious if it incorrectly sets itself to good in Step A or in Step B, or if it incorrectly sets its good neighbor(s) to bad in Step C.

The detailed algorithm is given below. Comments are given between two slashes "/".

Diagnostic algorithm for 3-majority linear arrays:

For each cell i do in parallel:

A. /Initialize/

1. Initialize status flags of all cells G_i =00 (unknown).
2. Send signature S_i to cells i-1 and i+1. /S_i is the signature of cell i./

   Receive signatures S_{i-1} and S_{i+1} from cells i-1 and i+1 respectively.
3. /signature comparison./

   If $S_i=S_{i-1}$, then E_i(i-1) = 0 else E_i(i-1) = 1

   If $S_i=S_{i+1}$, then E_i(i+1) = 0 else, E_i(i+1) = 1
4. Cell i sends E_i(i-1) and E_i(i+1) to cell i-1 and cell i+1 respectively.

   Cell i receives E_{i-1}(i) and E_{i+1}(i) from cells i-1 and i+1 respectively.

B. /Good cells set themselves to good./

If one of or both E_{i-1}(i) and E_{i+1}(i) are equal to 0, set G_i=01(good) .

C. /Good cells set bad neighbor to bad./

For each good cell, set the status of its neighbor with different signatures as faulty.

End.

For a n-cell 3-majority linear array, the number of faulty cells which can be diagnosed correctly by the above algorithms is $\lfloor n/3 \rfloor$.

The 3-majority diagnostic algorithm is quite simple since each good cell has at least one fault-free neighbor that sends the correct signature comparison result, and thus the cell can be immediately diagnosed. In the 5-majority diagnostic algorithm, however, a cell may have two faulty neighbors, making the diagnostic process more complicated.

Five-majority Arrays

The outline of the 5-majority algorithm is as follows. Following the outline, the 5-majority algorithm will be given.

A. Initialize.

   In Step A, each cell sets itself to unknown. Then each cell compares its signature with its both neighbors and exchanges the comparison results.

B. Good cells that belong to consecutive sequences of at least 3 good cells and their bad neighbors are diagnosed correctly.

   First, a good cell with two good neighbors is diagnosed as good in Step B.1. Then diagnosed good cells designate their neighbors as good cells in Step B.2. Finally, diagnosed good cells designate their bad neighbors as bad cells in Step B.3.

C. Good cells that belong to consecutive sequences of exactly two good cells and their bad neighbors are diagnosed correctly.

   In a 5-majority array, there can be consecutive sequences of at most two bad cells. When two bad cells are neighbors, each is at the end of a three or more consecutive sequence of good cells. Therefore, these bad cells were diagnosed as bad in Step B.3.

   If an unknown cell i receives $E_{i-1}(i)$=0 from its neighbor i-1, or $E_{i+1}(i)$=0 from its neighbor i+1, either i is actually good, or i and i-1 are both bad. But, if i and i-1 are both bad, then i can not be unknown in Step C. Therefore i is good. Cell i sets itself to good in Step C.1. Clearly, after C.1, the two consecutive good cells are marked as good.

   In Step C.2, the bad neighbors of all good cells are diagnosed as bad by a good neighbor.

D. Good cells with 2 bad neighbors are diagnosed correctly.

   Because of the 5-majority assumption, all bad cells are at the end of a consecutive sequence of 2 or more good cells. Therefore, all bad cells were already diagnosed as bad by the end of Step C. Thus the cells which are still unknown are those good cells with two bad neighbors. These cells are set to good in Step D.

The detailed 5-majority diagnostic algorithm is given below.

Diagnosis algorithm for 5-majority linear arrays:

For each cell i do in parallel:

A. Initialize.

   1. Initialize all G_i=00 (unknown).

   2. Send signature S_i to cells i-1 and i+1, and receive signatures S_{i-1} and S_{i+1} from cells i-1 and i+1 respectively.

   3. /Signature comparison./

      If $S_i = S_{i-1}$, then E_i(i-1) =0, else E_i(i-1)=1.

      /E_i(i-1)=0 (1) means that cell i considers that cell i-1 is good (bad)./

      If $S_i = S_{i+1}$, then E_i(i+1) =0, else E_i(i+1)=1.

4. Send E_i(i-1) to cell i-1, E_i(i+1) to cell i+1.

   Receive E_{i-1}(i) and E_{i+1}(i) from cells i-1 and i+1 respectively.

B.

1. / If cell i is bad, at least one neighbor is good and considers i as bad. Therefore, diagnose i as good if both neighbors consider it as good. /

   If both E_{i-1}(i) and E_{i+1}(i) are 0's, then cell i is good, G_i =01.

2. / Good cells designate their good neighbors as good. /

   For each good cell i (G_i=01) do:

   If there exists a neighbor x, and E_i(x) =0, set cell x as good (G_x=01).

   /cells x and i have the same signature./

3. / Good cells designate their bad neighbors as bad. /

   For each good cell i (G_i=01) do:

   If there exists a neighbor x and E_i(x) =1, set cell x as bad (G_x=10).

   /cell x has a different signature from cell i/,

C.

1. /Before this step is executed all two consecutive bad cells are correctly diagnosed. Therefore, a cell which is designated unknown and has the same signature as its neighbor must be a good cell./

   For each cell i which is unknown (G_i=00) do:

   If $E_{i+1}(i)$=0 or $E_{i-1}(i)$=0, then cell i sets itself as good (G_i=01).

2. For each good cell i, if there exists a neighbor x and E_i(x)=1, set x as bad (G_x=10).

D. / All bad cells were already diagnosed. Therefore, all cells which are unknown designate themselves as good. /

   Each unknown cell i, set itself as good (G_i=01).

End.

We illustrate the diagnostic algorithm for 5-majority arrays with the following example. Assume a linear ring array with good cells denoted by G and bad cells denoted by B as follows:

Cells	G	G	G	B	B	G	G	G	G	B	G	G	B	G	B	G

In Step A.1, all the cells are marked as unknown, denoted by U.

Cells	G	G	G	B	B	G	G	G	G	B	G	G	B	G	B	G
Status	U	U	U	U	U	U	U	U	U	U	U	U	U	U	U	U

In Step B.1, all good cells with two good neighbors are marked as good, denoted by status G.

Cells	G	G	G	B	B	G	G	G	G	B	G	G	B	G	B	G
Status	G	G	U	U	U	U	G	G	U	U	U	U	U	U	U	U

In Step B.2 and B.3, the diagnosed good cells mark their good neighbors as good and bad neighbors as bad, denoted by status B.

Cells	G	G	G	B	B	G	G	G	G	B	G	G	B	G	B	G
Status	G	G	G	B	B	G	G	G	G	B	U	U	U	U	B	G

In Step C.1, cells that belong to two consecutive good cells, set themselves as good.

Cells	G	G	G	B	B	G	G	G	G	B	G	G	B	G	B	G
Status	G	G	G	B	B	G	G	G	G	B	G	G	U	U	B	G

In Step C.2, diagnosed good cells mark their bad neighbors as bad.

Cells	G	G	G	B	B	G	G	G	G	B	G	G	B	G	B	G
Status	G	G	G	B	B	G	G	G	G	B	G	G	B	U	B	G

In Step D, cells designated unknown set themselves to good.

Cells	G	G	G	B	B	G	G	G	G	B	G	G	B	G	B	G
Status	G	G	G	B	B	G	G	G	G	B	G	G	B	G	B	G

For a n-cell 5-majority linear array, the number of faulty cells which can be identified correctly by the above algorithm is $\lfloor 2(n/5) \rfloor$.

CONCLUSIONS

In this paper, we employed a built-in self-test method for systolic/wavefront arrays to generate in parallel a signature for each cell of arrays. Thus, the time for signature generation is independent of the size of an array. The results of signature comparison of each cell with its neighbors are the basis of diagnosis in our approach.

The classical PMC system diagnostic approach is not suitable for diagnosing systolic/wavefront arrays because of low diagnosability for such arrays. But, for arrays with regular interconnection structure and uniform fault distribution (e.g. m-majority systolic arrays), we have shown that the diagnosability can be improved significantly. Similar to the PMC model, we assumed that the diagnostic units of faulty cells are unreliable, the comparison results of faulty cells are not to be trusted , and no faulty cell is allowed to make any diagnostic decision. Efficient algorithms for 3 and 5-majority linear arrays were developed and proved correct. The same approach can be extended to general m-majority arrays with little effort.

REFERENCES

1. H. T. Kung, "Why systolic architectures?," Computer, Jan. 1982, pp. 37-46.
2. S. Y. Kung, "On supercomputing with systolic/wavefront array processors", Proc. of IEEE, July 1984, pp. 867-884.
3. W. Moore, A. McCabe and R. Urquhert (editors), "Systolic Arrays," Adam Hilger, Bistol, England and Boston, MA, 1987.
4. IEEE Computer - special issue on systolic arrays, July 1987.

5. H. Fujiwara, "Logic testing and design for testability," The MIT press, Combridge, MA, 1985.

6. S. Y. H. Su, M. Wang and K. K. Saluja, "Fault-tolerant VLSI/WSI systolic architecture," Proc. 9th Int. Conf. on Fault-tolerant Systems and Diagnostics, Brno, Czechoslovia, June 1986, pp. 36-44.

7. S. Y. H. Su, M. Cutler, M. Wang and K. K. Saluja, "Self diagnosis of linear and mesh arrays by signature comparison," in *Systolic Arrays* (W. Moore et al. Editors), Adam Hilger, Bristol England and Boston, MA 1987.

8. S. Y. H. Su, M. Cutler and M. Wang, "Fault detection and location for tree array processors," Proc. of 11th Int. Conf. on Fault Tolerant Systems and Diagnostics, Suhl, German Demoncratic Republic, June 6-8, 1988. Summary in Proc. of CompEuro 87 Conf. on VLSI and Computers, Hamburg, W. Germany, May 11-15, 1987.

9. F. P. Preparata, G. Metze and R. T. Chien, "On the connection assignment of diagnosable systems," IEEE Trans. Electronic Computers, Dec. 1967, pp. 848-860.

10. A.T. Dahbura and G. M. Mason, "An $O(n^{2.5})$ fault identification algorithm for diagnosable systems", IEEE Trans. Compt. June 1984, pp.486-492.

11. A. T. Dahbura, K. K. Sabnani and L. King, "The comparison approach to multiprocessor fault diagnosis", Proc. of 15th Symp. on Fault-tolerant Computing, June 1985, pp. 260-265.

12. S. Hosseini, J. Kuhl and S. M. Reddy, "A diagnosis algorithm for distributed computing system with dynamic failure and repair", IEEE Trans. Compt. Mar. 1984, pp. 223-233.

13. S. E. Kreutzer and S. L. Hakimi, "Distributed diagnosis and the system user", IEEE Trans. Compt. Jan. 1988, pp. 71-78.

14. B. Konemann, J. Mucha and G. Zwiehoff, "Built-in logic block observation techniques," Digest of papers, 1977 Test Conf. pp. 37-41.

15. K. D. Wagner, C. K. Chin and E. J. MaCluskey, "Pseudorandom testing," IEEE Trans. Compt., March 1987, pp. 332-343.

16. E. B. Eichelberger and E. Lindbloom, "Random-pattern coverage enhancement and diagnosis for LSSD logic self-test," IBM J. Res. Dev., Mar. 1983, pp. 265-272.

17. J. E. Smith, "Measures of the effectiveness of fault signature analysis", IEEE Trans. Compt. June 1980, pp. 510-514.

18. K. E. Batcher, "Design of a massively parallel processors," IEEE Trans. Compt. Sept. 1980, pp. 836-840.

7 DEFECT- AND FAULT-TOLERANT PROCESSORS

DESIGNING FOR HIGH YIELD: THE NS32532 MICROPROCESSOR

B. Maytal, A. Danor, V. Karpati,
R. Nassrallah, Y. Sidi, and E. Shihadeh

National Semiconductor (Israel) Ltd.
Herzlia, Israel

ABSTRACT

Complex VLSI devices, like general purpose microprocessors, require novel design techniques to ensure high manufacturing yield. This paper suggests that device tolerance to electrical parameter variations is the major design yield contributor when nonstructured VLSI devices are involved. The hierarchical methodology employed to design the NS32532 microprocessor is presented. The methodology guidelines and the design stages are described in detail. This methodology's contribution to yield is presented.

INTRODUCTION

The NS32532 is a complex, highly integrated 32-bit microprocessor which incorporates 370,000 transistors (Alpert et al., 1987). The microprocessor incorporates a 4 stage pipeline, a Memory Management Unit, Instruction and Data Caches (Figure 1). It is manufactured in a double metal, 1.25u CMOS process. The NS32532 is specified to run at 30MHz in a voltage range of 4.5v to 5.5v.

The yield of any VLSI device, especially that of a complex microprocessor, depends on both the quality of the design and fabrication. These factors influence the yield independently and in combination. The main fabrication factors that effect the yield are:

(1) Electrical parameter variation throughout the fabrication process, e.g Leff, Vt values.

(2) The hard defect density, e.g metal bridging, missing contacts.

The design team effects the yield mainly by:

(1) Device tolerance to electrical parameter variations.

(2) The die size.

(3) Device tolerance to hard defects.

In order to meet the high degree of integration sought by the NS32532, the design team had to take advantage of most of the die size allowed by the available lithography technology. The NS32532 die size is large 460 mil * 560 mil at 1.25u. The chip incorporates many types of circuits , The biggest array of identical cells, the Data Cache, occupies less than 10% of the chip area, making fault tolerance techniques inefficient To increase the yield the design team had therefore to guarantee device parametric tolerance - a wide process parameter range of device functionality.

Wide process parameter range of device functionality is obtained by minimizing the number of circuit design flaws. In simple devices the number of such possible flaws is fairly small and may be detected by applying traditional circuit simulations of problematic hardware (Agarwal, 1983) In addition, the existence of such problems can be overcome by correct tuning of the fabrication line.

This is not the case for a complicated, big device like the NS32532. The number of possible paths and combinations is large and only a small portion of it can be simulated manually. The large number of engineers employed in such a project only complicates the issue since each engineer has a unique design style . Many design flaws can be caused due by faulty communication between the engineers.

Another approach is to use a verification CAD tool which will run on the entire data base, and detect faulty circuits. Such a tool is complicated to develop and use, and is very slow on big data bases.

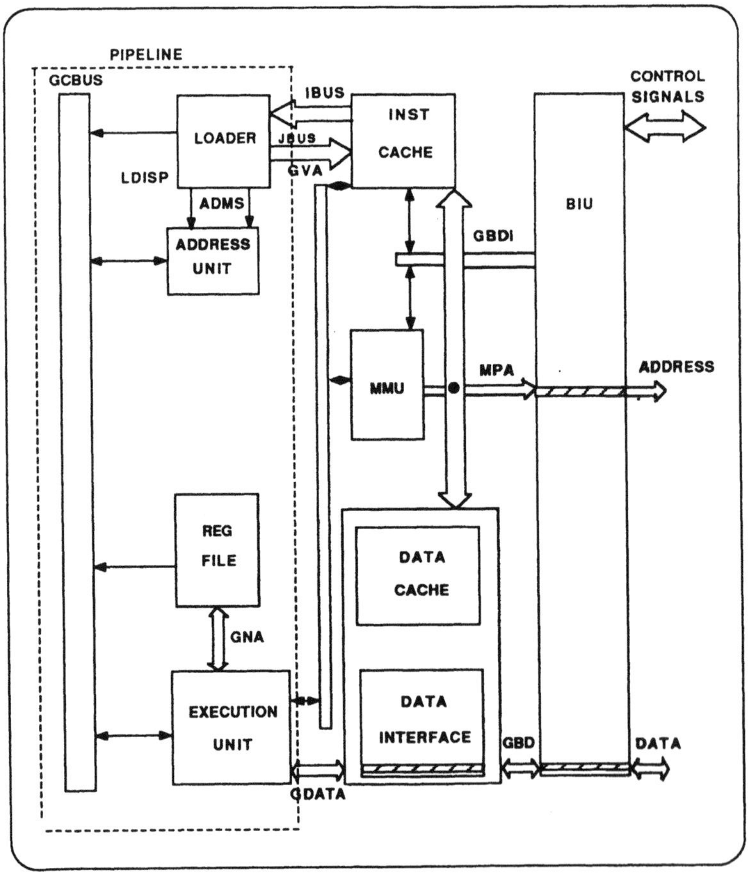

Figure 1. NS32532 Internal Organization

The development of an efficient design methodology, backed by appropriate CAD tools becomes therefore essential in such a project. The hierarchical design methodology developed for the design of the NS32532 and aimed at reducing the number of design flaws and increasing the yield is described in the following sections.

DESIGN FOR HIGH YIELD GUIDELINES

The design team goal of minimizing the number of design flaws found on silicon had to be achieved under the constraints of the fabrication process, the required level of integration, the tight schedule and the electrical specifications. The NS32532 was designed in a hierarchical manner (Weiser et al.,1987) The chip was divided into major sections and subsections on down to the transistor level. Each of the blocks in the hierarchy had its own electrical characteristics. This allowed for the development of relativly simple in-house CAD tools which were used to enforce the methodology.

The main guidelines of the methodology were aimed at reducing in a timely manner the number of design flaws found on silicon. These were to:

(1) Strictly define all internal signals.

(2) Standardize design in order to increase productivity and minimize the number of design styles.

(3) Debug and correct the yield problems (e.g. speed): a) early in the design process, and b) before the various functional blocks were completely integrated.

(4) Control paths correct by construction.

(5) Design for automatic shrink to 1.0u in order to achieve higher yield during mass production.

(6) Speedily debug and correct the yield problems in first silicon.

(7) Help the fabrication line target the process.

USING THE DESIGN METHODOLOGY TO DETECT DESIGN FLAWS

The NS32532 was implemented in four stages (Figure 2):
(1) High level design

(2) Low level design

(3) Layout

(4) Post silicon debug and corrections.

Design flaws which effected the yield were detected during each level of the design. The rest of this paper discusses the design flaws found on each level, and how they were detected.

High Level Design

The high level design of the NS32532 was done in two phases. During the first phase a behavioral description of the microprocessor's major blocks and their interconnections was given in an in-house HDL (I. Sprecher, 1985). In the second phase, the major block structures were described in the HDL language. Each major block incorporated a hierarchy of blocks and their interconnections down to the basic functional units. The basic functional units were defined to be implemented by either PLA, random logic or special structures like ALU or registers.

The main yield enhancement goal at this level was to detect potential circuit design flaws like speed or race problems in the block interfaces, and create a base of information that could be used for design and debug at a lower level.

Bug detection proceeded as follows: Each interface signal was assigned a timing class attribute which could be checked by the logic simulator. This attribute provided general information about the clock edge from which the signal was activated, and the clock edge at which the signal was guaranteed to be valid. A "1s" timing class , for example, meant that a signal was guaranteed to be valid during clock phase 1, while "1v" timing class meant that the signal was guaranteed to be valid only at the trailing edge of clock phase 1. This information ensured that the high level blocks interfaced correctly. This data was used at the low level design stages when more precise timing information was needed.

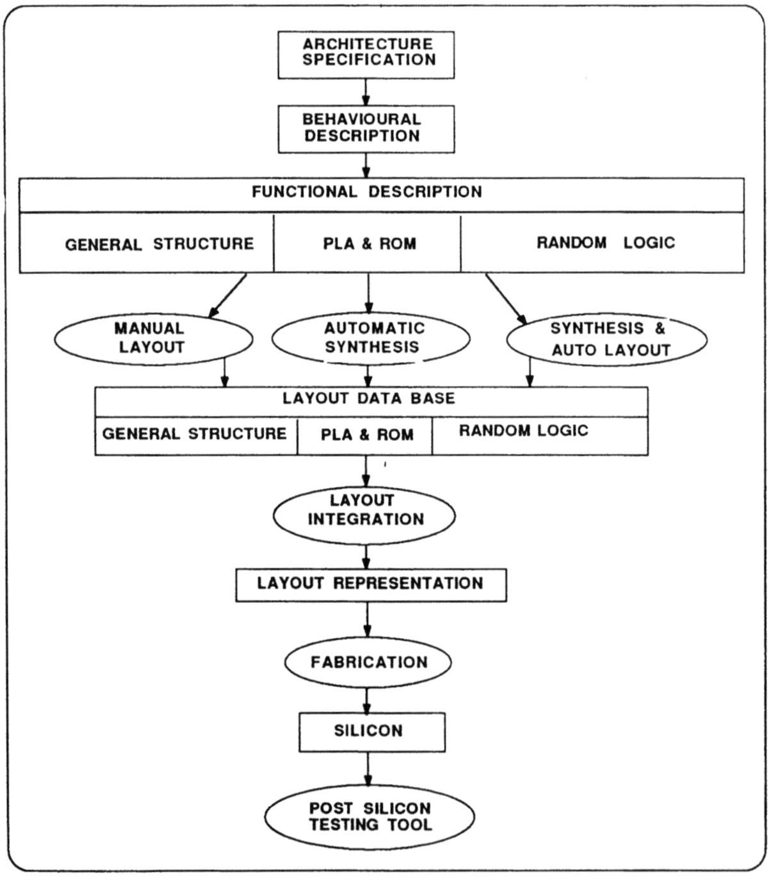

Figure 2. NS32532 Design Process

Low Level Design

The low level design is the functional description in transistors. As in the high level, the low level design stage was concerned with identifying and correcting bugs which affected yield. The low level design was broken down into a number of steps:

Step one identified the design restrictions for each basic block. Electrical characteristics were specified for each interface signal in the NS32532 . The setup time and input capacitance values were estimated for each block's input signals. Designers also estimated the valid time and capacitance load for each block's output signals. Interconnection capacitance was specified in higher hierarchy levels. In this way, the electrical specifications of the device were broken down to electrical specifications on each block in the internal hierarchy. Once the "mini" specifications for a given block were set, its design could proceed independent of other blocks.

A special CAD tool calculated the required capacitance load on each signal, and flagged contradictions between the source and destination specifications. These estimated timing and capacitance values were replaced by actual values after the layout was complete.

Step two was to translate the functional description to transistors. Block designs were executed in three styles:

* Special structures - Registers, ALU's, memories etc.

* PLAs and ROM.

* Random logic (Figure 3).

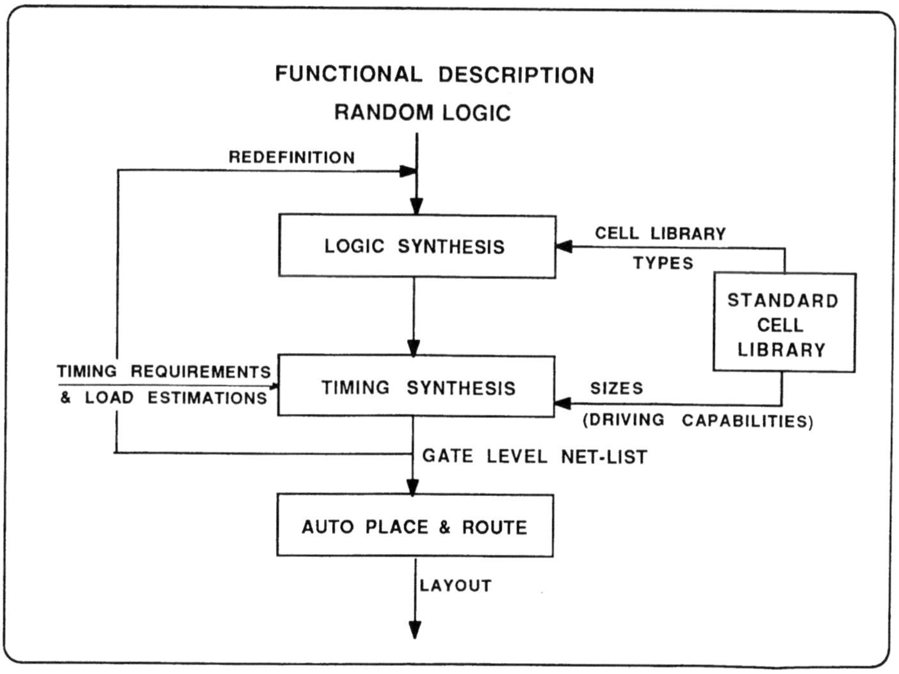

Figure 3. Random Logic Design Path

Special structures were designed in the traditional logic and circuit design approach. Worst case voltage, temperature and process were used for circuit simulations.

Special attention was given to control paths. They were implemented either by PLA's or random logic, where the possible number of circuit combinations is very big. Automated synthesis tools used the interface specifications and HDL for creating the PLA and random logic transistor descriptions. This was done by the use of a library of basic PLAs cells and basic logic cells. Automation ensured that no speed flaws entered the design and minimized other potential circuit problems. The tools would alert the user each time they could not meet a specified restriction. All paths were automatically checked to see if they met the speed requirements.

Step three took place after the design was complete. During this stage another CAD program checked that the design adhered to design methodology rules. Methodology violations (ex. race condition) were detected with information from the timing classes of the interface signals and the timing models of the standard logic cell. The timing models specified the allowed combinations of the gate input and output timing classes.

Layout

Random logic, PLAs, and on-chip memories were laid out automatically. Chip level routing was performed semi-automatically. The basic layout cells of all chip structures and data path structures were layed out manually to save chip size. The existence of 2 metal layers supported the hierarchical design well. M2 layer was dedicated to inter block connectivity and power supplies.

The design rules were set such that automatic shrink to 1u could take place. Metal bridging problems were minimized by setting non-aggressive design rules for Metal 1 and Metal 2 pitch. Specifically, Metal 1 was laid out at 3.9u while Metal 2 was laid out at 5u (for 1.25u process). In addition, spare Via's and contacts were put wherever possible. These measures gave some immunity to hard defects.

Post Production Debug and Correction

A production test analysis tool was developed to detect yield limiting circuit problems found in the first silicon. Each die was tested under a variety of conditions. The tool then identified the cycles when failures most frequently occured, as well as the effect of VCC, temperature, and burn-in on yield and failing mechanisms. This , together with the data available on the internal chip signals enabled engineers to detect problematic circuits within a few days.

A special tool was developed for identifying speed bugs. Cycles which caused a failure were identified by tracing backwards from the point of failure by widening one clock cycle at a time.

In order to provide guiding information to the fabrication line, Electrical Test (ET) results were recorded for each wafer. Statistical analysis was then performed on the effect of each ET parameter on the yield. Specific failure mechanisms were identified and a correlation was made between the failure mechanisms and the ET parametric results. This data enabled fabrication line process engineers to tune the line to achieve higher yield.

QUALITY OF THE DESIGN

Only seven problems which could effect yield have been identified since first silicon. These are two speed problems, two design sensitivity problems and three problems with the I/O buffers. The CPU has exhibited no race or charge split problems. The second version of the NS32532 ran at 30 MHz. The yield of the NS32532 was 6 times better than the yield of another device done in the same process but with traditional methodology.

CONCLUSIONS

The hierarchical design methodology enabled the design team to clearly identify the appropriate restrictions on each piece of hardware, to complete the design quickly, and later to verify the design cleanliness and to detect sensitive circuits which decrease the yield. The NS32532 design methodology contributed significantly to the final device yield.

ACKNOWLEDGEMENT

The authors wish to thank Jay Finkelstein for his editorial assistance and guidance in writing this paper.

REFERENCES

P. Agarwal, "Design Methodologies Applicable to Microprocessors", Proceedings IEEE international conference on Computer Design, October 1983.

D. Alpert, J. Levy, B. Maytal, "Architecture of the NS32532 Microprocessor", Proceedings IEEE International Conference on Computer Design, October 1987.

I. Sprecher, "Z: A Hierarchical Multi-level Interactive Logic Simulationtor", Conference Proceedings of the Israel Society for Computer Aided Design and Manufacturing, December 1985.

U. Weiser, Y. Sidi, L. Epstein, D. Biran, A. Kaminker, "Design of the NS32532 32-Bit Microprocessor", Proceedings IEEE International Conference on Computer Design, October 1987.

DEFECT TOLERANCE IN A 16-BIT MICROPROCESSOR

Regis Leveugle, Mohammad Soueidan,
and **Norbert Wehn***

Institut National Polytechnique de Grenoble
46, Avenue Felix Viallet
38031 Grenoble Cedex, France

***Technische Hochschule Darmstadt**
Schloßgartenstrasse, 8
6100 Darmstadt, F.R.G.

INTRODUCTION

The goal of this demonstrator of the ESPRIT project (824) was to prove the feasability of a high yield defect tolerant 16-bit microprocessor. Such a device is supposed to become the core of an application specific microprocessor-based system integrated on a single chip. A lot of effort was done to obtain a regular design allowing the introduction of standby elements at an adequate level. The area of redundant elements was limited to 30%.

GENERAL PRESENTATION OF HYETI

The HYETI microprocessor is dedicated to highly dependable real time control systems and its characteristics have been determined in cooperation with potential industrial customers.

Special emphasis has been given to on-line test. A compaction device allows to compact any information on the bus through a parallel LFSR with a programmable characteristic polynomial. Dedicated on-line test instructions perform the verification of instruction code sequences according to a "adjusted signature" method (Leveugle et al., 1988).

Software characteristics

The instruction set has been defined according to the application area (masked operations, boolean matrix processing, ...).

The instruction set is divided into 10 classes :

- Arithmetic instructions.
- Logical instructions, which can be masked or not.
- Shift and rotate instructions.
- Comparison instructions with or without mask.
- Transfer instructions with or without mask.
- Instructions with implicit operand.
- Branch and jump instructions.
- Stack instructions.
- Interrupt instructions.
- Test instructions (for on-line test).

The masked instructions allow the programmer to process selected bits in a 16-bit word.

Eight addressing modes are provided : relative, inherent, immediate, direct, register indirect, register indirect with post modification and register indirect with pre-modification. For the last modes, the indirection register is automatically modified by adding a programmable arithmetical offset. These modes were introduced for efficient matrix processing.

Hardware characteristics

HYETI is a 16-bit microprocessor with an addressing capability of 16 pages of 64 K words. The memory interface signals are compatible with those of Motorola family.

Two hardware interrupt levels are provided, as well as a halt control signal.

The circuit is implemented in the CMOS double metal 1.2μ technology of SGS-Thomson Microelectronics. Its general architecture is shown in Figure 1. The elements of the datapath include registers, the ALU and two special devices: a masking device for masked operations and a signature device for on-line test.

RECONFIGURATION PRINCIPLES

Underlying strategy

The goal of the reconfiguration strategy is the tolerance of a large number of defects in order to increase the yield and to be able to use the processor in a WSI system. Nevertheless, all defects are not necessarily tolerated : a defect tolerance strategy is also defined at the system level. The interconnection lines between the controller and the datapath are for example not reconfigurable. Some elements in the datapath are not reconfigured either because it is not possible to have them repairable without a large area overhead. The goal is to achieve a good compromise between area overhead and yield enhancement.

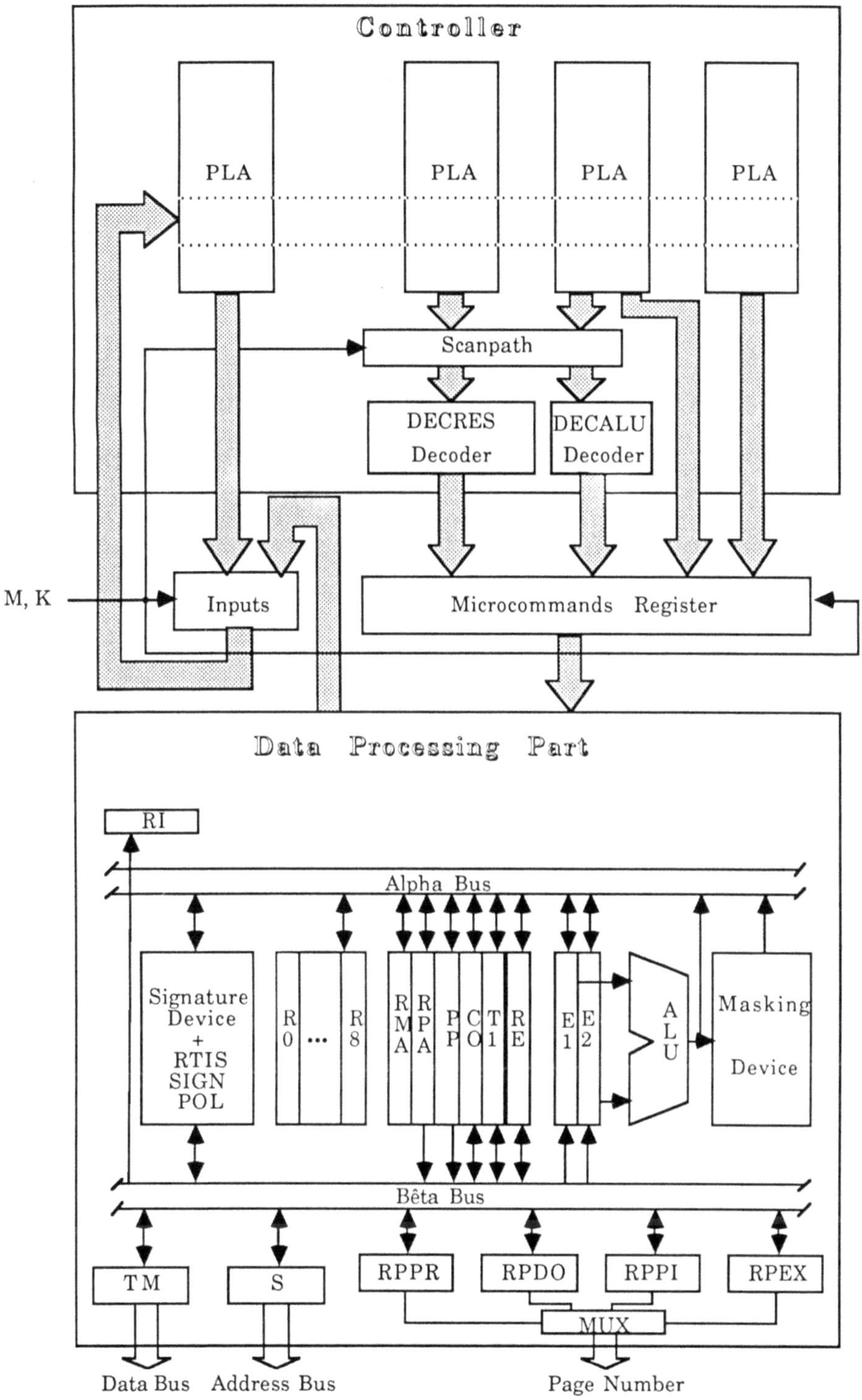

Figure 1 : General architecture of the HYETI microprocessor

In the yield analysis, we distinguish the elements of the circuit which are reconfigurable from those which have no redundancy. If we denote Aun the area in which a defect is fatal and Arec the area in which a defect can be repaired in most cases (some control lines may have no redundancy), the probability to have a circuit working after reconfiguration may be defined as

$$Pw = P(X=0 \; ; Aun) * P(X \leq d \; ; Arec) * Prec$$

where X is the number of defects,

P(X=i ; A) is the probability to have i defects in the area A,

P(X≤i ; A) is the probability to have less than or exactly i defects,

d is the number of defects which can be repaired,

Prec is the probability to have no fatal defect among those in the reconfigurable part.

The probabilities are computed using the negative binomial statistics and the technology parameters.

In our case, the microprocessor is divided into blocks where at least one defect can be corrected, except some fatal ones on the control signal lines. This partitioning respects the architecture: the datapath and the controller are considered separately.

A pessimistic yield formula is therefore

$$Pw = P(X=0 \; ; Aun) * \prod \left[P(X \leq 1 \; ; A_{block\ i}) * P_{rec\ i} \right]$$

since each block i of area $A_{block\ i}$ can tolerate at least one defect with the probability $P_{rec\ i}$.

This formula allows to determine the best size for a reconfigurable block, taking into account the estimated area of the chip and the estimated values of Aun and Arec. An optimum size for the blocks was found to be 7 to 8 mm^2 (Genestier, 1987).

Nevertheless, the definition of the blocks has to be closely associated to the architecture of the circuit. For example, it is impossible to consider that two PLAs are in the same block. They have obviously to be reconfigured separately. So, architectural and topological aspects of the design have to be considered during the partitioning.

Datapath reconfiguration

Generally speaking, the datapath of a microprocessor is made up of several functional units (registers, ALU, bus precharge logic, specialized operators) which are realized in a classical bit slice architecture.

The partitioning of the datapath can be either functional or structural (the bit slice is the basic block).

The functional partitioning consists in a separate reconfiguration of the different functional units. This partitioning may be necessary when there is no regular structure. The main drawback of this strategy is the complex connection network required between blocks in order to achieve the reconfiguration.

In the HYETI datapath, we take advantage of the bit slice structure. One spare slice is provided and can replace a defective one. Providing more spare slices would have led to an area increase superior to the limit fixed by the ESPRIT project. Furthermore, the area of the datapath is about 8 mm^2, which is about the optimal size determined by Genestier, (1987). Therefore, this part of the circuit can be considered as only one block.

The area increase for reconfiguration is about 20 to 25%. The non reconfigurable part of the datapath is about 20% of the area. Without reconfiguration, the expected yield of the data processing part of the circuit is

$$P(X=0 ; 6 \text{ mm}^2) \approx 0.67.$$

The expected yield with the reconfiguration devices is (with $P_{rec\ i} \approx 0.95$ according to the area of the non-reconfigurable lines in the reconfigurable part)

$$P(X=0 ; 2 \text{ mm}^2) . P(X \leq 1 ; 6 \text{ mm}^2) . 0.95 \approx 0.77.$$

Notice that this yield increase considers restrictive hypotheses. In fact, more than one defect can be corrected within the datapath, provided that they occur on reconfigurable devices and on the same slice.

Controller reconfiguration

A global lack of regularity can be noticed in the most commonly used controller structures. Several types of elements are necessary and the reconfiguration requires a lot of redundancy because each element is a particular case.

In the HYETI controller, only one type of element (PLAs) is used. An initial virtual PLA is partitioned into several ones, some of them being decoders. This organization in two stages allows to reduce the silicon area (a single PLA would be relatively sparse). Furthermore, the first stage is composed of several PLAs for topological reasons. A method has been defined in order to minimize the global size of the controller and the area needed for the interconnections (Leveugle and Soueidan, 1988). The final controller is composed of 6 PLAs in the first stage and 2 decoders.

The PLAs which are used in the HYETI microprocessor have clocked dynamic planes with switched grounded array transistors for the AND plane and directly grounded array transistors for the OR plane.

The initial PLA matrices, obtained after a complex synthesis process (state encoding, minimization ...) with the ASYL system (Saucier et al., 1987), are modified for testing in order to be able to localize a defect. The PLA testing strategy is based on the testing scheme presented by Bozorgui-Nesbat and McCluskey (1984) and improved later (Bozorgui-Nesbat and Khakbaz, 1985; Aarts et al., 1986). The basic idea of this scheme is the ability to activate an

Table 1 : Yield analysis and area of a PLA versus the number of redundant terms

Number of redundant terms	Area mm^2	Yield
0	2,75	0.81
1	2,77	0.93
2	2,78	0.94
3	2,79	0.95

individual product term by adding some additional inputs to the AND-plane of the PLA. With this scheme, the test procedure requires only the application of some test vectors; no register is needed to control the activation of product or input lines. A more detailed presentation of the method was made by Wehn et al. (1988).

Each PLA is considered as a block from the reconfiguration point of view. Except for the test inputs, the redundancy provided is composed only of programmable product terms. As a matter of fact, when taking into account the external interconnection overhead needed for redundant inputs and outputs, it is obvious that redundant product terms are a good compromise between area overhead and fault occurence. The yield analysis and area overhead is given in table 1 for a PLA with 30 inputs, 30 outputs and 322 product terms. Three additional inputs were added. The table demonstrates that an significant yield increase takes place with one redundant product term.

Input and output lines were also designed on two layers that are multiply connected. This technique ensures fault-tolerance with regard to open defects. On the other hand, decreased speed due to the enhanced capacitance must be accepted.

In the controller of HYETI, the area increase for redundancy is inferior to 20%. Furthermore, the size of each PLA is inferior to the one of the example in table 1. A yield superior to 0.95 is therefore expected.

Yield increase for the HYETI chip

In the present version, the prototype of the HYETI chip is composed of the microprocessor only. Pads are then connected in order to be able to test the chip. Nevertheless, these pads do not have to be considered in the yield analysis since the goal is to obtain a reconfigurable microprocessor for wafer scale integration.

.The useful area of the chip for the present version is about 20 mm^2, with about 4 mm^2 dedicated to reconfiguration devices and about 4 mm^2 with non-reconfigurable devices. The expected yield without any fault tolerance would therefore be about 0.35. After reconfiguration, the expected yield is at least 0.5. The prototype which was developped will allow to verify the theoretical results.

RECONFIGURATION DEVICES

Different types of switching devices

Reconfiguration involves disconnection of defective elements and connection of spare ones. Several methods, software or hardware, can be used. Three types of hardware switches have been used in a first version: laser fuses, antifuses and floating gate FETs.

Laser fuses and antifuses can be used when no reversibility is needed. Floating gate FETs can be used when several programming attempts are necessary.

Datapath modifications

When a bit slice partitioning is chosen, two problems have to be overcome: the bus connections on the good slices and the routing of signals between slices with the bypass of the defective one.

The first problem arises because some devices are not reconfigurable without incurring a large area overhead. For example, in the case of the status register, each bit has a specific meaning and is associated with some random logic. The datapath is therefore composed of three parts, according to Figure 2. Depending on the slices used in the reconfigurable parts, the buses have to be reconfigured in order to correctly connect the non-reconfigurable block.

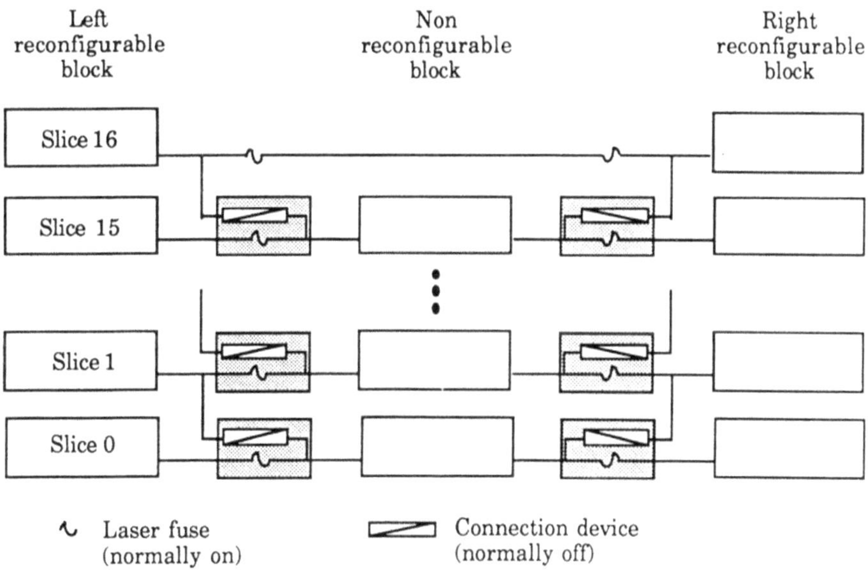

Figure 2 : General floorplan of the datapath

The initial connections of the buses are made with laser fuses. After test, the reconfiguration consists in cuting some fuses and connecting new lines. The connections may use antifuses. Nevertheless, it is useful from an economical point of view to minimize the number of different switches, namely the number of machines necessary during reconfiguration. Therefore, the connections are made by means of transfer gates. A special reconfiguration signal Crec is generated for each slice. It indicates if the slice must be bypassed and its logic level is determined by laser fuses. The Crec signal is generated on each slice by the circuit of Figure 3. On slice 0, the transfer gates are directly controlled by the Crec signal; on the other slices (1 to 15), the transfer gates are controlled by a logic combination of the Crec signals :

$$C_i = Crec_i \cdot C_{i-1},\ 1 \le i \le 15 \qquad \text{with} \quad C_0 = Crec_0$$

The connection circuit for these slices is shown in Figure 4. The Crec signals are also used to propagate correctly information between the slices according to Figure 5.

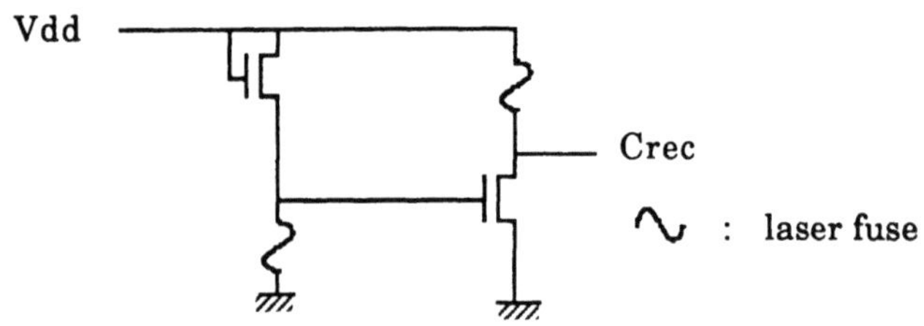

Figure 3 : Circuit used for the generation of the Crec signal

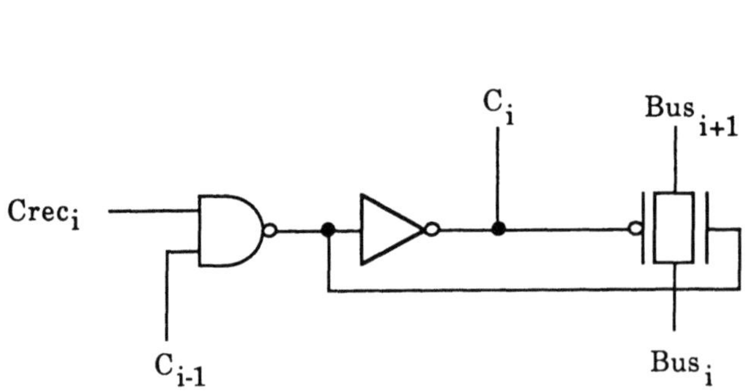

Figure 4 : Connection circuit for slice $1 \le i \le 15$

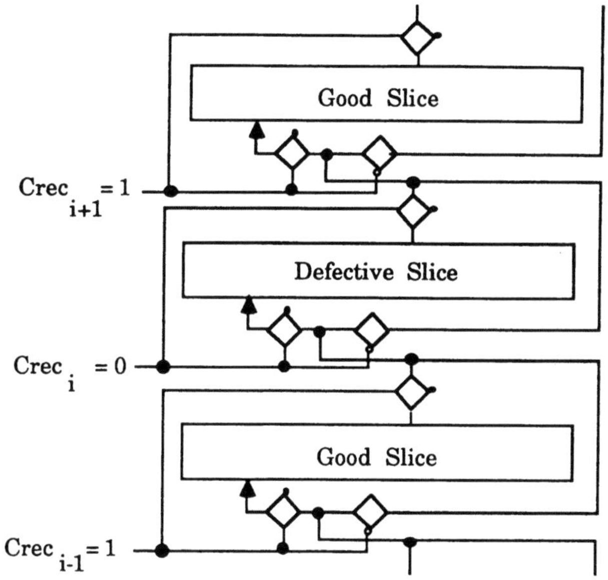

Figure 5 : Reconfiguration of signals propagated between slices

PLA modifications

As mentioned earlier, the reconfiguration of the PLAs is made by means of spare programmable product terms which can replace the defective ones. The programmation uses either laser fuses or N channel depleted floating gate FETs. In our case, fuses were used for three reasons : 1) reversibility is not essential, 2) fuses are already present in the datapath and 3) they are more reliable than floating gate FETs.

A normal product term is implemented as shown in Figure 6. A laser fuse allows to disconnect it from the OR array when it is proved to be defective. The programmable product terms are implemented as shown in Figure 7. These product terms have, at the end of manufacturing, no influence since transistors are connected in the AND array on all the input and complemented input lines. Some of these transistors are disconnected from the AND array when the redundant product term must be used. An example of reconfiguration is given in Figure 8. The programmable product terms can also be disconnected from the OR array when necessary.

OFF-LINE TEST AND RECONFIGURATION

At the end of manufacturing, a very efficient off-line test can be performed. This is due to the modifications introduced in the PLAs, to a special design of the ALU which has an ILA C-testable structure and to the implementation of scanpath registers. As a matter of fact, the microcommands register can be loaded or read serially in a test mode.

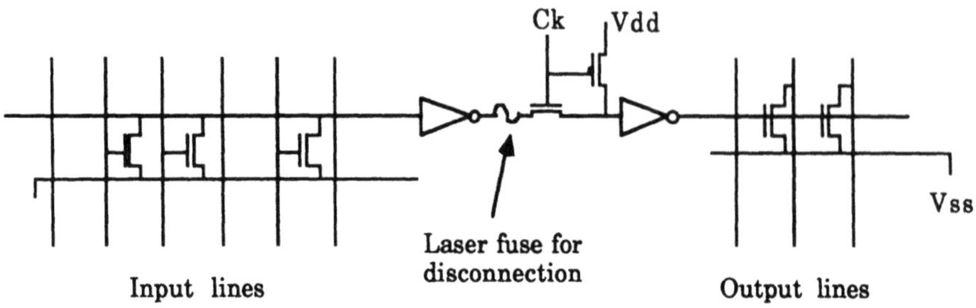

Figure 6 : Normal product term of a reconfigurable PLA

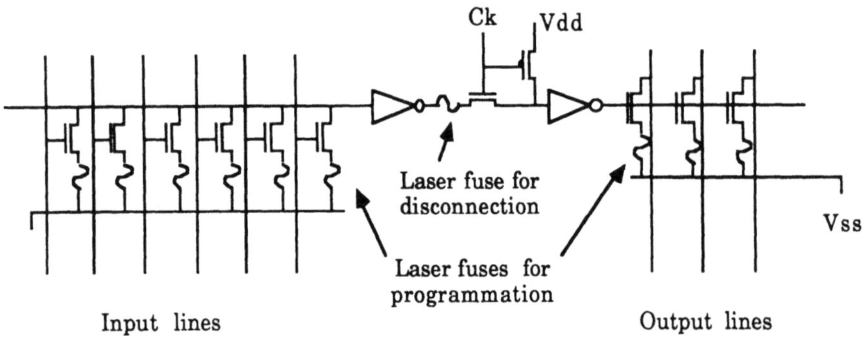

Figure 7 : Programmable product term of a reconfigurable PLA

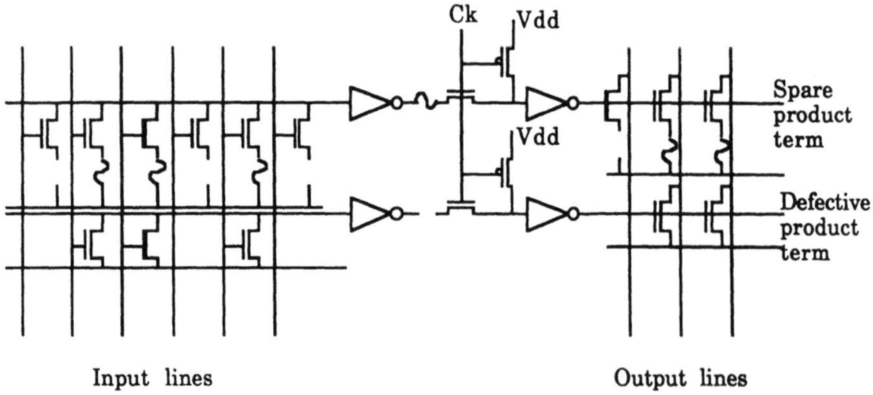

Figure 8 : Reconfiguration in a PLA

Furthermore, two scanpath registers are added onto the inputs of the controller and between the two stages of PLAs in the controller. Two signals M and K are provided to control the different scanpaths.

The test after manufacturing involves the following steps :

- Parametric test (general validation of the chip).
- Test of the 3 scanpath registers.
- Test of each PLA in the controller, using the scanpath registers.
- Test of the datapath elements, using the scanpath registers.
- Test of the whole microprocessor in working mode if all the other tests are successful.

During this phase, the test of the datapath is performed on the 17 slices at the same time according to a "start small" strategy.

When defects are detected in the reconfigurable parts of the processor, a reconfiguration phase takes place after this first test. In the different PLAs, the defective product terms are disconnected and replaced by spare ones which are programmed by laser cuts to implement the right function. In the datapath, the Crec signal of the defective slice D is set to 0 and all the fuses on the buses are cut on slices D ($0 \leq D \leq 15$) to 16. When no defect is detected or when the defective slice is the 17th one, only the fuses on this slice have to be cut.

When the reconfiguration is achieved, a second test is performed to verify the effectiveness of the corrections.

Notice that multiple consecutive test and reconfiguration runs are not allowed. A defect can be repaired only if it can be localized during the first test with the general test patterns. This restriction is economically motivated, since test and reconfiguration of a chip is a time-consuming process.

CONCLUSION

The method presented here may be applied to any type of circuit with a bit slice datapath and a controller. Nevertheless, some modifications are necessary during the design, especially for the complex dedicated operators. These techniques can be used to improve the productivity of VLSI circuits and may lead in the near future to the possible integration of whole systems on a single chip.

ACKNOWLEDGMENT

This project has been developped under the EEC ESPRIT 824 grant, under the leadership of Dr. J. Trilhe from SGS - Thomson Microelectronics. This work has been done under the responsability of Professor Saucier (I.N.P.G.) and Professor Glessner (T.H.D.). The authors thank also Philippe Genestier for his work on the reconfiguration strategies.

REFERENCES

Aarts, E.H.L., Beenker, F.P.M., and Ligthart, M.M., 1986, Design for testability of PLAs using statistical cooling, 23th DAC

Bozorgui-Nesbat, S., and McCluskey, E.J., 1984, Lower overhead design for testability of PLAs, Int. Test Conference

Bozorgui-Nesbat, S., and Khakbaz, J., 1985, Minimizing extra hardware for fully testable PLA design, ICCAD

Genestier, P., Jay, C., and Saucier, G., 1986, A reconfigurable microprocessor for Wafer Scale Integration, in : "Wafer Scale Integration", G. Saucier and J. Trilhe, ed., Elsevier Science Publishers, Amsterdam

Genestier, P., 1987, Conception de microprocesseurs à haut rendement, thèse de Docteur de l'INPG, Grenoble

Kuo, S-Y, and Fuchs, W.K., 1987, Fault diagnosis and spare allocation for yield enhancement in large reconfigurable PLAs, Int. Test Conference

Leveugle, R., and Soueidan, M., 1988, Design of an application specific microprocessor, International Workshop on Logic and Architecture Synthesis for Silicon Compilers, Grenoble

Leveugle, R., Soueidan, M., and Delord, X., 1988, HSURF : un microprocesseur à test intégré pour les applications à haute sûreté de fonctionnement, Sixième Colloque International de Fiabilité et de Maintenabilité, Strasbourg

Mangir, T.E., and Avizienis, A., 1982, Fault-tolerant design for VLSI : effect of interconnect requirements on yield improvement of VLSI designs, IEEE trans. on Computers, vol. C-31, n°. 7

Mangir, T.E., 1984, Sources of failures and yield improvement for VLSI and restructurable interconnects for RVLSI and WSI, Proc. of the IEEE, vol. 72, n°. 6

Saucier, G., Crastes de Paulet, M., and Sicard, P., 1987, ASYL: a rule-based system for controller synthesis, IEEE trans. on Computer-Aided Design, vol. CAD-6, n°. 6

Saluja, K.K., Kinoshita, K., and Fujiwara, H., 1983, An easily testable design of PLAs for multiple faults, IEEE trans. on Computers, vol. C-32, n°. 11

Somenzi, F., and Gai, S., 1986, Fault detection in Programmable Arrays, Proc. of the IEEE, vol. 74, n°. 5

Wehn, N., Glesner, M., Caesar, K., Mann, P., Roth, A., 1988, A defect-tolerant and fully testable PLA, 25th DAC

Wey, C-L, 1987, On the design of a redundant Programmable Logic Array, IEEE Journal of Solid-state Circuits, vol. SC-22, n°. 1

Wey, C-L, 1988, On yield consideration for the design of redundant Programmable Logic Arrays, IEEE trans. on C.A.D., vol. 7, n°. 4

DESIGN TECHNIQUES FOR A SELF-CHECKING SELF-EXERCISING PROCESSOR

Savio Chau* and David Rennels*

University of California
Department of Computer Science
Los Angeles, CA 90024

1.0 INTRODUCTION

Dormant faults and latent errors can impair recovery of fault tolerant systems. For some, a relatively long time may occur before detection. This can jeopardize recovery because hardware faults or transient errors can accumulate. Since most fault recovery mechanisms are only designed to cope with a single fault, accumulated faults during the recovery process may cause system failure. Several studies have shown that latent faults cannot be ignored in highly reliable systems [ShMc 75], [ShMc 76], [Chil 86], [Shin 86] and [Swer 87]. This is especially a problem in some aerospace systems which must operate in very severe environments in which high rates of transient errors are expected and external disturbances may cause multiple faults. This paper investigates VLSI design techniques which can search for these latent faults actively to allow rapid recovery before multiple errors build up. The basic principle of these techniques is to implement a system with concurrent error detection. Very short test cycles are then inserted into the system periodically as it performs its normal program execution in order to expose and detect latent errors and faults. The rate of inserting these test cycles is chosen (typically every 100 execution cycles) such that the entire system can be exercised in a fraction of a second. The goal is to achieve self-exercising without an excessive hardware overhead (beyond the initial design for concurrent fault-detection which is needed for fault-tolerance anyway) or significantly degrading the system performance. These techniques are also used to provide rapid hardware diagnosis, to simplify initial testing, (a major expense in space programs) and to provide the possibility of on- line acceptance testing (i.e. testing a circuit by just running applications programs.) A previous paper described how self-checking, self-exercising design could be effectively employed in memory [Renn 86]. This paper extends the approach to the processor.

2.0 DESIGN METHODOLOGY

In this paper we have focused on generic VLSI building-block components which can be used to implement the data section and control structure of a processor. First, an overall strategy is developed for implementing error detection and then redundancy at a system level, and then internal checkers are then added to the component logic structures to provide concurrent error detection. Second, the self-checking design is modified so that all of its internal logic structures can be more easily exercised by external inputs. (The designs involve careful consideration of VLSI layout issues and use both logic and circuit design techniques.) Third, test cycles are designed to cause faults and errors to be detectable by the internal checkers, and they are interleaved with normal sys-

* Also with the Jet Propulsion Laboratory, Pasadena, CA 91103

tem operation to test each component logic structure. Each test cycle is a small portion of the complete test of the component. Hence, the test cycles are very short and can be applied at a relatively high rate (e.g. one every 100us). If the components are easily testable, a complete test only requires a small number of test cycles. The fewer the number of test cycles needed, the quicker is the detection of the latent faults. The test steps also take advantage of the checkers for concurrent error detection to perform the test verification. Once the overhead of concurrent fault detection is paid for, only a modest amount of additional hardware is needed to provide self-exercising to expose latent errors.

3.0 A SELF-CHECKING SELF-EXERCISING DATA PATH

The data portion of a processor was first chosen to illustrate the self-checking exercising design techniques. It employs a 16-bit dual bus architecture and consists of three major function blocks: a register file, a shifter and an ALU (Figure 1). The VLSI architecture of this data path is similar to the OM2 data path chip [MeCo 78] developed by Caltech. The register file is a 16 x 16 dual port memory. The shifter performs shifting and rotation by extracting a 16 bit segment from the two buses. The ALU consists of two input registers, two output registers and a Manchester adder which uses three general function blocks at each bit position to produce generate, propagate and stop signals. The general function blocks can be programmed to performed addition, subtraction, logic operations and one-bit left shift. Figure 1 also shows in a general way, where additional logic is added for concurrent error detection and self-exercising. This will be described below.

Concurrent Error Detection. Two ways were considered for implementing concurrent error detection. The first is to duplicate the complete data path on-chip and compare the outputs of both copies. This approach was recommended by Univac, and design methodologies for layout with complementary logic signals in the duplicated copies have been developed [SEDM 80, TAMI 84]. The second approach is to use error detecting codes in a single copy of the data path as the primary means for error detection. It is well known that these codes preserve code properties under arithmetic operations and provide low-cost checking [AVIZ 71]).

The second approach was chosen for this design. Two check bits are added to the data path and are a check symbol of the modulo 3 inverse residue code [Aviz 71]. (Future designs may extend this to mod 7 or 15 for improved multiple error detection.) Registers, buses, arithmetic operations, and shifting (with augmented masking circuitry) are checked

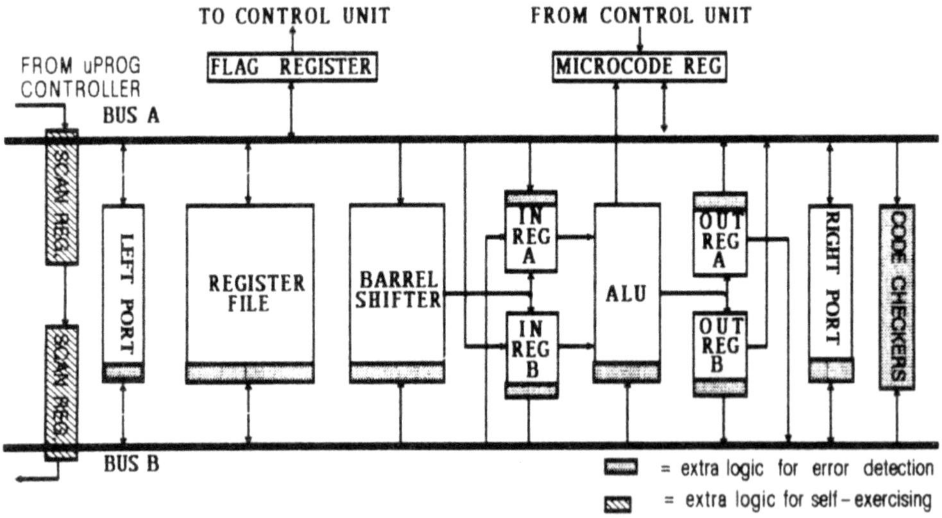

Fig. 1. Model of Data Path

using this code and logical operations are checked using duplication and comparison. This choice was made to optimize area (less logic than duplication) at the probable expense of performance (larger layouts and longer critical paths).

Self-Exercising Features. The next step is to modify the self-checking design so that its components can easily be exercised. Special scan registers are placed "under" the internal buses and at the input of other combinational circuits not accessible from the buses. These registers can be loaded or unloaded serially while normal computing is taking place. After they are loaded, their contents are output during a test cycle to generate a test vector into their associated circuitry. The test vectors are designed to cause latent faults to generate errors which are detectable by the built-in error checkers. (As an added benefit, the scan registers can also sample bus information and serialize it out off of the chip. This provides an access capability similar to LSSD for off-line localization of a fault.)

There are two types of test cycles, one for the sequential logic and one for the combinational logic. The combinational logic test cycles and the sequential logic test cycles are separate but interleaved with each other. The exercise sequence is illustrated in the following.

TIME	OPERATION
A.	NORMAL OPERATION (100 cycles)
B.	SELF-EXERCISE CYCLE: SEQUENTIAL LOGIC
C.	NORMAL OPERATION (100 cycles)
D.	SELF-EXERCISE CYCLE: COMBINATIONAL LOGIC
E.	NORMAL OPERATION (100 cycles)
F.	SELF-EXERCISE CYCLE: SEQUENTIAL LOGIC
.	
.	etc.

In sequential logic test cycles, sequential elements are isolated from each other so that they can be tested simultaneously. These elements are tested by toggling their contents, and no external test inputs are required. Also, the state of the data path will not be destroyed by the testing. Information in sequential elements is protected by error detecting codes, and they are specially designed to allow all of the registers to be code-checked during a few check cycles. The combinational elements must usually be tested individually because they share the same buses for test data input, and they share the same code checkers.

During the self-exercise cycle, all control signals are routed directly (with multiplexers) from the input pins and thus bypass any control register. Similarly, all flag signals are routed directly to the output pins and bypass the flag register. This is to ensure that the machine state is preserved so that normal operations can be resumed immediately after self-exercise.

The following sections describe the self-exercising features of each component.

3.1 A Self-Checking Self-Exercising Two Port Register File

The two port register file in the data path is protected by a two dimensional self-checking approach. The residue code is used to detect errors in the horizontal dimension (rows) during readout, and a special self-checking register cell design is used to detect any transient errors in the vertical dimension (columns). This specially designed register cell contains a static sub-cell (two inverters), and a dynamic sub-cell which is part of a built-in comparator (Figure 2). The CMOS pass transistors (D,E) store the (complement) value of the static cell as charge on gates B and C of the comparator. When a register cell is writ-

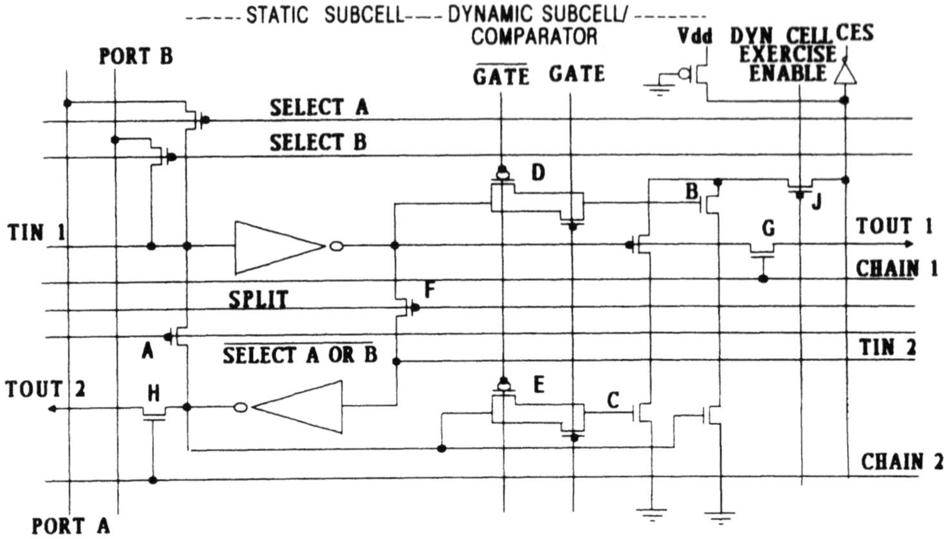

Fig. 2. Self-checking Self-exercising Register Cell

ten, the data is stored in both sub-cells, then the pass transistors are turned off. The sub-cells are compared to each other by the built-in comparator. The comparator output is connected to a signal line (called Column Error Signal or CES) shared by all the registers cells in the same column. A transient fault which flips either sub-cell of a register cell in a column can be detected instantly by the CES. The dynamic sub-cells are periodically refreshed in the test cycles.

Recovery from transient errors is also possible using this approach. When a CES is detected, all the words in the register file will be read out one by one. The faulty word can be detected by the residue code check and the faulty bit position is indicated by the CES. Hence, the error can be corrected by reversing the faulty bit.

This self-checking register file is not capable of detecting latent stuck-at faults. A conventional way to detect these faults requires (1) reading out one word at a time, (2) checking its parity, (3) reversing all the bits, (4) writing back the word, (5) reading and checking the inverted word, and (6) restoring the word by reversing the bits again. This approach will take too many test cycles. Therefore, it is necessary to modify the self-checking design to make the whole register file fully testable in a few test cycles.

Transistors F,G,H, and J are added to provide self-exercising features. Transistors A and F allow the two inverters of the static cell to be split, and transistors G, and H allow the upper and lower inverters in the static cells in a given row (word) to be connected together in a chain. In a test cycle, the two inverters of the static sub-cell are isolated from each other so that the data is stored dynamically at the gates of both inverters. Then one of the inverters is chained to the corresponding (upper or lower) inverters in neighboring cells so that a long chain of inverters is formed in each row. After that, a 1 and a 0 is applied alternatively to the input end of all the inverter chains. Any stuck-at fault will prevent the output end of the chains to alter. After testing, data can be restored from the untested inverter. Therefore, all the static sub-cells can be tested in two test cycles, regardless the size of the register file.

Since any shorted transistor in the dynamic sub-cell can easily be detected by the comparator, it is only necessary to check for stuck- open transistors in the dynamic sub-cell in the test cycle. This can be done by enabling the transistor J of the comparators in one and only one row of register cells while the chaining test (above) takes place. Since one of the inverters of the static sub-cells takes on both 1 and 0 and the trapped values at transistor B and C are not altered, one of the pull down paths in the selected comparators should be turned on and pull the CES low once in the test cycle. Stuck-open transistors are detected by non-toggling CES signal. The dynamic subcells are refreshed when the static sub- cells restore their values.

It is worth noting that this augmented register design can also provide delayed commitment on writes by temporarily splitting the two inverters of the static cell. New data can be loaded into the upper inverter only of a static cell (dynamically) while maintaining the old value dynamically in the lower inverter. This can be useful in some cases of checking delays. If the newly loaded data proves to be faulty, the lower inverter can be used to restore the old data. In any case, after a cycle or two either the new or old data is stored statically by opening the feedback pass transistors (A and F) in the appropriate sequence.

This design has been verified by SPICE simulation. The layout of the self-checking self-exercising cell is about 12% larger than a self-checking duplex static register cell.

3.2 A Self-Checking Self-Exercising Barrel Shifter

The barrel shifter in this data path performs three functions: left shift, right shift and rotation. A typical non-self-checking barrel shifter is shown in Figure 3 [WeKa 85].

Designing a barrel shifter with concurrent error detection is difficult. One method is to duplicate the shifter and compare the outputs. Unfortunately, since the shifter has a large number of input and output signals, this method requires too much routing area for VLSI design. (The total area overhead could be up to 500%.) A usual way to reduce the routing problem in duplex design is spatially interleaving the bits of the duplicate copies so that the comparison can be done locally at each bit output. However, this technique is impractical for shifters because the internal connection between bits is very difficult to lay out.

The approach taken here is to again make use the inverse residue code. Although the code is not invariant under general shift operations, it is under the rotate operation (if the data length is an integral multiple of the number of check bits). If the data is rotated N positions, the checksum will be correct if it is rotated N-modulo i positions in the same direction (where i is the number of check bits) [AVIZ 71]. Thus we always use the barrel shifter as a rotator and check it using the residue code. The VLSI implementation of the mod 3 inverse residue code checker is shown in Figure 4. It is relatively simple and small compared to the shifter.

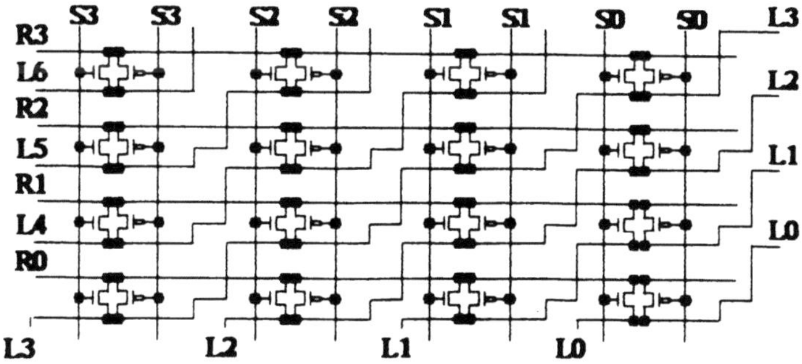

Fig. 3. CMOS Barrel Shifter

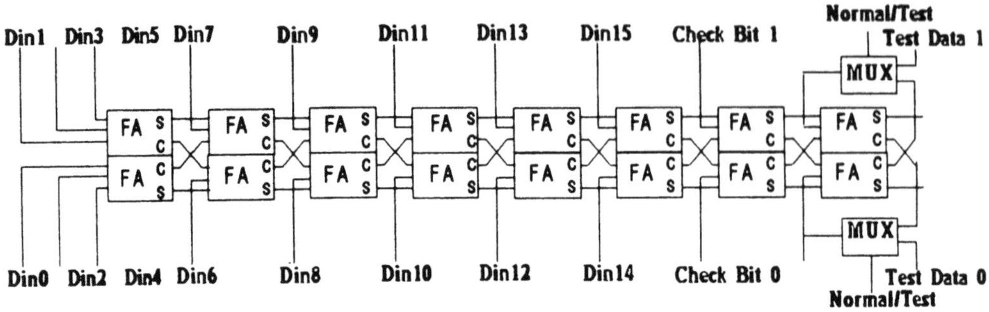

Fig. 4. MOD 3 Inverse Residue Code Checker

Based on this observation, a scheme was devised to check the shift operation by breaking it up into two steps. The first step is to rotate the input data by the number of bit positions to be shifted. Then the shift operation is achieved in a second step by masking the higher order or lower order bits of the rotated data with zeros or ones. The masking pattern is derived from decoding the shift distance with a modified 4-to-16 decoder.

Example 1: Left shift by 5 bits

Data		Left Rotate 5 bits		Mask Last 5 bits
01110010	≥	01001110	≥	01000000

Hence, the rotate operation can be checked by the inverse residue code as mentioned above. The mask operation can be checked by comparing the input and output of the masker. At output of the masker, the unmasked bits are compared with the input bits and the masked output bits are compared with zeros. Figure 5 shows a block diagram of this scheme.

Based on a transistor level circuit design, this self-checking design only requires twice area as the non-self-checking shifter. Therefore, this design is very practical in comparison to the other approaches such as the straight forward duplicate-and-compare. Since the barrel shifter is directly accessible from the buses, very little additional hardware is needed to make it testable by external test inputs. The only required extra logic is a multiplexer which connects the control signals of the shifter directly to the input pins during test cycles. A complete test sequence for stuck-at faults has been derived which consists of 22 test cycles.

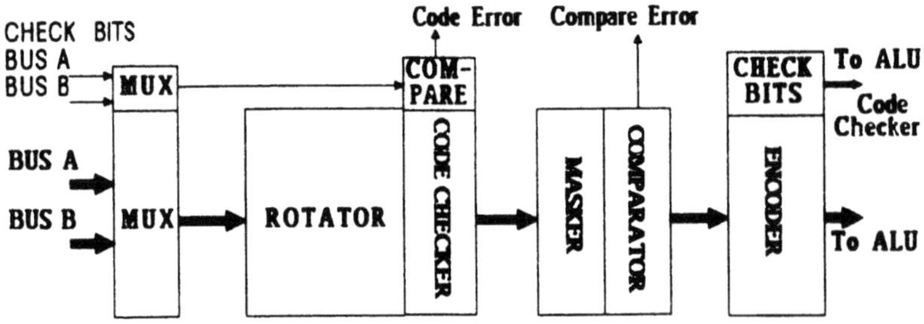

Fig. 5. Self-checking Rotator

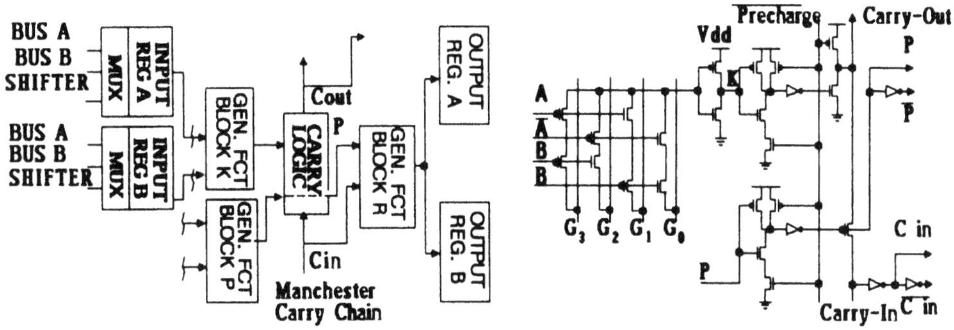

(a) Block Diagram (b) CMOS Circuit Design

Fig. 6. Basic Design of the ALU

3.3 Self-Checking Self-Exercising ALU

Figure 6a shows the basic design of the ALU in the data path. This ALU can perform three types of functions: arithmetic operations, logic operations and one bit left shift. It consists of two input and two output registers, three general function blocks (K, P and R) and a Manchester carry chain [MeCo 78]. The input registers can select either one of the buses or the shifter as their inputs. Their outputs are fed to the function blocks which can be programmed to perform any logic function on the two input data. In arithmetic operations, the function block P is programmed to generate a half- sum and the general function block K is programmed to generate a carry kill signal. The output of these two blocks are fed into the carry chain logic to generate the carry signal. Then the general function block R uses the half-sum and the carry signal to generate the final result. The result can be stored in either one of the output registers. The Manchester carry chain can also be used to perform the one-bit left shift operation. The CMOS implementation of the general function blocks and the Manchester carry chain is shown in Figure 6b.

The inverse residue code is the primary technique of checking the ALU for arithmetic operations. They are preserved under "ones complement" arithmetic operations (i.e. independently executed operations on data and check symbols preserve the code for add, and subtract. Since this is a "two's complement" machine, a code adjustment must be made for carries out.

For logic operations, duplication and comparison is used for error detection, and results are re-encoded before being sent to the bus. A straightforward duplicate-and-compare approach would require about 200% of area overhead, which includes duplicating the ALU, the area for the comparator and, most importantly, routing the ALU outputs to the comparator. Therefore, an alternative technique was found which does not require excessive overhead.

It was observed that each bit slice of the ALU is composed of three identical arrays of general function blocks [MeCo 78] and each of them can be programmed to perform any function for two Boolean variables. For arithmetic operations all three of the function blocks are used but in logic operations, only one of these function blocks is used and the other two function blocks are redundant. Thus, to accomplish self-checking in logic operations, it was only necessary to program the first and second function blocks to perform the logic operation and then program the third function block to compare their results. However, the input registers to the function blocks are duplicated to ensure that the function blocks are independent. Some additional logic is also needed to change the routing of the output signals of the functional blocks. The circuit is being laid out and the preliminary result indicates that the area overhead of this approach is about 80%, in contrast to nearly 200% of a conventional duplex design.

The exercise of the ALU consists of both sequential logic test cycles and combinational logic test cycles. The input registers to the ALU are duplicated, so they are be toggled and compared to detect stuck-at faults. The check is done by the same comparison circuitry for checking logic operations. This approach, which preserves the contents of the registers after testing, is also taken for the output registers. The combinational logic of the ALU are exercised in the combinational logic test cycles. A set of 1156 test vectors has been derived in [Chau 88].

In order to preserve the contents of the input registers while the combination logic of the ALU is being exercised, additional multiplexers are placed at the input of the general functional blocks. Hence, the general function blocks can get the test data directly from the buses without disrupting the input registers. A preliminary layout indicates that the self-exercising features add about 20% in area to an ALU which already uses inverse residue codes for concurrent error detection.

4.0 SELF-CHECKING SELF-EXERCISING CONTROL

A generic structure of a control unit is shown in Figure 7. It generates the regular control signals for the other modules of the system. In addition, test vectors are read from the microcode ROM into a parallel-to-serial register and sent to the various system modules via the Serial Test Vector Bus. During typical operation, the Control Unit loads a set of test vectors into the parallel- serial registers, and then executes normal microprogram steps (typically 50-100 cycles) while the test vectors are being transferred serially into selected external units. It then transfers to a short sequence of test microinstructions which execute the a series of tests using the current vectors, and then it initiates sending the next set of test vectors (for the next test). The process repeats with execution of the next burst of normal microcode followed by another test sequence, etc.

A set of read addresses to sequential locations in the ROM are also interleaved with normal execution and test sequences to scrub the ROM and provide self exercising as described below. To do this there are three interleaved microprogram counters -- one for normal processing, another for controlling test sequences, and a third for control memory (ROM) scrubbing. A timer causes the controller to periodically switch between the three microprogram counters.

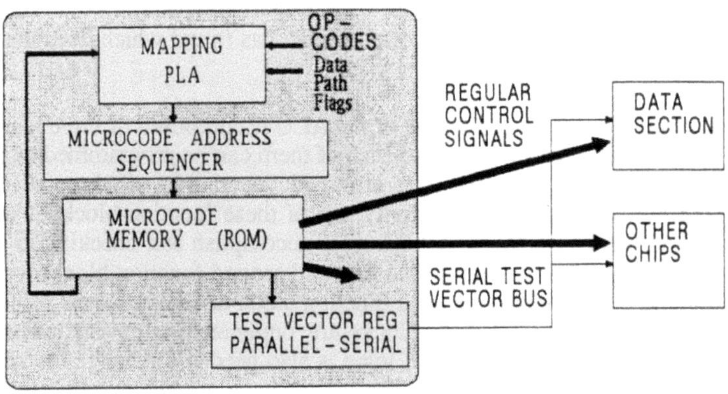

Fig. 7. A Strawman Control Unit

4.1 Self-Checking Self-Exercising of the Components

The Microcode ROM. For the microcode ROM, concurrent error detection is provided using an inverse residue code, and self- exercising is done by scrubbing. Single cycles are interleaved with testing and normal execution to sequentially read out each word and check its code. For these special scrubbing cycles, microcode output is only code-checked. It is not sent to external units so as not to affect the state of the rest of the machine. There are three types of microprogram control words: (1) Regular control words are sent out to the memory and data section to control normal program execution, (2) Test Control words are also sent out to control exercise tests, and (3) Load-Test words are used to load test vectors to be serialized and sent to the data section.

The Mapping PLA. This design takes advantage of two facts. First, it is a large PLA used for mapping op-code and condition signals to generate microprogram addresses and can thus afford the overhead of a modest amount of special testing circuitry. Second, for this mapping function it is possible to program the PLA so only one AND term is active at a time. (A general controller design was presented in Renn 81 a PLA with non-concurrent AND terms.) Previous work has shown that non-concurrent AND terms are necessary to achieve self-checking using coding techniuqes [WaAv 79]. This design has the constraint that a separable code be used for checking outputs (i.e. the data and check bits must be distinct) because the output is used directly as a microprogram address.

The approach we have taken for concurrent error detection is summarized below. First, the PLA inputs are received in encoded form with Hamming distance two. (This is part of the error protection in the rest of the system, fields of the OP-code are encoded in a separable (hamming or inverse residue) code and condition-code inputs are duplicated.) Minterms in the AND plane include the redundant checking bits. This means any non-selected AND term line will have at least two physically separated transistors pulling down on it. Single stuck-at faults in the input register, input select lines, and AND plane transistors, will result in zero AND-term lines being selected.

It is possible that shorting faults can result in more than one AND term being selected. To detect this condition the VLSI design of the PLA is augmented by adding a parity checking circuit over the AND terms that signals an error if an even number of terms is active. With this check in place, the remaining undetectable single stuck-at or bridging faults in the OR plane will only affect one output bit. An inverse residue code is added to the OR plane to detect there errors. This code is then carried through the microcode address sequencer to protect its arithmetic operations.

The principal problem in generating interleaved test cycles for self-exercising the PLA is that the AND plane of a PLA is not a full decoder. In other words, the number of minterms is usually a tiny fraction of the number of all possible input addresses. Hence, the use of a counter to generate the test address is very inefficient. Our approach uses a scheme that was proposed by Fujiwara [Fuji 81] and Hong [Hong 80]. Their designs use a shift register to activate the minterms individually. Therefore, each word in the OR plane can be called out disregarding the personality of the AND plane. Then the output words can easily be checked using the residue code.

The input register is similarly modified with a shift register for testing so that the input lines of the AND plane can also be activated individually. One additional AND term is added which is programmed to make an odd number of AND lines high for each input line. Thus, the program points on each input line can be checked by the parity checker over the AND lines. The PLA design is summarized in Figure 8.

The Microcode Address Sequencer. A Microcode Address Sequencer is shown in Figure 9. It contains three microprogram counters (for controlling normal program execu-

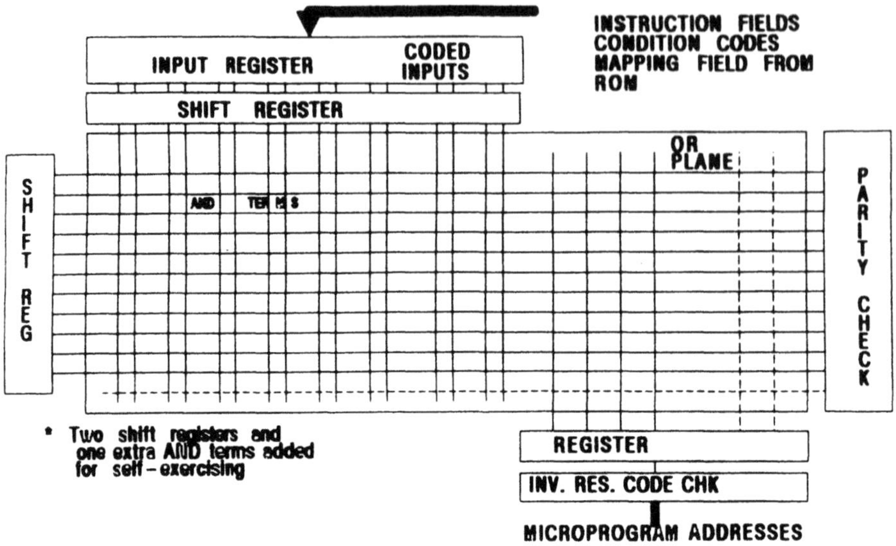

Fig. 8. The Mapping PLA in the Control Section

tion, scrubbing of the microcode ROM, and sequencing test vectors for self-exercising in external modules), a multiplexer, and an incrementer. Since the primary functions provided in this block of logic is arithmetic, inverse residue codes are used for error detection.

A novel technique for self-exercising is employed. The three program counters are periodically shifted through the three registers (R1, R2, R3) shown in Figure 9. When the real-time counter overflows, the microprogram counters are rotated in the three registers (microprogram, test, and scrub as shown in the figure). The value in the Active Register determines the microprogram addresses currently being executed. Thus the active register will sometimes contain the regular microprogram addresses. At other times it will contain

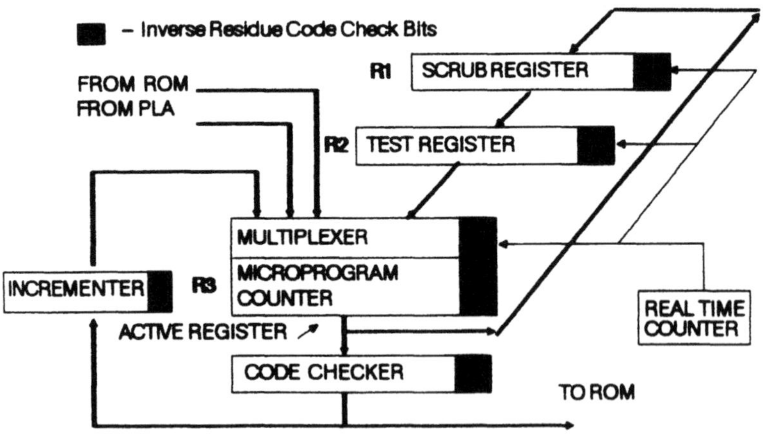

Fig. 9. Self-checking Self-exercising Microcode Address Sequencer

the address of the test microprogram, and at other times the address of the scrubbing of the ROM. Since the counter which sweep-tests the microcode ROM generates all possible addresses, it provides exhaustive test vectors for the registers, incrementer and code checker in the sequencer. (For every one of its values, it will reside in all three registers, and be incremented by the incrementer and be checked by the checker.) The Real Time Counter is duplicated and compared with a self-checking comparator.

5.0 CONCLUSIONS

This paper has presented a methodology and some techniques to design a self-checking self-exercising system which is capable of detecting active faults when they occur and latent faults concurrently with normal program execution in a fraction of a second. The work is still in progress, and some of the more complex issues of testing control circuits is still under study, but we feel that the preliminary results are promising. It is intended as an example to show that synthesis of architecture, concurrent fault detection, specialized VLSI layout, and design for test techniques can lead to new and interesting fault-tolerant designs.

References

[Aviz 71] Avizienis A., "Arithmetic Error Codes: Cost and Effectiveness Studies for Application in Digital System Design", *IEEE Trans. Comp.,* Vol C-20, No. 11, Nov 1971, pp. 1322-1331.

[BoMc 80] Bozorgui-Nesbat s., McCluskey E., "Design for Autonomous Test", *Proc. 1980 IEEE Test Conf.,* pp. 15-21.

[Chau 88] Chau S., "Self-Exercising in Self-Checking Fault Tolerant Computers", PhD Dissertation, UCLA, in preparation.

[Chil 86] Chillarege R., Iyer R., "Fault Latency in the Memory - An Experimental Study on VAX 11/780", *Digest 16th FTCS,* Vienna, Austria, July 1986.

[Fuji 81] Fujiwara, H., Kinoshita, K., "A Design of Programmable Logic Arrays with Universal Tests," *IEEE Trans. Computers,* Nov. 1981, pp. 823-828.

[MeCo 78] Mead C., Conway L., *Introduction to VLSI Systems,* Addison-Wesley. 1979.

[RENN 86] Rennels, D. and S. Chau, ''A Self-Exercising Self- Checking Memory Design,'' *Dig. Int. Symp. Fault-Tolerant Computing,* Vienna, June 1986, pp. 358-363

[SEDM 80] Sedmack, R., and Liebergot, H., "Fault Tolerance of a General Purpose Computer Implemented by Very Large Scale Integration, ", *IEEE Trans. Computers,* Vol C-20, No. 6, June 1980, pp. 492-500.

[Shin 86] Shin K., Lee Y-H., "Measurement and Application of Fault Latency", *IEEE Tran. Comp.,* Vol C-35, pp. 370-375, Apr. 1986.

[ShMc 75] Shedletshy J., McClusky E., "The Error Latency of a Fault in a Combinational Digital Circuit", *Digest 5th FTCS,* Paris, France, June 1975, pp. 210-214.

[ShMc 76] Shedletshy J., McClusky E., "The Error Latency of a Fault in a Sequential Circuit", *IEEE Trans. Comp.,* June 1976, pp. 655-658.

[Swer 87] Swern F. et el, "The Effects of Latent Faults on Highly Reliable Computer Systems", *IEEE Tran. Comp.,* Aug. 1987, pp. 1000-1005.

[TAMI 84] Tamir, Y. and SéQuin, C. H., "Reducing Common Mode Failures in Duplicate Modules," *International Conference on Computer Design,* Port Chester, NY, pp. 302-307 (October 1984

[WaAv 79] Wang, S., and Avizienis, A., "The Design of Totally Self- Checking Circuits using Programmable Logic Arrays," *Proc. 1979 Int. Symp. on Fault-Tolerant Computing,* Madison, WI, June 1979, pp. 173-180.

[WeKa 85] Weste N., Eshraghian K., *Principles of CMOS VLSI Design: A Systems Perspective,* Addison-Wesley, 1985.

[Will 82] Williams T., Parker K., "Design for Testability - A Survey", *IEEE Trans. Comp.,* Vol C-31, No 1, Jan. 1982.

CACHE MEMORY ORGANIZATION TO ENHANCE THE YIELD OF HIGH-PERFORMANCE VLSI PROCESSORS†

G.S. Sohi

University of Wisconsin-Madison
Department of Computer Sciences
Madison, WI 53706

Abstract

High-performance VLSI processors make extensive use of on-chip cache memories to sustain the memory-bandwidth demands of the CPU. As the amount of chip area devoted to on-chip caches increases, we can expect a substantial portion of the defects/faults to occur in the cache portion of a VLSI processor chip.

This paper studies the tolerance of defects/faults in cache memories. We argue that, even though the major components of a cache are linear RAMs, traditional techniques used for fault/defect tolerance in RAMs may neither be appropriate nor necessary for cache memories. We suggest a scheme that allows a cache to continue operation in the presence of defective/faulty blocks. Then we present the results of a trace-driven simulation analysis that evaluates the performance degradation of a cache due to defective blocks. From the results we see that the on-chip caches of VLSI processors can be organized such that the performance degradation due to a few defective/faulty blocks is negligible. We conclude that by tolerating such defects without a noticeable performance degradation, the yield of VLSI processors can be enhanced considerably.

1. Introduction

Advances in semiconductor technology have led to the development of high-performance single chip VLSI processors. For such processors, an increase in CPU speed must be coupled with an increase in memory bandwidth. By far the most popular technique for improving memory bandwidth in general purpose processors is the use of cache memories. For a single chip processor, this translates into the use of on-chip cache memories. Many recent single-chip processors use some form of on-chip cache to provide adequate memory bandwidth and reduced memory latency for the CPU [1, 7]. For example, the MIPS-X processor devotes more than half of its chip area to an on-chip instruction cache [5]. We expect that, in the future, most single-chip VLSI processors will devote a sizeable fraction of their chip resources to cache memories.

An increase in the circuit density of single chip processors is coupled with an increase in defects and, with a sizeable fraction of the chip resources devoted to cache memories, we can expect a large fraction of defects/faults in the processor chip to be present in the cache memory portion. If such defects can be tolerated without substantial performance degradation and/or increase in cost, the yield of single-chip VLSI processors could be enhanced considerably [6].

The crux of defect- and fault-tolerance techniques is the use of redundancy. Generally, some form of redundancy is provided explicitly for defect-/fault-tolerance. Sometimes, redundancy may be introduced in a processor not for defect-/fault-tolerance but for performance enhancement. A cache memory is an example of such a "redundant" resource. Cache memory is "redundant" because

†This work was supported in part by NSF Grant CCR-8706722.

the correctness of processor operation is not dependent upon the cache. A processor can still operate correctly, albeit with severly degraded performance, in the absence of an architecturally-invisible cache memory. When "redundant" components are present in a processor, two important and interesting questions arise: (i) what defect-/fault-tolerance technique should be used for "redundant" portions of a processor such as the cache memory and (ii) what are the performance implications of such a scheme? The focus of this paper is to evaluate the effects of defects/faults in a cache memory and provide answers to the above questions in the context of a cache memory.

The outline of this paper is as follows. We start off by discussing the nature of defects in cache memories and see how they affect cache operation. We discuss the use of techniques that use additional redundancy and see why their use may not be a good choice for cache memories, especially for the on-chip caches of VLSI processors. We present a technique that allows a cache to continue operation even though some of its blocks may be defective. Then we evaluate the performance of cache memories to determine if defective blocks cause any appreciable loss in performance. Next, we discuss a sector cache organization. Finally, we present some concluding remarks.

2. Defects in Cache Memories and Their Tolerance

A cache memory consists of several *blocks* or *lines* of data. Each cache block is occupied by data elements from a block of the memory. A block consists of several contiguous bytes of memory. As data is referenced by the processor, it is brought from the memory into the cache. Data from a memory block is present in the *data memory* or *data array* portion of the cache. Each block in the cache has an associated tag which is kept in the *tag memory* or *tag array* portion of the cache. The tag is used to distinguish between one of several memory blocks that map onto a cache block. A simplified description of cache operation follows; a detailed description can be found in any text on computer architecture or in a survey paper by Smith [12].

When the CPU generates a memory request, a portion of the address is used as a tag and is compared with the tag(s) stored in appropriate locations in the tag array. If a match results we have a cache *hit* and the data is accessed from the corresponding location in the data array. If no match results we have a cache *miss*. On a cache miss, the entire memory block is transferred from the memory to the cache and is then accessed from the cache. In more sophisticated *sector cache* organizations, a block could be sub-divided into several *transfer blocks* [12]. To simplify the discussion, we shall initially assume a non-sector cache organization (indeed, this is the more common case). Then, in Section 4, we shall discuss a sector cache organization.

2.1. Types of Cache Defects

Components of a processor such as registers, buses, control logic, and the ALU, are critical to the functioning of the processor. Defects in such components are *critical defects* because the defect will lead to incorrect processor operation unless some action is taken to tolerate and/or correct such defects. Consider, for example, a defect in a register. Instructions that utilize the defective register have no alternate modes of operation without violating the architectural definition of the instruction and will fail unless means are provided to tolerate the defect. Likewise, an ability to tolerate defects in the main memory must also be provided. As mentioned earlier, cache memory is not an "essential" component of the processor as far as correct operation is concerned. Cache memory is present in a processor mainly for performance reasons. The processor will be able to operate in a correct but degraded fashion if parts (or all) of the cache memory are unavailable and if alternate means are provided to recover and access correct data. If data cannot be accessed from a defective cache block, it can always be recovered from the memory without violating the architectural definition of the instruction. We call defects in non-critical components such as the cache, *non-critical defects*.

A majority of fabrication defects can also be classified as random *spot defects* [13]. Our defect-model assumes random spot defects. We also assume that the defective area is small enough so that a single defect affects only one block of the cache. If the defect occurs in the tag array of the cache, we call it a *cache tag defect* and if it occurs in the cache data array, we call it a *cache data defect*. A cache tag defect will not pollute the data stored in the cache, i.e., it will not pollute the contents of the cache data array, but it will affect the cache *hit* operation. Examples of incorrect operation due to a cache tag defect include: (i) a miss indication even though data for the block is present in the cache, (ii) a hit indication even though the block is not present in the cache and (iii) a "multiple" hit resulting from several tags matching. A cache data defect does pollute the data in the cache data array but does not affect the tag array. Such a defect does not affect the cache hit operation but results in the access of incorrect data.

2.2. Use of Redundancy To Tolerate Defects

Since the use of redundancy is a popular way of enhancing the yield and reliability in several contexts, one might be tempted to use redundancy in the cache portion of the VLSI processor to enhance the yield. Both the major portions of the cache, i.e., the tag and the data arrays are linear RAMs and redundancy techniques that are useful for RAMs could easily be applied to a cache. These techniques fall into two broad categories: (i) spare resources and a reconfiguration mechanism to switch out the defective resource and (ii) use of error checking and correction (ECC) codes to mask out defects within a resource. Below, we discuss both of these options in some more detail.

2.2.1. Spare Cache Blocks and Reconfiguration

A cache memory could be designed with spare cache blocks in the data and tag arrays. If a block is defective, it can be switched out and a spare block substituted in its place. The overhead for doing so includes the additional chip area for the spare blocks and the additional logic needed to implement the reconfiguration.

While this overhead is not very significant, there is no reason to have a "full" cache in order to ensure correct operation. A "full" cache is a cache with the same number of defect-free blocks available for use in the caching operation as a completely defect-free cache. If the cache can be designed to operate in the presence of defective blocks with a negligible performance degradation, the use of spare cache blocks is wasteful.

2.2.2. Error Checking and Correction (ECC)

ECC techniques have been used widely to tolerate faults in memory systems [11]. A typical memory system uses a Single Error Correcting Double Error Detecting (SECDED) Hamming code to correct single errors and detect double errors in the memory system. To carry out the detection and correction process, redundancy in the form of check bits must be incorporated into the memory data word. The fault-tolerance capability of an ECC technique is determined by the number of check bits used.

ECC techniques can also be used to enhance the yield of memories by masking out defective bits [8]. However, ECC techniques have two forms of overhead: (i) the time penalty introduced by the ECC logic and (ii) the additional RAM required to store the check bits. Let us consider the implications of these overheads.

Since the degradation in memory access time is a good indicator of the degradation in performance of a VLSI processor let us consider the degradation in memory access time due to the ECC logic. A typical processing system that uses a high-performance VLSI processor would have at least 3 levels in its memory hierarchy (excluding backing store). These levels are: (i) the on-chip or level 1 cache that serves to reduce the latency of CPU requests, (ii) an off-chip or level 2 cache that serves both to reduce the latency of off-chip memory requests and, in the case of a multiprocessor, to reduce the traffic on the interconnect[4] and (iii) the main memory. For such a 3-level memory hierarchy, the effective memory access time as seen by the CPU is:

$$T = h_1 t_1 + (1-h_1) h_2 t_2 + (1-h_1)(1-h_2) t_m \qquad (1)$$

where h_1 is the hit ratio and t_1 is the access time of the level 1 cache, h_2 is the hit ratio and t_2 is the access time of the level 2 cache and t_m is the access time of the main memory.

Using equation (1), let us see how the use of ECC at various levels in the memory hierarchy affects the overall effective memory access time. Let us consider a processing system built using high-performance VLSI processors in which the relative access times of the level 1 cache, the level 2 cache and the main memory are 1, 3 and 10 time units, respectively. These values are typical of processing systems built using single-chip processors such as the NS32532 processor [3]. To determine the time penalties due to ECC at various levels, we carried out a VLSI layout of an on-chip cache and a paper design of a level 2 cache and a main memory. A timing analysis indicated that SECDED ECC degraded the the access time of the level 1 cache, the level 2 cache and the memory by 20%, 15% and 10%, respectively. Assuming these degradations in access times to be representative of a large class of high-performance processing systems, we computed the average memory access time as seen by the CPU.

Table 1: Average Memory Access Times For Varying ECC Usage

ECC Usage	$h_1=0.6; h_2=0.95$	$h_1=0.8; h_2=0.95$	$h_1=0.9; h_2=0.95$
No ECC	1.940	1.470	1.235
ECC only in Main Memory	1.960	1.480	1.240
ECC in Memory and Level 2 Cache	2.131	1.566	1.283
ECC at all levels	2.251	1.726	1.463

Table 1 presents the average access time as seen by the CPU for three cases: (i) a small level 1 cache (h_1=0.6), (ii) a medium level 1 cache (h_1=0.8) and (iii) a relatively large level 1 cache (h_1=0.9). In all cases, the on-chip level 1 cache is backed up by a typical medium-sized off-chip level 2 cache (h_2=0.95). The results in Table 1 are presented for varying degrees of ECC usage.

From Table 1 we can see that, in all cases, the use of ECC in the main memory does not affect the overall memory access time to any appreciable extent (a degradation of about 1%). If ECC is used in the level 2 cache, the overall memory access time is degraded slightly (3.9%-9.8% over the no-ECC case) but the degradation is less severe if the level 1 cache is larger and has a higher hit ratio. Note that the level 2 cache and the main memory are built from several chips and ECC would be necessary for fault-tolerance. Indeed, the use of ECC for the main memory is very desirable. The use of ECC is also desirable for the level 2 cache especially if the level 2 cache is a copy-back cache and is built from DRAMs.

However, the use of ECC in the level 1 cache degrades the overall memory access time significantly (17%-18% over the no-ECC case). Therefore, the use of ECC in the on-chip cache for yield enhancement does not seem to be an attractive option for high-performance VLSI processors. Furthermore, in the absence of adequate cache-coherence algorithms for on-chip caches, the on-chip caches are generally used to cache read-only information (such as instructions) or are write-through caches. For a read-only or write-through cache, correct information always exits elsewhere in the system (the level 2 cache or the memory) at all times. Therefore, a simple error-detection capability is all that is needed even for fault-tolerant operation of the level 1 cache.

It is possible that the degradation in memory access time due to ECC could be reduced for the on-chip cache [9]. For example, data could be read from the on-chip cache assuming that no error exists and supplied directly to the CPU. The ECC computation could be carried out in parallel with the CPUs use of the data. If the ECC computation indicates an error, the CPU would be informed and the computation aborted. However, the additional RAM overhead still exists.

Ideally, ECC must be provided on the smallest writeable unit [9]. Since the smallest writeable unit in most processors is a byte, this implies the use of 1 parity bit for single error detection, 4 check bits for single error correction and 5 check bits for SECDED for each byte in the cache. This per bit overhead can be reduced by maintaining ECC check bits at the word (16-bit) or double word (32-bit) level. However, doing so can complicate the access of data in the cache when only a byte needs to be accessed since ECC information must be computed for more than a byte [9].

Because of the time and space overheads associated with it, ECC techniques to tolerate defects in an on-chip cache may be of limited utility. Indeed, if a defect affects more than a single bit in a cache block (as our defect model allows), the RAM overhead for storing the check bits for a multiple-error-correcting ECC code can be very large. Even if the overheads are tolerable, we would like to know if they are worthwhile. Therefore, we would like to see how a cache can operate in the presence of defective blocks and how its performance would be degraded in such a case.

2.3. Operation With Defective Blocks

To operate in the presence of defective blocks, the cache control logic must be able to distinguish between defective and defect-free blocks. To do so, we append to each block of the cache an *availability bit*. This bit is similar to the *fault tolerance* bit proposed for the RISC-II instruction cache [10]. When the cache is tested, the availability bit for a block is set if the block is free of defects and is reset if a defect exists in the block. The defect can either be a cache data defect or a cache tag defect (in Section 4, we shall see how these bits and their setting change for a sector cache organization). The cache control logic makes use of the availability bit when it makes decisions during cache operation. The defective block is excluded from cache operation, i.e., is never chosen as the target block for the cache placement algorithm. If a reference maps on to a defective block, it is treated as a miss.

When data is fetched from the memory, it is normally supplied to the CPU through the cache. If the cache is set associative and there is at least one defect-free block in each set of the cache, data transfer between the CPU and the memory can be carried out through a defect-free block. However, if all the blocks in a particular set are defective, data references that map on to the defective set cannot be carried out through the cache. This problem is particularly significant in a direct mapped cache where there is only one block in each set. To overcome this problem, selective bypass of the cache for must be possible. We believe that this selective bypass capability is not a significant problem. Most processors have an option to turn off the on-chip cache thereby bypassing the cache for all memory references. Since the basic data paths already exist, extending the capabilities to allow selective bypass is straightforward.

The approach of using an availability bit to allow cache operation has little overhead - a single bit for each cache block. The cache can continue operation in the presence of cache tag and cache data defects. However, correct operation cannot be guaranteed in the rare case in which the defect exists in an availability bit. If the cache can be organized so that the degradation due to defective blocks is negligible, this approach can be used profitably to enhance the yield of a VLSI processor with an on-chip cache.

3. Cache Performance Under Defect Conditions

In this section, we evaluate the performance of various cache organizations in the presence of defective blocks. First, we see how sensitive a cache organization is to a defective or missing block and then we carry out a detailed performance evaluation using extensive trace-driven simulations.

3.1. The Sensitivity of a Cache Organization to Defective Blocks

Let us suppose that the memory consists of M blocks, the cache consists of C blocks and the set-associativity (number of blocks per set) of the cache is S. For this organization, there are $(M \times S)/C$ blocks in an *equivalence* or *congruence* class. All blocks from the same equivalence class are mapped on to the same set of the cache, i.e., the $(M \times S)/C$ blocks of an equivalence class are mapped on to one of S blocks in the cache.

In a direct mapped cache, the set-associativity is one ($S=1$) and, if a cache block is defective, M/C memory blocks are excluded from the cache. Consider, for example, the cache-memory system of Figure 1. The cache has 4 blocks ($C=4$) and the memory has 16 blocks ($M=16$). If the cache were direct mapped ($S=1$), under normal operation 4 memory blocks, namely $\{M_0, M_4, M_8, M_{12}\}$, map onto cache block C_0. If cache block C_0 is defective, 4 memory blocks $\{M_0, M_4, M_8, M_{12}\}$ will be excluded from the cache.

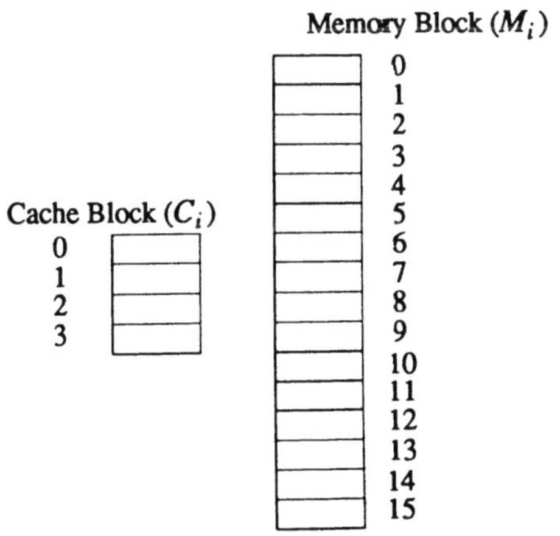

Figure 1: An Example Cache-Memory System

Since memory blocks $\{M_0, M_4, M_8, M_{12}\}$ cannot be present in the cache, references to these blocks must be serviced by the CPU-memory interface directly without passing the data through the cache. Therefore, any reference to these blocks would be a miss. In general, if there are D defective blocks in the direct mapped cache, $(D \times M)/C$ memory blocks would be excluded from the cache. Therefore, we can expect the miss ratio of a direct mapped cache to degrade linearly with the number of defective blocks if each cache block is equally likely to be referenced.

A set associative cache is less restrictive. A single defective block does not automatically exclude any memory block from the cache. In fact, as long as every set in the cache has at least one defect-free block, no memory block is excluded from the cache. Blocks from a congruence class are excluded only if all the cache blocks of the corresponding set are defective. However, the miss ratio will degrade because the probability of interference among the blocks that map on to a set with defective blocks increases. For example, suppose that the cache of Figure 1 were 2-way set associative (S=2) and cache blocks $\{C_0, C_1\}$ comprised set 0 of the cache. Under normal operation, memory blocks $\{M_0, M_2, M_4, M_6, M_8, M_{10}, M_{12}, M_{14}\}$ could be present in either cache blocks C_0 or C_1. A defect in cache block C_0 will not exclude any memory block from the cache completely, however, the probability of interference amongst the memory blocks that map onto set 0 of the cache will increase. Since no memory block is excluded from the cache unless all blocks in a set are defective, we can expect the degradation in miss ratio for a set-associative cache to be less than the degradation in miss ratio for an equivalent direct mapped cache.

A fully associative cache (S=C) always allows every memory block to be cached (unless the entire cache is defective). Furthermore, all memory blocks are treated uniformly and no set of memory blocks experiences a greater interference than another set. The degradation in the miss ratio, therefore, will be solely due to the increased probability of interference. The probability of interference is small, especially for larger cache sizes. Therefore, we expect a fully associative cache to have little degradation in miss ratio because of defective blocks, especially if the number of available defect-free blocks is large.

In summary, based on our understanding of cache operation in the presence of defective blocks, we expect that a direct mapped cache will be more sensitive to faults than a set-associative cache with the same cache and block size. Furthermore, we also expect a cache with a larger number of blocks to be less sensitive to defective blocks than a cache with fewer number of blocks.

3.2. Simulation Methodology

To get an accurate estimate of degradation in cache performance (as measured by the miss ratio) due to defective blocks, we carried out an extensive trace-driven simulation analysis. Trace-driven simulation is the most popular way of evaluating cache memory performance. We simulated three different cache sizes: (i) a 256 byte cache which is a typical on-chip cache size for VLSI processors of the early- to mid-1980's (such as the Motorola 68020), (ii) a 1K byte cache which is a typical on-chip cache size for VLSI processors of the mid- to late-1980's and (iii) an 8K byte cache which we expect be a typical on-chip cache size for high-performance VLSI processors of the near future. A direct mapped, a 2-way set associative and a fully associative organization were simulated for each cache size. An LRU replacement strategy was used for the set and fully associative organizations. The block size was also varied for each cache.

The benchmark programs used to simulate the caches were taken from the widely-used traces generated for a VAX-11/780 using the ATUM trace technique [2]. Each cache organization was simulated for approximately 1 million references. The caches were unified instruction and data caches. We assume that the defects in the cache occur randomly, i.e., there is no clustering of defects. A defect has the effect of preventing any data from being cached in the defective block. Since various blocks of the cache are not accessed precisely in the same fashion, two different caches with the same number of defective blocks (but different defective blocks) may differ slightly in performance. In order to overcome this problem, we simulated each cache organization several times for the same number of defective blocks but with a different set of defective blocks for each run and averaged out the miss ratios.

The results of our simulation are presented in Figures 2, 3 and 4. The figures plot the cache miss ratio (averaged over all traces) versus the percentage of blocks that are defective for direct mapped (DM), 2-way set associative (TW) and fully associative (FA) caches with various block sizes (in bytes). For the 256 byte and 1K byte caches, we have plotted the complete range of defective blocks. For the 8K byte cache, we have truncated the curves at 50% defective blocks to allow for a better look at the miss ratio degradation, especially for a fully associative cache.

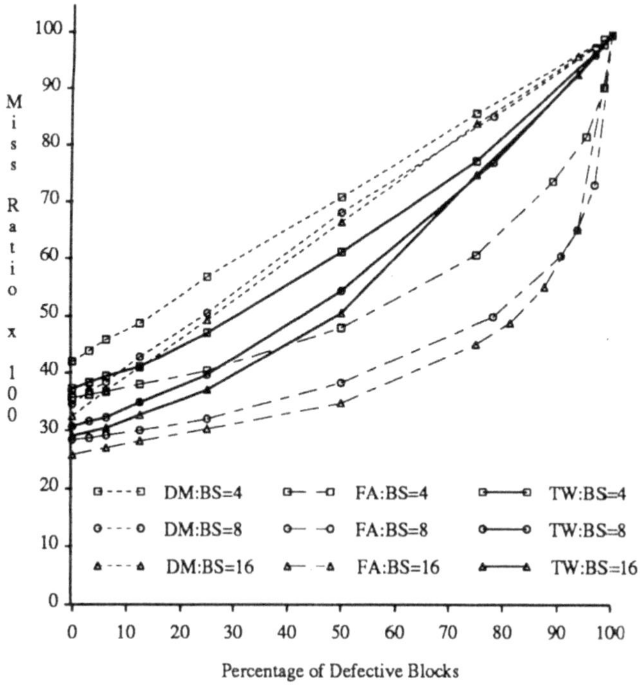

Figure 2: Miss Ratios; Cache Size = 256 bytes.

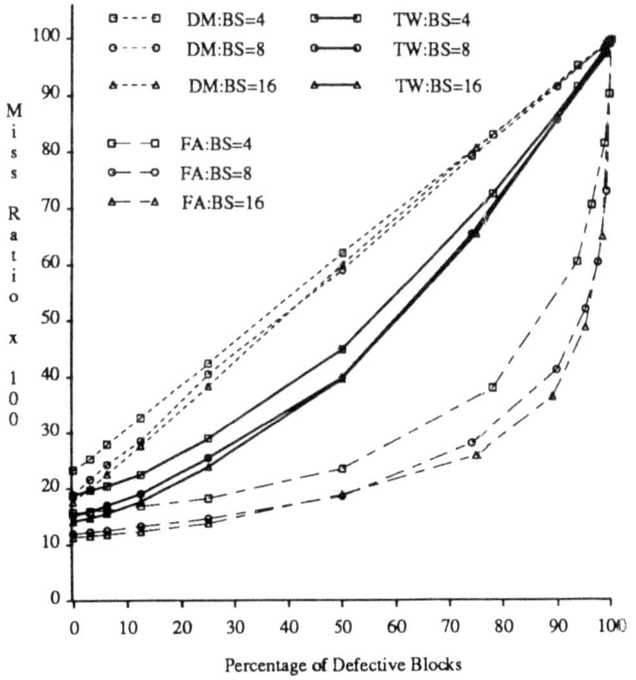

Figure 3: Miss Ratios; Cache Size = 1K bytes.

3.3. Discussion of the Simulation Results

Consider the results for a 256 byte cache (see Figure 2). If the cache is organized as a direct mapped cache, the miss ratio would degrade almost linearly with the number of defective blocks. This is indeed what we had expected. Thus, if the block size was 16 bytes (BS=16), four defective blocks would imply that 25% of the blocks were defective and the miss ratio would degrade from about 0.325 to about 0.493. If, on the other hand, the cache were organized as a fully associative cache, the miss ratio would degrade only from about 0.259 to about 0.304. The degradation in miss ratio for other set associative organizations would be in between the two limits. For a 2-way set associative organization, the miss ratio would degrade from about 0.307 to about 0.397. If the block size is small (4 bytes) and there is only a single defective block (1.56% of all blocks are defective), the miss ratio would degrade by only 0.002 (from 0.357 to 0.359) for a fully associative cache and by 0.01 (from 0.42 to 0.43) for a direct mapped cache.

The results for cache sizes of 1K and 8K bytes (Figures 3 and 4) follow a similar pattern. However, because of a large number of total blocks, more defect-free blocks are available for caching operation and the degradation in miss ratio is much smaller. For example, in a direct mapped 8K byte cache with a block size of 8 bytes, 4 defective blocks (1.56% of all blocks) would degrade the miss ratio by only 0.0037 (from 0.0689 to 0.0726). If the 8K byte cache is 2-way set associative with 8 byte blocks, the degradation in miss ratio due to 4 defective blocks would have only been 0.001. For a fully associative organization, the degradation is negligibly small.

An interesting point to note from Figure 4 is that for a fully associative 8K byte cache, a loss of 50% of its blocks would only degrade the miss ratio from 0.054 to 0.064 if the block size is 8 bytes. Since a fully associative 8K byte cache with 50% of its blocks defective is essentially the same as a fault-free, fully associative 4K byte cache, the reader might be tempted to conclude that a 4K byte fully associative cache is adequate; additional chip resources utilized in going to an 8K byte cache might be better utilized elsewhere. However, we would like to point out that the results for the fully associative cache have been presented as an upper bound. In many cases, fully associative caches are prohibitively expensive to implement and most practical caches employ a set associative or a direct mapped organization.

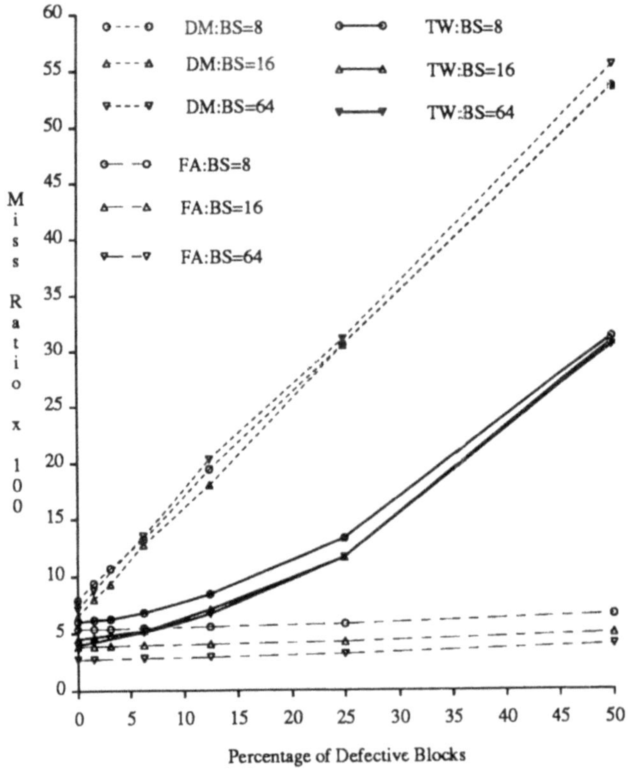

Figure 4: Miss Ratios; Cache Size = 8K bytes.

Table 2: Relative Memory Access Time In the Presence of Defects

Block Size (bytes)	Number of Defective Blocks	Direct Mapped Cache Size (bytes)			2-way Set Associative Cache Size (bytes)			Fully Associative Cache Size (bytes)		
		256	1K	8K	256	1K	8K	256	1K	8K
8	0	1.812	1.443	1.185	1.722	1.356	1.142	1.669	1.281	1.126
	1	1.864	1.458	1.188	1.743	1.360	1.142	1.676	1.282	1.126
	2	1.901	1.475	1.190	1.759	1.363	1.142	1.686	1.284	1.126
	4	2.007	1.507	1.194	1.820	1.372	1.142	1.706	1.287	1.126
16	0	1.764	1.413	1.154	1.684	1.332	1.106	1.608	1.266	1.090
	1	1.866	1.427	1.158	1.717	1.336	1.106	1.637	1.267	1.090
	2	1.963	1.441	1.162	1.769	1.340	1.106	1.686	1.268	1.090
	4	2.158	1.469	1.171	1.872	1.346	1.106	1.713	1.271	1.090

Since the results presented in Figures 2-4 do not indicate the degradation in the memory access time as seen by the processor, we converted the degradation in the miss ratio of the level 1 cache to a degradation in the relative memory access time as seen by the processor. We assumed the same system parameters as in Section 2.2.2, i.e., the level 2 cache has a hit ratio of 0.95 and a relative access time of 3 and the main memory has a relative access time of 10. The results for block sizes of 8 and 16 bytes and various cache organizations are presented in Table 2.

From the results presented in Figures 2-4 and Table 2, we can make two observations about cache behavior in the presence of defective blocks. First, for an arbitrary cache organization, the performance degradation due to a defect is larger if the fraction of cache resources that it affects is larger. For example, the degradation due to a single defective block in a cache with a block size of 8 bytes is smaller in a 1K byte cache (0.78% defective blocks) than in a 256 byte cache (3.125% defective blocks). The fraction of cache blocks that are unavailable due to defects can be reduced by increasing the cache size and/or decreasing the block size. However, smaller block sizes mean more tags and consequently more chip area. As we shall see in Section 4, a *sector cache* organization provides a good middle ground.

Second, for an arbitrary cache size, the performance degradation is smaller if the cache has a higher degree of set associativity. Increasing set associativity not only improves defect-free cache performance, it also enhances the ability of the cache to tolerate defects independent of or in conjunction with the cache size and the block size.

4. Sector Caches

The cache organizations that we have discussed so far have a single block size and fetch the entire block into the cache on a miss. In general, large block sizes are preferable because they reduce the number of tags required (and consequently the size of the tag array RAM). However, large block sizes have two problems that limit their use: (i) additional traffic is generated on the cache-memory interconnection and (ii) the time penalty incurred on a miss to fetch the block into the cache is larger. To alleviate these problems, a sector cache organization can be used.

In a sector cache, an *address block* is divided into several *transfer blocks* [12]. Tags for an entire address block are maintained in the tag array. An additional bit called a *presence bit* is maintained for each transfer block in the data array. A reference is a miss either if the address block or if the desired transfer block is not present in the cache. On a miss, only the desired transfer block is brought into the cache. By having large address blocks and small transfer blocks, we can reduce the cache-memory traffic as well as minimize the penalty incurred on a miss.

A sector cache can be made defect-tolerant in the same way as a regular cache - by the use of a single availability bit for each address block. The availability bit is reset if a defect exists in the tag portion (address block) or in either of the data portions (transfer blocks) of a particular block. However, if a cache data defect is confined to a small number of transfer blocks from a particular address block, the performance degradation can be reduced even further by associating an availability bit with each transfer block in the data array. In this case, there are a total of T+1 availability bits where T is the number of transfer blocks in an address block. Now, in case of a single cache tag defect, an entire address block (T transfer blocks) is unavailable for use and the performance degradation that we can expect would be similar to the performance degradation for a non-sector cache. However, in case of a single cache data defect, only a single transfer block is unavailable for use (the other $T-1$ transfer blocks of the address block can still be used for caching). Since the "scope" of the defect is much smaller, the degradation in performance will also be smaller.

5. Conclusions

To achieve high performance in VLSI processors, the use of an on-chip cache memory is highly desirable. Since the on-chip cache is expected to consume a large portion of the chip area, techniques to enhance the yield of the cache can profitably be used to enhance the yield of VLSI processors. In this paper, we investigated techniques to enhance the yield of on-chip cache memories of VLSI processors. We saw that traditional techniques that use additional redundancy may neither be appropriate nor necessary since a cache can continue operation in the presence of defects. Then we suggested a scheme that allows a cache to continue operation in the presence of defective blocks.

We evaluated the degradation in cache performance due to defects using an extensive trace-driven simulation analysis. Our evaluation indicates that the performance degradation due to defective blocks is small in cases where: (i) the cache is large, (ii) the block size is small and (iii) the set-associativity is high. In many cases, the performance degradation is insignificant.

Two major conclusions can be drawn from the work presented in this paper: (i) by choosing an appropriate cache organization, the need for additional redundancy in a cache to allow defect-/fault-tolerant operation can be avoided and (ii) defects in the on-chip cache portion of a VLSI processor can be tolerated with a minimum performance loss and very small area overhead and consequently, the yield of high-performance VLSI processors can be enhanced considerably.

References

[1] *IEEE Journal of Solid-State Circuits, Special Issue on Logic and Memory*, October 1987.

[2] A. Agarwal, R. L. Sites, and M. Horowitz, ''ATUM: A New Technique for Capturing Address Traces Using Microcode,'' in *Proc. 13th Annual Symposium on Computer Architecture*, Tokyo, Japan, pp. 119-127, June 1986.

[3] D. Alpert, J. Levy, and B. Maytal, ''Architecture of the NS32532 Microprocessor,'' *Proceedings IEEE International Conference on Computer Design*, October 1987.

[4] J. R. Goodman, ''Using Cache Memory to Reduce Processor-Memory Traffic,'' *Proc. 10th Annual Symposium on Computer Architecture*, pp. 124-131, June 1983.

[5] M. Horowitz, et al, ''MIPS-X: A 20-MIPS Peak, 32-bit Microprocessor with On-Chip Cache ,'' *IEEE Journal of Solid-State Circuits*, vol. SC-22, pp. 790-799, October 1987.

[6] I. Koren and D. K. Pradhan, ''Modeling the Effect of Redundancy on Yield and Performance of VLSI Systems,'' *IEEE Transactions on Computers*, vol. C-36, pp. 344-355, March 1987.

[7] D. MacGregor, D. Mothersole, and B. Moyer, ''The Motorola MC68020,'' *IEEE Micro*, pp. 101-118, August 1984.

[8] W. R. Moore, ''A Review of Fault-Tolerant Techniques for the Enhancement of Intergated Circuit Yield,'' *Proceedings of the IEEE*, vol. 74, pp. 684-697, May 1986.

[9] D. A. Patterson and C. H. Sequin, ''Design Considerations for Single-Chip Computers of the Future,'' *IEEE Transactions on Computers*, vol. C-29, pp. 108-116, February 1980.

[10] D. A. Patterson, P. Garrison, M. Hill, D. Lioupis, C. Nyberg, T. Sippel, and K. Van Dyke, ''Architecture of a VLSI Instruction Cache for a RISC,'' *Proc. 10th Annual Symposium on Computer Architecture*, pp. 108-115, June 1983.

[11] D. K. Pradhan, *Fault Tolerant Computing: Theory and Techniques*. Englewood Cliffs, New Jersey: Prentice Hall, 1986.

[12] A. J. Smith, ''Cache Memories,'' *ACM Computing Surveys*, vol. 14, pp. 473-530, Sept. 1982.

[13] C. H. Stapper, F. M. Armstrong, and K. Saji, ''Integrated Circuit Yield Statistics,'' *Proceedings of the IEEE*, vol. 71, pp. 453-470, April 1983.

8 DEFECT- AND FAULT-TOLERANT MEMORIES

DIAGNOSIS AND REPAIR OF LARGE MEMORIES: A CRITICAL REVIEW AND RECENT RESULTS

W. Kent Fuchs and **Ming-Feng Chang**

University of Illinois
Coordinated Science Laboratory
Urbana, IL 61801

INTRODUCTION

As semiconductor device sizes continue to scale down, the size of random access memory (RAM) quadruples every two or three years. Dense RAM chips are susceptible to manufacturing defects with the result that the manufacturing yield may become unacceptably low . Due to the regular array structure of RAMs, one important method of increasing yield is through design for restructuring, which adds on-chip redundancy to replace faulty memory cells [1]. The most popular approach is to add spare rows and columns of bit-cells. Each spare row and column may have a dedicated programmable decoder, which can activate the spare row or column and disable the row decoder or column multiplexer of the bit-cell array. In this way any faulty memory cell can be replaced either by a spare row or by a spare column. An alternative implementation is one in which each row or column can be disconnected from the decoder or multiplexer by electrical or laser fuses [1]. The repair of a RAM chip with redundancy can be divided into two phases: a diagnosis phase to detect and locate all faulty cells and a repair phase to allocate spares for all faulty cells.

Although there has been extensive published research on testing memory chips, there is little published work on efficient algorithms for diagnosis (location) of coupling faults in memory. In the last two years there have been several papers concerning algorithms for repairing memories with stuck-at faults [2-9], which has been shown by Kuo and Fuchs to be an NP-complete problem [5]. This paper presents a critical review of the state of the art in memory diagnosis and algorithms for memory repair. The basic approaches of the algorithms are compared and recent results are presented.

MEMORY FAULTS AND DIAGNOSIS

Fault Model

The complexity of a sequential functional test that can detect all possible faults in a memory array has been shown to be $O(2^m)$ [10], where m is the total number of bit-cells. As a result, most testing procedures restrict themselves to a subset of all possible faults, and choose a

Acknowledgment: This research was supported by the Semiconductor Research Corporation under contract SRC 87-DP-109.

more feasible fault model. The restricted fault models used in these testing procedures include stuck-at faults, coupling faults and pattern-sensitive faults [11]. Most existing efficient sequential test procedures, whose complexity are $O(m)$, are targeted to detecting stuck-at faults and coupling faults [11-14].

A RAM chip typically consists of four functional blocks: bit-cell array(s), row decoder(s), column multiplexer(s) and read/write circuits. It has been shown that decoder faults can be modeled as stuck-at and coupling faults in memory cells as long as the faulty decoder remains a combinational circuit [12]. Moreover, these kinds of decoder faults can be repaired by row or column redundancy [15]. Stuck-at faults and shorts or capacitive coupling faults for read/write circuits also have been modeled as stuck-at and coupling faults of memory cells [12]. The authors are not aware of any published research concerning diagnosis and repair of arbitrary pattern sensitive memory faults. In this paper we restrict ourselves to diagnosis and repair of stuck-at and coupling faults. The diagnosis and repair procedures vary according to the type of coupling fault and therefore we distinguish between two classes of faults.

An *active coupling fault* is present if a read or write (transition or nontransition) operation at address i forces a write of the same value at another address j. In this case, address i is the *coupling address* and address j is the *coupled address*. An active coupling fault is fully specified if its coupled address and coupling address are located.

A *passive coupling fault* is present if the read or write operation at address i does not result in the proper output or transition in the presence of a specific bit value in the cell at address j. In this case, address i is the *coupled* address and address j is the *coupling* address.

Since a read operation may also cause an error due to an active coupling fault, the content of any bit-cell may change when a read operation is executed. As a comparison, the coupling fault model used by Nair, Thatte and Abraham [12], specified that the coupled cell not necessarily make the same transition when a write is made to the coupling address. However, they also assumed that the read operations will not change the content of any cells. Multiple access faults [14] caused by a faulty decoder are a special case of the active coupling faults considered here.

Repair Requirements

We outline here the requirements for repairing stuck-at and coupling faults.

Stuck-at faults : Any cell that has a stuck-at fault may be repaired by replacing the row or the column which contains the cell. This is the classical repair requirement assumed in most papers on algorithms for repairing memory.

Active coupling faults : In order to repair an active coupling fault, either the coupled cell or the coupling cell may be replaced. If we replace the coupling cell then the coupled cell is no longer affected by the fault and hence becomes functional. However, the repair must ensure that the coupling cell is not accessed, even internally to the chip. In most DRAMs this can be done only by disabling the row (word line), not by disabling the column (bit line). On the other hand since the coupling cell is functional, we can also replace the faulty coupled cell. As a result, an active coupling fault is repaired if the coupled cell is replaced by a spare row or column, or the coupling cell is replaced by a spare row (word line).

Passive coupling faults : For a passive coupling fault, the coupled cell must be replaced. If the coupling cell is replaced then the coupling cell's content becomes uncontrollable and unknown, as a result read/write operations at the coupled address still may not be functional.

Diagnosis Procedures

There has been extensive published research on memory testing as described in the survey paper on functional testing of memories by Abadir and Reghbati [11]. A highly efficient testing procedure for stuck-at faults only is the MATS (Modified Algorithmic Testing Sequence) algorithm by Nair [13]. The complexity of MATS is $4m$, where m is the size of the memory. MATS has been shown to cover all stuck-at faults for RAMs with wired-OR logic behavior and arbitrary decoder design [13]. MATS is a modification of the ATS procedure by Knaizuk and Hartmann [16]. Efficient testing procedures for stuck-at and coupling faults have been developed by Nair, Thatte and Abraham (30m complexity) [12] and by Suk and Reddy (16m complexity) [14].

Testing procedures are typically targeted to detecting and not locating faults. Therefore they usually locate the coupled cells without locating the coupling cell. However, by locating both the coupled cell and the coupling cell of a coupling fault the probability of repairing a memory can be increased. A diagnosis procedure that can locate the coupled cells and the coupling cells has been developed by Chang, Fuchs, and Patel [17].

The read/write operations performed on a RAM at any address under test are denoted as:
$R_d(i) \equiv$ a read operation at address i with an output of d expected.
$W_d(i) \equiv$ an operation to write d into the cell at address i.

The diagnosis procedure consists of three subprocedures:

Identification of stuck-at faults and coupled cells for passive coupling faults

Diagnosis-1 : Assume there are m total addresses $(0 : m-1)$
step 1: For i=0 to $m-1$, $W_1(i)$.
step 2: For i=0 to $m-1$, $W_0(i)$, $R_0(i)$, $W_1(i)$, $R_1(i)$.
step 3: For i=0 to $m-1$, $W_0(i)$.
step 4: For i=0 to $m-1$, $W_1(i)$, $R_1(i)$, $W_0(i)$, $R_0(i)$. □

In *Diagnosis-1*, we assume that a cell cannot be the coupling cell of one passive coupling fault and the coupled cell of another passive coupling fault. All the cells that have stuck-at faults cannot pass *Diagnosis-1*, because they would fail to make 0-1 or 1-0 transitions. Moreover, all the coupled cells of passive coupling faults cannot pass *Diagnosis-1*, because they would fail to make a proper transition, either in *step 2* when their coupling cell's content is 1, or in *step 4* when their coupling cell's content is 0. Those addresses which cannot pass *Diagnosis-1* should be repaired, and are no longer tested in the remaining diagnosis procedure. All the addresses that pass *Diagnosis-1* can make proper 1-0 and 0-1 transitions independent of the contents of cells at other addresses.

Identification of the coupled cells of active coupling faults

Diagnosis-2 :
step 1: For i=0 to $m-1$ and i passes *Diagnosis-1*, $W_1(i)$
step 2: For i=0 to $m-1$ and i passes *Diagnosis-1*,
$R_1(i)$, $W_0(i)$, $R_0(i)$
if $R_1(i)$ fails, record cell i in list *LAC0*
step 3: For i=0 to $m-1$ and i passes *Diagnosis-1*, $W_1(i)$
step 4: For i=$m-1$ to 0 and i passes *Diagnosis-1*,
$R_1(i)$, $W_0(i)$, $R_0(i)$
if $R_1(i)$ fails, record cell i in list *HAC0*
step 5: Repeat *step 1* to *step 2* exchanging 0 and 1 for all read and write operations to create list *LAC1*
step 6: Repeat *step 3* to *step 4* exchanging 0 and 1 for all read and write operations to create list *HAC1* □

The active coupling fault that we have assumed forces the *same value* to be written in the coupled cell as the value being read or written in the coupling cell. All the coupled addresses of active coupling faults are located in *Diagnosis-2*, since the addresses that are coupled by a lower address would be detected in *step 2* and *step 5*. Those that are coupled by a higher address are detected in *step 4* and *step 6*. Thus *LAC0* is the list of Lower Address Coupled cells with forced value 0 and *HAC0* is the list of Higher Address Coupled cells with forced value 0. Similarly *LAC1* and *HAC1* are lists for Lower and Higher address coupled cells with forced value 1.

Coupling fault pattern diagnosis

Diagnosis-3:

step 1: For each $j \in LAC\,0$

begin

$W_1(j)$

For each i passing *Diagnosis-1* and $i<j$ **do**

begin

$W_0(i)$, $R_0(i)$, $R_1(j)$

if $R_1(j)$ fails

then begin

record the coupling fault $(i \rightarrow j)$

$W_1(j)$

end

end

end

step 2: For each $j \in LAC\,1$, repeat *step 1* but exchange 0 and 1 for all read and write operations

step 3: For each $j \in HAC\,0$, repeat *step 1* but replace the condition $i<j$ by $i>j$

step 4: For each $j \in HAC\,1$, repeat *step 3* but exchange 0 and 1 for all read and write operations □

Diagnosis-3 locates all the coupling addresses whose read or write operations change the content of the cells at other addresses. $i \rightarrow j$ represents the condition when an active coupling fault exists between the coupling address i and the coupled address j.

Let *SCD* be the set of all coupled addresses detected in *Diagnosis-2*. The total number of test operations for *Diagnosis-1* is $10m$, for *Diagnosis-2* it is $16m$ and for *Diagnosis-3* it is $((\frac{3}{2}m+1)|SCD|+c)$, where m is the total number of memory addresses and c is the total number of active coupling faults. We believe there exist more efficient procedures if *Diagnosis-1* and *Diagnosis-2* are combined. However, the two procedures are separated in order to distinguish the cells that must be replaced and those that may be used. If $|SCD|$ is very large, *Diagnosis-3* may represent a large overhead in the overall diagnosis procedure. An example of a small memory with two stuck-at faults and six active coupling faults is shown in Fig. 1. For this example, it takes 360 read/write operations for *Diagnosis-1*, 576 read/write operations for *Diagnosis-2* and 405 read/write operations for *Diagnosis-3*.

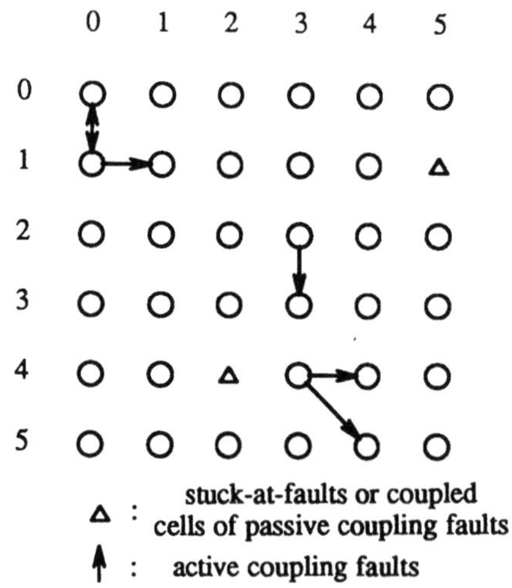

Fig. 1. A memory array with stuck-at and coupling faults.

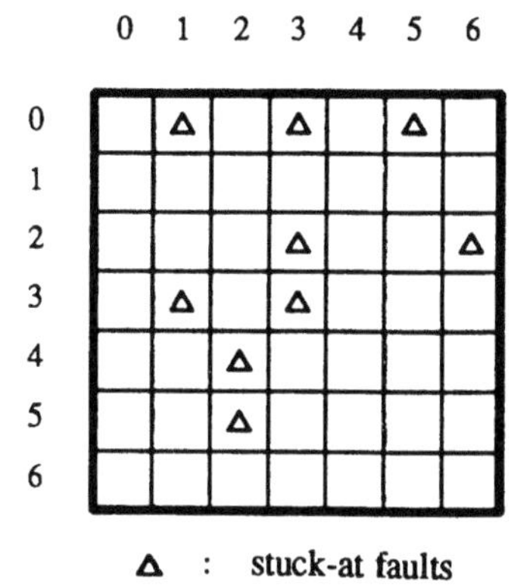

Fig. 2. A 7x7 memory with stuck-at faults.

MEMORY REPAIR

Problem Complexity

Given a memory with *TR* rows, *TC* columns, *ASR* spare rows, *ASC* spare columns and a set of faulty cells, the repair problem for stuck-at faults can be formally described as a vertex cover problem for a bipartite graph $BG = (A,B,E)$, where each vertex in A represents a row, each vertex in B represents a column and an edge $\{r,c\}$ is in E if cell (r,c) is faulty. A simple example is shown in Fig. 2 and its bipartite graph model is shown is Fig. 3. The repair problem is to find a subset R of A and a subset C of B, such that $|R| \leq ASR$, $|C| \leq ASC$ and every edge in E is incident to a vertex in R or a vertex in C. Kuo and Fuchs demonstrated the repair problem to be NP-complete by transforming the Clique problem to the repair problem in polynomial time [5]. All existing repair algorithms are either heuristic algorithms which take polynomial time and may not find a solution even if a solution does exist, or are searching algorithms which may take exponential time to find a solution, if one exists. Note should be made that the bipartite graph has also been used to model other repair problems such as arrays using Singh's interstitial redundancy scheme [18].

If only redundant rows (columns) are available, the complexity of the repair problem for stuck-at-faults only is $O(F)$ where F is the number of faulty cells. However, this is not the case for coupling faults. The authors have shown that the repair problem for coupling faults, even when only spare rows are available, is NP-complete [17].

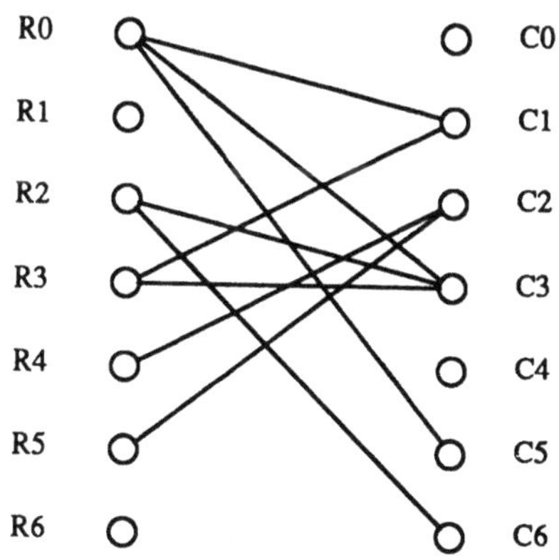

Fig. 3. Bipartite description of Figure 2.

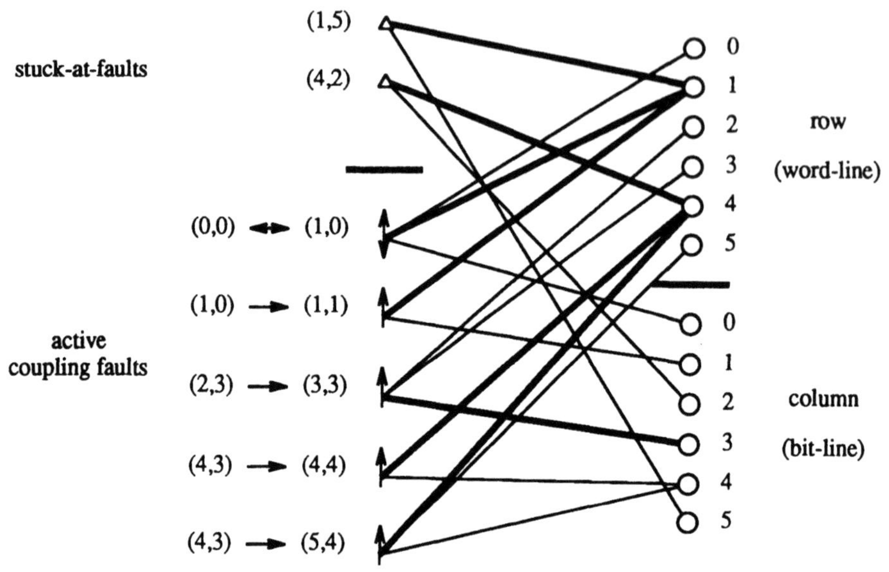

Fig. 4. Repair model for stuck-at and active coupling faults.

A bipartite graph can also be used to represent the repair problem for coupling faults [17]. Instead of vertices representing the columns and rows containing faulty cells, the repair model for memory with coupling faults is represented by a bipartite graph $BG = (F,R,E)$, where F is a set of vertices corresponding to the faults in the array and R is another set of vertices corresponding to the rows and/or the columns of the array. There is an edge $\{f,r\}$ in E between vertex f in F and vertex r in R, if the fault corresponding to f is repaired by replacing the row or column represented by r. The problem is to find a subset S of R such that for any f in F, there is a r in S and $\{f,r\}$ in E. Moreover the number of rows (columns) in S must not be greater than ASR (ASC). The repair graph model for the example in Fig. 1 is shown in Fig. 4. Fig. 4 also indicates by the bold edges a solution to the repair problem utilizing two rows and one column.

Repair Heuristics

Several key heuristics have been developed to guide the process of searching for a solution to the repair problem. These heuristics can be classified into two categories: row/column selection and unrepairability checking. Given a memory with faulty cells, a repair solution which utilizes the least number of rows and columns is typically considered to be the minimum solution. Note should be made however that in some cases the cost of utilizing a spare row (column) may be higher than that for a spare column (row), which would effect what is considered to be a minimum cost solution. Two faults will be considered to be *dependent* in the following discussion if they can be repaired by replacing a single row or column.

Row/column selection heuristics

Must-repair [2]: A row (column) which contains more than ASC (ASR) faults must be replaced, since all the faults in the row (column) cannot be repaired by using column (row) redundancy only.

Repair-most [2]: While there are unrepaired faults and spares available, repair the row or column which contains the largest number of faults. Note that this heuristic will often demonstrate greedy behavior resulting in not finding a solution or in not finding a minimum solution.

Effective-repair [8]: The effective coefficient $F(B)$ of a row (column) B is defined as: $F(B) = N_0+N_1-N_2$. N_0 is the number of faulty cells in B. N_1 is the number of faulty cells in B which have no dependent faulty cells not in B. N_2 is the number of faulty cells in B which have only one dependent faulty cell not in B, and the dependent faulty cell has no other dependent faulty cell. While there are faults unrepaired and spares available, repair the row or column which has the largest effective coefficient.

Critical-set-repair [9]: In some cases a critical set can be found, which is a set of rows and columns that must be replaced for all minimum solutions. For stuck-at faults, the critical set is unique and can be found in polynomial time.

Unrepairability heuristics

Two faults are *independent* if they cannot be repaired by replacing a single row or column. An independent fault set is defined to be a set of faults which are pairwise independent. Let MIF be the largest (maximum) independent fault set of a memory, then $|MIF|$ is the lower bound for the number of rows and columns required to repair the memory.

Diagonal-test [19]: Since all the faulty cells on a diagonal line of a memory are independent, if the number of faulty cells on a diagonal line is greater than the total number of spare rows and columns, the memory is unrepairable.

Maximum-matching [5]: One way to find MIF is to obtain a maximum matching for the bipartite graph representing the memory faults (Fig. 3). If the size of the maximum matching is greater than $ASR+ASC$, the memory is unrepairable.

Total-faults [2]: The maximum number of faults that can be repaired by ASR spare rows and ASC spare columns is $N = TR \times ASC + TC \times ASR - ASR \times ASC$. If a memory has more than N faults, it is unrepairable.

Fault-count after Must-repair [3]: The total number of unrepaired faults that can be repaired after Must-repair is $2 \times ASR \times ASC$, since there can be no more than ASC faults on a row and no more than ASR faults on a column.

Polynomial Time Heuristic Repair Algorithms

Evans' algorithm [2]:

S1- Apply Diagonal-test. If the number of faulty cells is greater than $ASC+ASR$, abort the test and the memory is unrepairable.

S2- Count the number of faulty cells while testing for stuck-at faults. The test is aborted and the memory is unrepairable if the number of faulty cells is greater than $TR \times ASC + TC \times ASR - ASC \times ASR$.

S3- Must-repair.

S4- Apply Repair-most to repair the unrepaired faulty cells. □

Evans' algorithm is an efficient algorithm that can find a solution for most repairable memories, though the solution may not utilize the minimum number of spare rows and columns. The algorithm also reduces the diagnosis time by aborting the diagnosis when the faulty cells diagnosed are unrepairable. However, the unrepairability condition is not tight enough to discard many unrepairable memories. Another drawback in Evans' algorithm is that it may not find a solution for some repairable memories. An example given by Evans is shown in Fig. 5.

Tarr, Boudreau, and Murphy's algorithm [3]

S1- Must-repair

S2- If the number of unrepaired faulty cells is greater than $2 \times ASR \times ASC$, the memory is unrepairable.

S3- Use a greedy algorithm to perform final analysis in allocating spare rows and columns. □

The authors did not state the details of their final analysis algorithm, however, they do indicate that for their test cases less than 0.1% of repairable devices are discarded by the algorithm. This algorithm is more efficient than Evans' algorithm in that a tighter unrepairability condition is utilized.

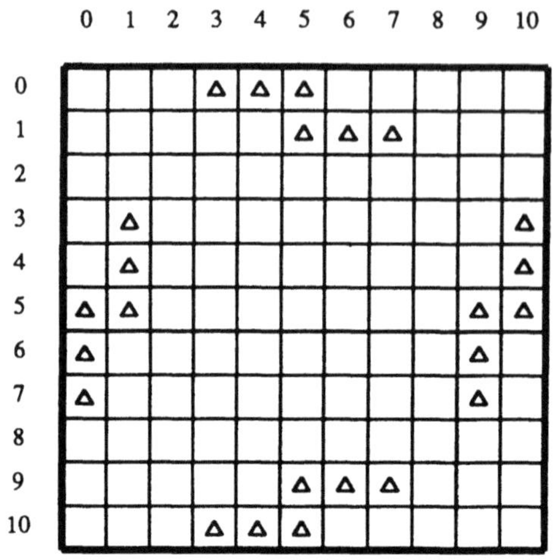

$ASR = ASC = 4$

Fig. 5. A faulty memory that cannot be repaired by some heuristic algorithms.

Kuo and Fuchs' heuristic algorithm [5]

S1- Must-repair.

S2- For each row (column) v, let $cc(v)=W(v)$ and $dc(v)=D(v)$, where $D(v)$ is the number of faulty cells in v, $W(v)=8$ for a row and $W(v)=15$ for a column (as an example). Let the cost to repair a row (column) be $cc(v)/dc(v)$.
While there are unrepaired faults and spares available, replace the row or column with least cost. Let C be the cost of the replaced row (column). Adjust the cost of each vertex u which is adjacent to the replaced row (column) in the bipartite graph as follows: $cc(u)=cc(u)-C$, $dc(u)=dc(u)-1$. □

This heuristic algorithm is a modification of Clarkson's approach to vertex covering for general graphs [20]. As with the other heuristic approaches, there are repairable arrays for which it will not find a solution.

Wey and Lombardi's algorithm [6]

S1- The leading element is the first faulty cell in the first faulty row of the memory. Once a leading element is found, all the faulty cells in the same row or column as the leading element are assumed to be repaired. The repair set S is a set of all the leading elements of the memory. If $|S| > ASR + ASC$, the memory is unrepairable.

S2- Determine the number of elements in S satisfying the following conditions: There exists a faulty cell i in the same row as the element (say k), a faulty cell j in the same column as the element, and both i and j are independent from every element in S except k. Let d be the number of such elements. If $|S| + d > ASR + ASC$, the memory is unrepairable.

S3- Generate the repair solution according to S by assigning a spare row and/or a spare column to each element in S. □

S is an independent fault set, but may not be the maximum independent fault set. In procedure S2, S is meant to be extended to a maximum independent set. Therefore, the prescreening condition in S2 is as tight as the prescreening condition in Kuo and Fuchs' algorithm. The drawbacks for this algorithm are that it may not find a solution for some repairable memories and it may not find the minimum solution. A repairable memory that cannot be repaired by this algorithm is shown in Fig. 6. The authors did not state how to assign spares to each element in S, which is typically the most difficult part of memory repair.

Lombardi and Huang's LECA [8]

The effective coefficient $F(B)$ of a row (column) B is defined as: $F(B) = N_0+N_1-N_2$. N_0 is the number of faulty cells in B. N_1 is the number of faulty cells in B which have no dependent faulty cell not in B. N_2 is the number of faulty cells in B which has only one dependent faulty cell not in B, and the dependent faulty cell has no other dependent faulty cell.

S1- Replace the row or column with maximum F.

S2- If $ASR=0$ then replace all the unrepaired faulty columns, else if $ASC=0$ then replace all the unrepaired faulty rows. If a solution is not found and spares are available, go to S1. □

This algorithm is an improvement over Repair-most, since the effective coefficient more accurately estimates the merit of replacing a row or column. However, if the faulty cells are highly clustered, N_1 and N_2 tend to be zero and this algorithm behaves similarly to the Repair-most algorithm. The example shown in Fig. 5 which cannot be repaired by Repair-most, cannot be repaired by this algorithm either.

Searching Algorithms

Day's algorithm [4]

S1- If the number of faulty cells is greater than $TR \times ASC + TC \times ASR - ASR \times ASC$, the memory is unrepairable.

S2- Must-repair. If the number of unrepaired faulty cells is greater than $2 \times ASR \times ASC$, the memory is unrepairable.

S3- Link all the unrepaired faulty cells in a list L. Given the first partial solution obtained from Must-repair, do a breadth-first search by expanding every partial solution to repair another unrepaired faulty cell in L either by a spare row or by a spare column. □

Day's algorithm is an exhaustive tree-search procedure and can find all solutions for repairable memories. However, the number of partial solutions that are generated by the algorithm increases exponentially as the number of faults increases. For unrepairable memories that pass the prescreening condition, it takes $O(2^F)$ time to determine that the chip is unrepairable, where F is the number of faults.

Kuo and Fuchs' branch-and-bound algorithm [5]

S1- Must-repair.

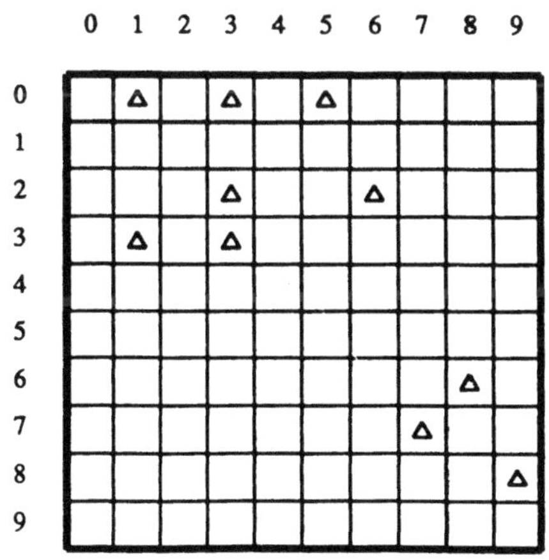

$ASR = ASC = 3$

Fig. 6. A second example that cannot be repaired by some heuristic algorithms.

S2- Find a maximum matching for the BG representing all the faults (Fig. 3). If $|MIF|$ $> ASR + ASC$ then the memory is unrepairable.

S3- Link all the unrepaired faulty cells in a list L. For each partial solution, the cost is $RC \times R + CC \times C$ where R (C) is the number of rows (columns) that are replaced and RC (CC) is the cost associated with allocating a row (column). Given the first partial solution obtained from Must-repair, do the best-first search by expanding the partial solution with the least cost to repair faulty cells in L either by a spare row or by a spare column. □

The unrepairability condition which checks the size of the maximum independent fault set in this algorithm is a very tight condition. Based on experimental results most of unrepairable memories will be screened out in S2. This algorithm is better than Day's algorithm, because the user can specify the cost for each partial solution and only the partial solution with least cost is expanded. However, if an unrepairable memory passes the tight prescreen checking, the number of partial solutions expanded by this algorithm will be as large as Day's algorithm.

Lombardi and Huang's FLCA [8]

S1- Must-repair. if the number of unrepaired faults is greater than $2 \times ASC \times ASR$, the memory is unrepairable.

S2- Given the first partial solution obtained by Must-repair, for every existing partial solution select the row or column which covers the most unrepaired faults. If the row or column selected covers only one fault, do not expand the partial solution, else expand the partial solution by repairing all the faults in the selected row (column) by one row (column) or by multiple columns (rows).

S3- If no solution is generated in S2, the memory is unrepairable, else select a minimum solution from the solutions generated in S2. □

The searching algorithm in S2 runs faster than Kuo and Fuchs' algorithm, because a faulty row or column instead of a faulty cell is considered in each step. However, the prescreening condition is not as tight. The unrepairable memories that pass the prescreen condition in S2 have to finish the exhaustive search in S3 to be designated as unrepairable.

Hasan and Liu's algorithm [9]

Hasan and Liu demonstrated that at least one solution utilizing $|MIF|$ rows and columns can be found in polynomial time, but the number of rows (columns) required may be greater than ASR (ASC). They also demonstrated that the critical set, a set of rows and columns that must be replaced for all minimum solutions, is unique and can be found in polynomial time.

S1- Find a maximum matching for the BG representing all the faults (Fig. 3). If $|MIF|$ $> ASR + ASC$ than the memory is unrepairable.

S2- Find the critical set CS [9]. If the number of rows in CS is less than ASR and the number of columns in CS is less than ASC, then a minimum solution can be constructed by replacing all the elements in CS and randomly assigning spare rows or columns to the uncovered matching edges in BG.

S3- If no solution is found in S2, a best-first search algorithm similar to FLCA is performed with the elements in CS selected first. □

Most of the unrepairable memories will be screened out in S1. For many of the repairable memories, the minimum solution will be found in S2 which requires only polynomial time. Even if a memory cannot be repaired in S2, the critical set obtained in S2 is useful in the best-first search in S3.

Concurrent Diagnosis and Repair Algorithms

Haddad and Dahbura's algorithm [7]

S1- While the memory is under test, execute the following on-line algorithms in parallel: Must-repair, Maximum-matching and Branch-and-bound search. If the memory is determined to be unrepairable the test is aborted. □

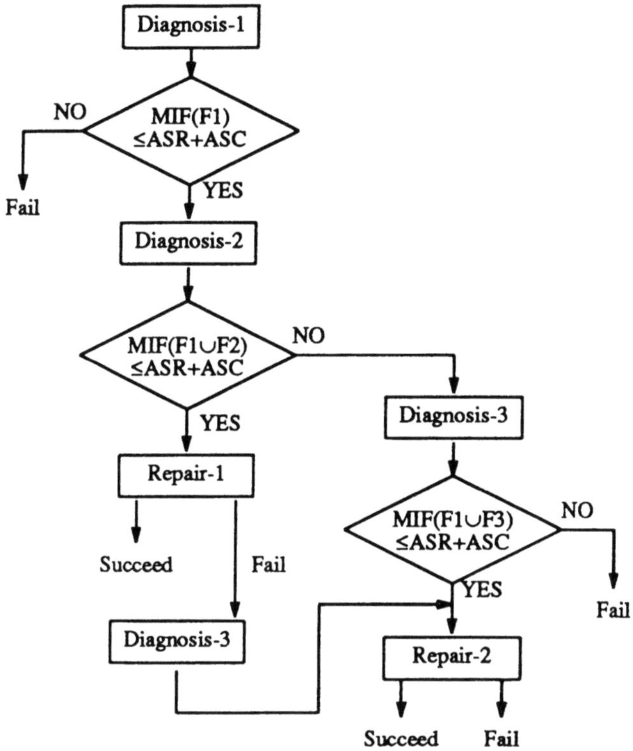

Fig. 7. Flowchart of the Diagnosis-and-Repair process.

Haddad and Dahbura's repair algorithm is similar to Kuo and Fuchs' branch-and-bound algorithm except that the algorithms are executed concurrently as the memory is tested in order to increase the throughput of test, analyze and repair. If a memory is determined to be unrepairable by the on-line Must-repair or Maximum-matching algorithm, the test is aborted. This approach can increase the throughput of memory testing significantly for large memories.

Chang, Fuchs, and Patel's diagnosis and repair procedure [17]

Chang, Fuchs, and Patel's algorithm merges the diagnosis and repair in order to increase the throughput. The flow chart of the algorithm is shown is Fig. 7. Since all the stuck-at faults must be repaired, if they are not repairable then the procedure is aborted. Since a coupling fault is repaired if the coupling cell is replaced by a spare row or the coupled cell is replaced, if all the stuck-at faults and coupled cells can be repaired, the memory is repaired and the coupling cells need not to be located. Otherwise, the coupling cells are located to increase the probability of successful repair.

APPLICATION OF DIAGNOSIS AND REPAIR ALGORITHMS

In general a combined memory test and repair system includes a tester with wafer probe, a repair analysis system and a laser repair system. The time required to test/diagnose and repair a memory die includes: wafer load and alignment time, test/diagnosis and verify time, repair analysis time and repair time. Since there are often more than one hundred dies in a wafer, the test/diagnosis, repair analysis and repair can be pipelined to increase the throughput. If only stuck-at faults are considered, the test/diagnosis time is fixed and the repair time is bounded for each die. Therefore, for an integrated system, the average repair analysis time should be chosen to be approximately the same as the test/diagnosis time to increase throughput. An example of

Table 1. An example of wafer test and repair

TEST RESULT	TEST TIME (sec.)	NUMBER OF DEVICES/WAFER	TIME/WAFER (sec.)
Good	4	12	48
Repairable	4	58	232
Reject	2.5	48	120
Quick-reject	0.1	48	5
TOTAL TEST TIME/WAFER (sec.)			405
TOTAL REPAIR TIME (sec.)			113
TOTAL NUMBER OF DIES/WAFER			166
AVERAGE TEST TIME/DIE (sec.)			2.44
AVERAGE REPAIR TIME/REPAIRABLE DIE (sec.)			1.9

test and repair time for 256k DRAM is shown in Table 1 [21]. The average test time per die is 2.44 seconds. The average repair analysis time can be expressed as follows:

$$T = P + p \times R$$

where P is average prescreen execution time, R is the average repair analysis time, and p is the probability that a memory die passes the prescreen. A tighter prescreen condition can reduce the total repair analysis time. Both P and R depend on the fault distribution and the number of spares available. Therefore, the repair algorithm should be chosen in the circuit design stage. For the example shown in Table 1, the average laser repair time is less than the average test time. In some cases, it may be necessary to find the minimum solution if the laser repair time is the bottle-neck of the test and repair system.

REFERENCES

[1] W. R. Moore, "A Review of Fault-Tolerant Techniques for the Enhancement of Integrated Circuit Yield," *Proc. IEEE*, vol. 74, pp. 684-697, May 1986.

[2] R. C. Evans, "Testing Repairable RAMs and Mostly Good Memories," *Proc. Intl. Test Conf.*, pp. 49-55, Oct. 1981.

[3] M. Tarr, D. Boudreau, and R. Murphy, "Defect Analysis System Speeds Test and Repair of Redundant Memories," *Electronics*, pp. 175-179, Jan. 12, 1984.

[4] J. Day, "A Fault-Driven Comprehensive Redundancy Algorithm," *IEEE Design and Test*, vol. 2, pp. 35-44, June 1985.

[5] S. Y. Kuo and W. K. Fuchs, "Efficient Spare Allocation in Reconfigurable Arrays," *IEEE Design and Test*, vol. 4, pp. 15-22, Feb. 1987.

[6] C. L. Wey and F. Lombardi, "On the Repair of Redundant RAM's," *IEEE Trans. on Computer-Aided Design*, vol. CAD-6, No. 2, pp. 222-231, March 1987.

[7] R. W. Haddad and A. T. Dahbura, "Increased Throughput for the Testing and Repair of RAMs with Redundancy," *Proc. IEEE Intl. Conf. on Computer-Aided Design*, pp. 230-233, 1987.

[8] F. Lombardi and W. K. Huang, "Approaches for the Repair of VLSI/WSI Repairable RAMs by Row/Column Deletion," *Proc. IEEE Intl. Fault-Tolerant Computing Symposium*, pp. 342-347, June 1988.

[9] N. Hasan and C. L. Liu, "Minimum Fault Coverage in Reconfigurable Arrays," *Proc. IEEE Intl. Fault-Tolerant Computing Symposium*, pp. 348-353, June 1988.

[10] J. P. Hayes, "Detection of Pattern-Sensitive Faults in Random-Access Memories," *IEEE Trans. on Comp.*, vol. C-24, pp. 150-157, Feb. 1975.

[11] M. S. Abadir and H. K. Reghbati, "Functional Testing of Semiconductor Random Access Memories," *ACM Computing Surveys*, vol. 15, pp. 175-198, Sep. 1983.

[12] R. Nair, S. M. Thatte, and J. A. Abraham, ''Efficient Algorithms for Testing Semiconductor Random-Access Memories,'' *IEEE Trans. on Comp.*, vol. C-27, pp. 572-576, June 1978.

[13] R. Nair, ''Comments on 'An Optimal Algorithm for Testing Stuck-at Faults in Random Access Memories','' *IEEE Trans. on Comp.*, vol. C-28, pp. 258-261, March 1979.

[14] D. S. Suk and S. M. Reddy, ''A March Test for Functional Faults in Semiconductor Random Access Memories,'' *IEEE Trans. on Comp.*, vol. C-30, pp. 982-985, Dec. 1981.

[15] Y. Hayasaka, K. Shimotori, and K. Okada, ''Testing System for Redundant Memory,'' *Proc. Intl. Test Conf.*, pp. 240-244, Nov. 1982.

[16] J. Knaizuk, Jr. and C. R. P. Hartmann, ''An Algorithm for Testing Random Access Memories,'' *IEEE Trans. Comp.*, vol. C-26, pp. 414-416, Apr. 1977.

[17] M.-F. Chang, W. K. Fuchs, and J. H. Patel, ''Diagnosis and Repair of Memory with Coupling Faults,'' *Proc. IEEE Intl. Conf. on Computer-Aided Design*, pp. 524-527, Nov. 1988.

[18] A. D. Singh, ''An Area Efficient Redundancy Scheme for Wafer Scale Processor Arrays,'' *Proc. IEEE Intl. Conf. on Computer Design*, pp. 505-509, Oct. 1985.

[19] J. F. M. Bindels, J. D. Chlipala, F. H. Fischer, T. F. Mantz, R. G. Nelson, and R. T. Smith, ''Cost-Effective Yield Improvement in Fault-Tolerant VLSI Memory,'' *Proc. IEEE Intl. Solid-State Circuits Conf.*, pp. 82-83, Feb. 1981.

[20] K. Clarkson, ''A Modification of the Greedy Algorithm for Vertex Cover,'' *Info. Proc. Letters*, vol. 16, pp. 23-25, Jan. 1983.

[21] D. M. Stewart, ''Production Test and Repair of 256k Dynamic Rams with Redundancy,'' *Proc. Intl. Test Conf.*, pp. 471-474, Oct. 1983.

A RECONFIGURABLE SRAM 4.5 MBIT WSI MEMORY

B. Nasreddine, E-F Kouka, Y. Wang,
D. Marron*, and J. Trilhe*

Institut National Polytechnique de Grenoble
46, Avenue Felix Viallet
38013 Grenoble Cedex, France

*SGS-Thomson Microelectronics
France

Introduction

Memory chips are always at the leading edge of integration and their size gives permanently the state of the art of the technology. As the design is very repetitive, no other factor such as architecture choices or innovative layout is very decisive. The state of the art has led to 1 Mbit static RAMs in CMOS technology (0.8 µm) with 25 ns access time. This state of the art is summarized in table I in appendix 1.

Even if the chip sizes grow rapidly, the demand of the customers is never satisfied as extra large computers, especially, for large scientific computations require "infinite" memory storage. An original and innovative way to satisfy this need is to worry about chip size increase and to investigate the "upper limit" which is Wafer Scale Integration [9]. This would allow to win at each instant a factor 10 with respect to the state of the art of the chip. The idea is very simple but the difficulties are numerous. The first problem is the yield problem. A maximum number of largest available memory chips, the design of which is available at a given instant, is implemented on a wafer. Depending on the manufacturing yield, a certain number of good chips is identified. The problem is then to collect the good chips in order to create or configure the definitive memory. If this architectural approach is solved in a closer manner, some other problems showed up such as capacitance of long lines, packaging, etc.

As far as Wafer Scale Integration prototypes are concerned, three companies have to be mentioned. NTT has made a 4 Mbit ROM on a three inches wafer in 1980 for word recognition purposes [1] and a 1.5 Mbit SRAM on a four inches wafer in 1984 for image storage [2], [3]. The strategy for both circuits is the same. The elementary chips are organized in pairs. A self test is performed by parity checking in each chip. The pair is used if, of course, at least one chip of the pair is fault free. This leads to an area cost comparable to conventional duplicated fault tolerant architecture.

This reconfiguration technique is very attractive because it is active each time the device is powered up, such that it is possible to correct not only end of manufacturing defects but also defects that occur during the life time of the circuit. Inspection and shift functions can be automatically completed within about 1.5 µs. The memory is not very fast, access time being about 520 ns.

A second pioneer company in this field is SINCLAIR, the activity of which in the domain has been transferred to ANARMARTICS and is presently shared with TANDEM and FUJITSU. The goal of their approach is the design of serial memory of large capacity.

The third compagny is INOVA microelectronics. Their architecture consists of implementing chips in the wafer around a centralized decoder (central column of the wafer). After external test, faulty elements are disconnected and good elements are connected to create the final memory. Connection and disconnection of chips are done by laser techniques. This company has announced a 55 ns 8 Mbit SRAM on a five inches wafer for application where speed is premium, such as microprocessor cache memories [7], [8]. The drawback of this approach is a non regularity of the layout due to the central decoder. Therefore INOVA needs twice more reticles than in a standard process. This has a drastic importance for production.

The approach proposed in this paper has been studied within the european ESPRIT project. Elementary memory chips are implemented on the wafer with a regular layout allowing step and repeat manufacturing. This overcomes the drawback of the INOVA approach. After an external test, the cartography of the good chips is established. An original automated configuration algorithm finds out an optimized global architecture. Corresponding connection and disconnection of memory chips are then realized by laser techniques. To illustrate the feasibility of our approach, a demonstrator of a 4.5 Mbit Static RAM using 64k elementary chips has been realized within the ESPRIT project. This memory is faster than the prototypes of NTT and ANAMARTIC. Furthermore, it can be produced economically in medium volume by using a process closer to a standard one than INOVA.

Part I gives some brief information on the memory architecture based upon basic cells which are 64k memories [10]. The yield prediction is summarized in part II. The third part gives some experimental details on the switches manufactured and tested within this project. The fourth part presents more details on the extra hardware to cope with defect tolerance. Part IV summarizes the configuration strategy to build the target memory with the defect free resources.

I. Architecture of the Memory

One goal of this project was to design a defect tolerant 4.5 Mbit static RAM, organized, by request of potential customers, as 256k words by 18 bits. The basic cell used to build it is a 64k word by 1 bit memory from ST which is scheduled for the open market in 1989 under the logo 4187 [10]. It is organized in two planes, separated by the line decoder. The main logical functions are implemented as close to the pads as possible in order to achieve an access time of 35 ns.

There is already built-in redundancy in the 4187. If a faulty column is detected during test, a dedicated pad is powered to blow a fuse. By doing so, the faulty column under test is discarded and replaced by a spare one. Up to 4 defects can be corrected in a 64k cell memory. A second level of redundancy allows to connect only good 64k to the interconnection network.

At the end of manufacturing, a test program checks the location of defective cells or connections on the wafer. Then a 256 kword by 1 bit block, is constructed using a set of four 64 Kword by 1 bit active cells chosen as close as possible. The four cells of one block are connected to the same bit line which is assigned to an I/O pad (figure 1). Typically, 18 blocks of 256k*1 bit can be found on a four inches wafer, i.e. a 4.5 Mbit device. The device is easily

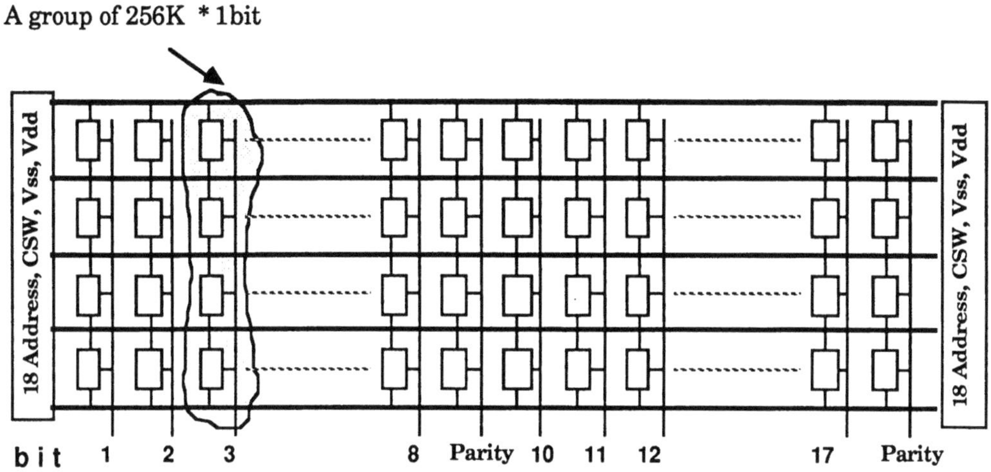

Figure 1 : Virtual organization of the WSI-memory

programmable to meet different user requirements, for example, 256k words by 16 bits + 2 code bits, 2*(256k words by 8 bits + 1 parity bit).

The construction of a 256k word by 1bit block out of four 64k words by 1bit cells requires the addition of a decoder (1 among 4) to each 64k basic cell. This decoder has to be programmable in order to be independent of the cartography of good cells on the wafer. The architecture of a block of 256k words by 1 bit is given in figure 2.

This decoder activates only the quarter of the 4.5 Mbit (i.e. 64k * 18 bits). So static and dynamic consumptions are reduced. The decoders of the four 64k of the same block are all different and are programmed with the switches which may be floating gates or laser antifuses. Both are implemented presently in parallel. A few hundred ones are implemented on the wafer.

II. Yield prediction and redundancy implementation

II.1 Probabilistic models

Redundancy has been implemented within the 64kbit cell to improve its yield. A minimum yield of 53% is required to embed the 4.5 Mbit memory on our four inches wafer. Following are the probabilistic models we used for the yield prediction.

Occurrence of defects on a given circuit area is considered as a random variable with a Poisson distribution. The parameter of this distribution

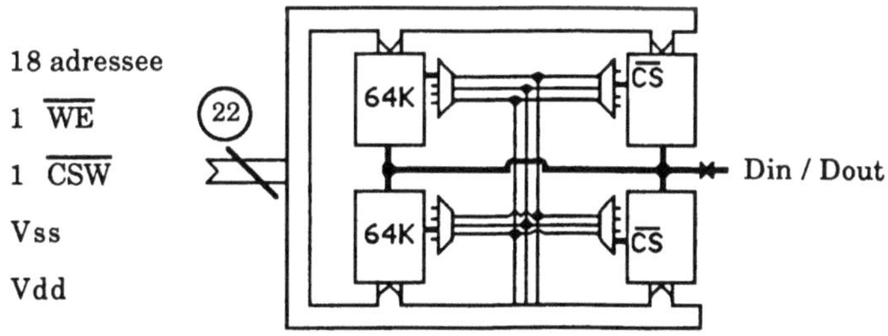

Figure 2 : Decoders and block of 256K 1bit*

depends on the area A and the expected density of defects D. This last parameter represents the quality of a given step of the fabrication process. So the following formula gives the probability P_k of the occurrence of k defects (k =0, 1, 2 ...) on an area A during the ith step of the fabrication process (represented by the defect density D_i) :

$$P_k(D_i) = \Sigma_{j=1,\ k} (A{*}D_i)^j {*}e^{-S{*}D_i}$$

The yield is defined as the probability of no occurrence of defect at this step, that is :

$$P_0(D_i) = (A{*}D_i)^0 {*}e^{-A{*}D_i} = e^{-A{*}D_i}$$

As we have no prior information on the densities of defects, D_i, these have been considered as independent random variables with the same density function F(D). Thus we have the following formula :

$$\Lambda F(D){*}d(D) = 1$$

Then the expected probability of no occurrence of defects at any stage of the fabrication process can be assessed as follows :

$$P_i = \Lambda P_i(D)F(D){*}d(D) = \Lambda\ e^{-A{*}D_i} F(D_i){*}d(D_i)$$

As the occurrences of defects are assumed to be independent from one step of the manufacturing process to another, then after n fabrication steps the mean probabibilty of no occurrence of defects on the overall circuit is estimated as follows :

$$P = \P_{i=1,n}\ P_i = e^{-A{*}n{*}D}$$

Then the yield can be seen as the probability of no occurrence of defects on a circuit with an area A; with an expected critical density of defect D associated with each step of the fabrication process. That is $P = P_0(A{*}n{*}D)$.

Different models are proposed in the literature [14,15] for the yield prediction. Depending on the density function F(D), the resulting formulas are obtained for the yield :

$$P_i(D) = (1 - e^{-A{*}2{*}n{*}D})/A{*}2{*}n{*}D \quad \text{(Murphy)}$$

$$P_i(D) = e^{-@A{*}n{*}D} \quad \text{(Seeds)}$$

The model to be used must fit the manufacturing process. Concerning the 1.2µm technology, experimental trials have shown that the best fit was obtained as follows :

- a model based on the mean value of the predictions given by the Murphy model and the Seeds model is used for low values of $A{*}D$.
- the Bose-Einstein model is performed for large values of $A{*}D$.

We therefore used the following balanced formula for the yield prediction :

Yield(n, S, D) = Min (Murphy-Seeds, Bose-Einstein).

II.2 Study of the theoritical yield of the 64 kbits memory

This section shows the yield enhancement resulting from the redundancy. The total area of the memory Ta consists in an area without redundancy NRa and the complementary area Ra containing redundancy. The redundant area is divided into p identical areas Pa, each one containing k reparable elements and r redundant elements. Each element occupies an elementary area Ea. The following equations summarize the above assertions :

Ta = Ra + NRa ;
Ra = p*Pa ;
Pa = (k+r)*Ea;

We get the following equation :

P(Ta) = P(NRa)*[P(Ra) + P(RDRa)] = a memory yield

where

P(NRa) is the probability of a defect free and not repairable area,
P(Ra) is the probability of a defect free and repairable area,
P(RDRa) is the probability of repairing a fault occuring in the repairable area.

Let P(REa) be the probability of replacing a faulty element by a redundant one, and let P(A) = Yield (A, n, D) be the probability of a defect free area A according to the previously presented formula.

As Ra = p*Pa, using the Bernouilli scheme, P(RDRa) can be computed as follows :

P(RDRa) = $\&_{i=1,\,p}\,{}^{(p)}$ prob[i defective and repairable parts in RA]*prob[p - i defect free parts] ;

Let P(DRPa) be the probability of repairing a defect occuring in Pa. As Pa = (k+r)*Ea, we get :

P(DRPa) = $\&_{j=1,\,k+r}\,{}^{(k+r)}$prob[j defective and repairable elements in Pa]*[k+r-j defect free elements in Pa].

Then $P(DRPa) = \Sigma({}^{k+r})P(DREa)^{j}*(1-P(DREa))^{k+r-j}$.

where P(DREa) is the probability of repairing a defect occurring in an elementary area Ea.

Due to relaxed design rules and the low interconnection density, the density of defects on a non repairable surface is smaller than that one associated with the redundant surface. So we get the following formula :

P(NRa) = yield(NRa, n, DANR) with (DANR < 1 is the density of defects on ANR).

From a program computing all these parameters we can draw the variations of the yield depending on the number of redundant elements (see figure 3).

III TECHNOLOGY ASPECT

We have rejected massive redundancy (duplication and triplication) because they are too much area consuming. We don't believe that discretionary wiring

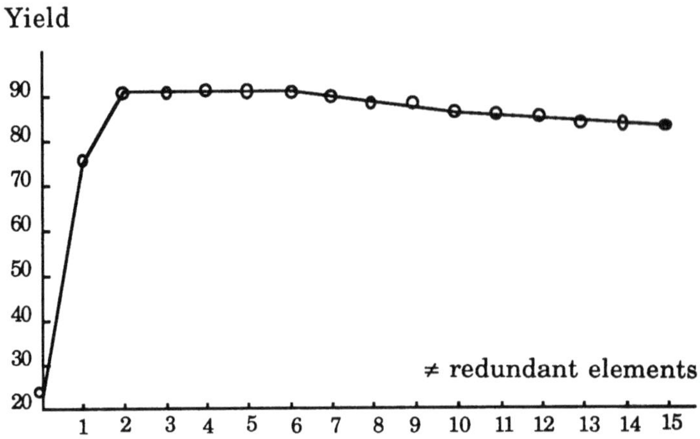

Figure 3 : The predicted yield versus the number of redundant elements.

can be an efficient solution, due to yield problem, for medium volume production we are concerned with. So we have focus our studies on switches to discard faulty elements and replace them by spares.

We only consider here the technology aspect, so we not deal with soft switches. The remaining alternative is between physically and electrically programmed switches. Electrically programmed switches (EPROM, EEPROM) have been further rejected due to their poor compatibility with SRAM process, that is why we have studied only laser and e-beam programmable switches.

The basic process is 1.2 µm CMOS, two levels aluminium technology with minor adds-on which will be described in this part. We will first present the switches and then the copper track which is used to drive power and global data.

III-1 Switch technology

If the technique to disconnect with laser is well established, it is not the case of the technique to connect with a low resistance. The connection technique, or "antifuse" has many solutions.

A first class contains the antifuses done with a single level of metal. Lateral links can be formed by a laser shot between two adjacent aluminium lines; this technique is called also laser melted gap [16]. An intrinsic polysilicon film surrounded by two N^+ diffusions can be made conductive if annealed by a laser shot which causes redistribution of doping in the polysilicon resistor [17]. An extra aluminium level can be selectively deposited on a wafer by laser enhanced Chemical Vapour Deposition [18]. Laser lithography can be used to pattern an aluminium level in order to make the required interconnections.

A second class concerns the connection between two different levels of metal. It is the most popular antifuse; it consists of an amorphous polysilicon sandwitches between two layers of aluminium; the laser shot induces conduction between the two layers of aluminium [19].

We have chosen to develop two switches for connection : the laser lithography which is a good compromise for the resistance and the maturity of the technique, and the floating gate FET which is the only one to be reversible.

A first batch was issued in March 1987 to study the feasibility, the compatibility with the process, the yield and the long term reliability of laser fuses, laser lithography antifuses and floating gate FETs. We will present now more significant results .

III-1-1 Laser Fuses

Laser fuse was already well integrated in the process. An array of fuses with different design rules, organized in a 16 kbit ROM was designed. Data for several million of shots were thus obtained. The results were pretty good, for 1.8 μm wide aluminium fuses, the yield (percentage of successfull shots) was 99.8 %. The good 16 kbit ROMs have been packaged and studied in temperature enhanced long term reliability. We stopped the experiment after 2000 hours and obtained then a yield of 99.98 %. We are presently analyzing failure statistics on the fuses which failed during the reliability study.

III-1-2 Antifuses by laser lithography

To achieve a sufficient yield, it appears necessary to first planarize the wafer with SOG. Then we evaluate the definition of the antifuse either by standard etching or by lift-off. Exposure being done in both cases by direct writing on the wafer with an Argon laser. Our target was 5 μm antifuse in order to drive a few mA. The yield obtained is 99.98 %. Reliability studies are in progress to evidence any electromigration problem.

III-1-3 Floating gate FETs

E.beam switches are based on the charge in a scanning electron microscope of the gate of a floating gate FET. We have previously seen that floating gate FETs do not need wiring for control and they are reversible. Their drawback is the need of a transfer gate in order to drive medium range current and an impossibility to drive large current (size and yield constraints).

Deposit only of negative charge on N-channel depletion device has been obtained with a sufficient yield. That means that a transistor "ON" at end of manufacturing, has become "OFF" by charging with e-beam. A N-channel depletion transistor requires one more than usually in our standard CMOS process.

Erasure of the programming has been performed routinely with UV, like for EPROM. We have also demonstrated [20] the possibility to erase selectively in the scanning electron microscope by increasing the voltage of secondary electron detector.

Several thousands of floating gate FETs have been handled by now, the yield is quite satisfactory when a dose of 10 exp-3 Ccm^2 is deposited at 3 keV through 4,000Å deposited oxyde. For these conditions, we have evidenced no significant loss of charge after 2000 hours at room temperature or at 200°C, so the results are far better than those published by D. Shaver [21].

III-2 Copper tracking

To drive power and data across a whole wafer, it is very attractive to deposit on the 1μm thick aluminium conductor a 3μm thick copper track. Doing so, the resistance of the line will decrease by a factor of 4 without any significant change of the line capacitance.

On passivated wafers, the process steps are as follows :

- opening of the passivation where electrical contact to copper is required (conventional lithography).
- deposition of Intermediate Layer Metallurgy (ILM).
- spin coat with 3 μm thick liquid resist.
- conventional 1:1 aligner to define copper tracks.
- electroplating of copper.

- stripping photoresist.
- wet or sputter etching of ILM not covered with copper.

Copper tracks are typically 10µm wide with a pitch of 20µm but the technology has been demonstrated down to 3 µm line. With such critical dimensions, it becomes possible to use copper track. This is a good alternative to antifuse. When the high flexibility of laser exposure is used, it can be a way to deposit a metal on a break of an aluminium 2 line, in order to repair.

IV LAYOUT

The building of the 18 blocks of 4x64 k requires routing for the lines and facilities to connect/disconnect the memory cells. Consequently the following features are implemented on the wafer :

- an interconnection network.
- 138 chips, each one made up of :
  . one memory cell (64 k word by 1 bit).
  . a set of switches.

IV-1 The chip (figure 4)

The 64k are isolated from the lattice by transfer gates which are non conducting at end of manufacturing. So it is possible to get the cartography of good 64k using standard probe, the good 64k are then connected to the interconnection network by charging the floating gate 1 on figure 3 which controls all the upper transfer gates. The floating gate 1' has the same use for the lower part, the role of this second floating gate FET is only to allow the suppression of a long metal 2 wire between the top and the bottom of the reticle. The efficiency of the connection of the 64k to the lattice is monitored using the power comsuption of the power supply. Then the decoder is programmed.

In each 256k word by 1 bit block, the four decoders need to have the four possible different settings. We have provided in the first run of 4.5 Mbit a duplication of e-beam switch with laser switch in order to program the appropriate setting of the decoders. The first possibility is to charge two among four floating gate FETs (one after each stage of the decoder of figure 2). An equivalent solution, in case of failure of the first one or if the device has to work in the military range of temperature, is to blow the fuse in serie with the floating gate charged in the previous scenario and deposit the antifuse in parallel of the floating gate to be programmed.

Let us mention that, due to these decoders, only a quarter of the 4.5 Mbit is activated (i.e. 64 kword by 18 bits); doing so, static and dynamic consumptions are reduced.

In case of failure of the transfer gate or of their control logic, fuses have been implemented in order to disconnect a defective 64k. Then a test of the 4.5 Mbit from the periphery of the wafer can be performed, but at low frequency (2.5 MHZ at Vcc = 5 +/- 0.5 V) due to the high Ron resistance of the transfer gate on power and I/O. If the result of test is bad, the floating gates are erased with UV, and a new configuration is tried, avoiding the area where defects have been evidenced. If the result of test is good, the transfer gates 2 and 2' are bypassed with a laser antifuse. In that case, there is more high Ron resistance and a test at full speed can be performed.

IV-2 The reticle

Manufacturing constraints (step and repeat) impose the use of repetitive pattern (reticle). The reticle, presented in figure 5, is made up of 2 chips. The

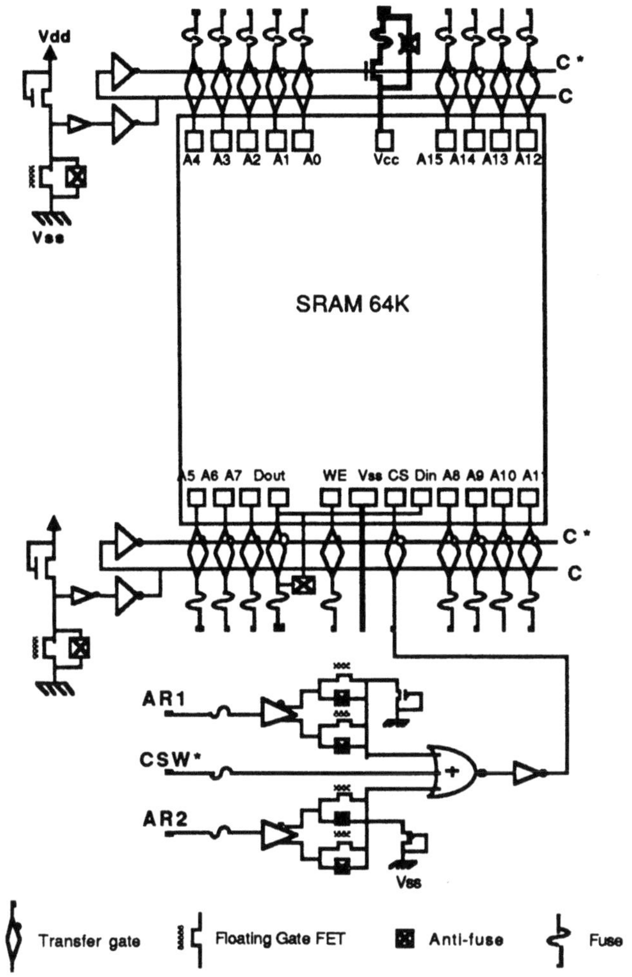

Figure 4: A 64 k word cell, a row decoder and transfer gates.

size of the basic 64k is 3.876 x 7.055 mm^2 leading to a reticle size of 8.8 x 8.7 mm^2 including test structures.

Although yield data can't be divulged, a theoretical approach for redundancy implementation is presented in the next section.

V CONFIGURATION STRATEGY

V-1 Preliminaries

After the test phase, the cartography of good elements is given as input data to a configuration program. The output gives the physical organization of the WSI memory and that is given as input to the laser or e-beam equipment to fix the definitive structure.

The configuration purpose is to determine 18 groups of 4 cells each (a cell is a 64 kword memory) in order to achieve the aimed 4.5 Mbit memory.

Each group must be assigned to a peripheral I/O pad. Due to communication delays, it is recommended to associate a group with its nearest I/O pad. Each pad is linked to a horizontal data line. Hence, four cells of a group have to be connected to the same data line. We have not implemented a bus of 18 data lines between each row of 64k in order to save area. With our usual model of defect distribution, four data lines seem enough between two rows of 64k; that is to say that occurrence of more than 4 x 4 good 64k in a row is very low. When a multiple of four cells can't be found in a line, communication from one line to another is provided, but through two vias and one *metal 1* line, so with a high penalty on the conductance. This has been taken into account in the optimization of the embedding of the 4.5 Mbit with the cost function detailed in the following section.

V-2 The Cost function

For monotoring the configuration algorithms, a cost function is defined on cells, groups and configuration. The cost of a cell, relatively to an I/O pad, is defined as the length of a line linking this cell to the I/O pad. Each element (cell, I/O pad) is assigned two coordinates (x, y) relatively to an arbitrary origine. So connecting cell **I** with coordinates (x_i, y_i) to I/O pad **J** with coordinates (x_j, y_j) involves a cost $\mathbf{C_{ij} = /x_i - x_j/ + w{*}/y_i - y_j/}$.

However the cell **I** is not directly connected in fact to the I/O pad, but actually to the data line associated with the pad, the cost C_{ij} can be considered as the bottleneck induced by the cell when accessing its group from the I/O pad. The parameter **w** is a weight which penalizes the use of vertical connection lines. The cost of a group is defined as the cost of the most expensive cell in that group plus the number of vertical connections necessary to interconnect cells within the group. A group which crosses the vertical median line of the wafer is penalized by twice the cost of the most expensive cell in the group. Similary, the cost of a configuration (a set of 18 mutually exclusive groups) is defined as the cost of the most expensive group in the configuration. Hence the configuration

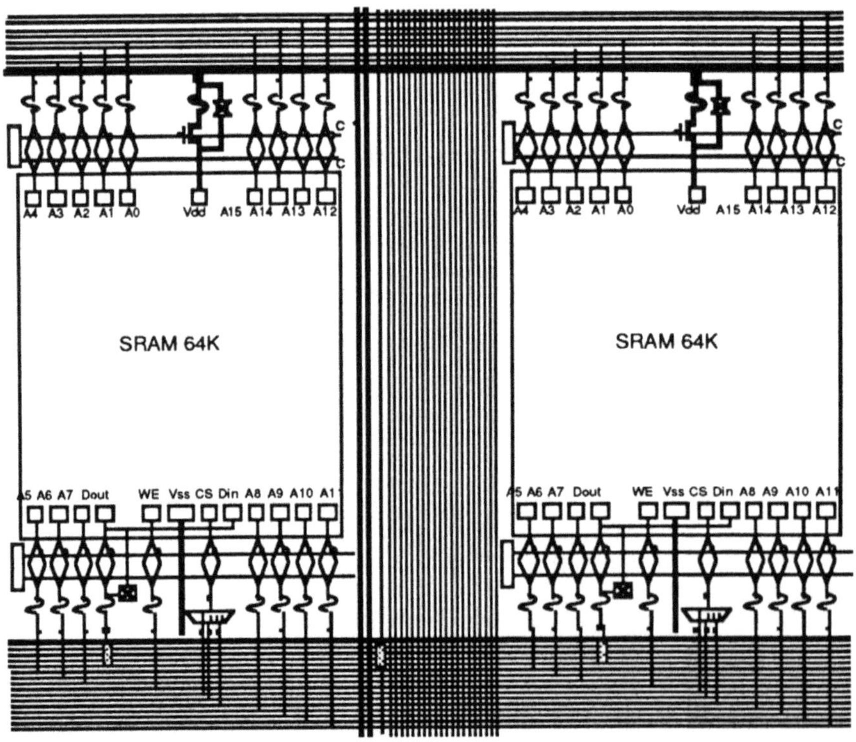

Figure 5: The reticle including two 64k cell memories.

problem aims at minimizing the following bottleneck problem :

$$\text{Min} (\text{Max } F(Gi))$$
$i = 1, \ldots 18$ (Number of groups)
Cardinal(Gi) = 4 if $i \neq j$ (Gi contains 4 interconnected good cells)
$Gi \cap Gj = \emptyset$ (Groups are mutually exclusive).

IV-3 Algorithms

Two configuration algorithms have been developed. The first is based on a simple linear scanning of the wafer. The algorithm tries to create groups which do not extend beyond a row. The second is based on a sorting of the cell costs relatively to the I/O pads. The algorithm tries to give globally the best configuration of 18 groups .

The algorithms operate from a cartography issued from a test program previously performed on the wafer. Therefore the location of defective cells, defective I/O pads and defective data lines, are known.

V-3-1 Algorithm 1

The main idea is to create groups which do not involve vertical connection lines. A group is then fully contained in a row. The wafer is bottom-up scanned row after row . A group is formed as soon as a collection of four good cells is met during the scanning. Each group is associated with its nearest I/O pad and its cost is then computed. At the end, when all the rows have been visited, we keep the cheapest groups for the configuaration.

This algorithm has been exercised on several instances of wafer cartographies. Cartographies were randomly generated from a yield model. An example of result is presented in figure 6.

The linear approach presented above leads to an O(N-D) algorithm where N is the total number of cells on the wafer and D the number of defective cells. A configuration is quickly obtained. On the other hand, avoiding systematically vertical connection lines is very constraining and that makes the algorithm performance decrease in the presence of low yields. The following method overcomes this limitation.

V-3-2 Algorithm 2

This is a sorting based configuration approach. The algorithm operates as follows.

At the first step we compute the cost of each good cell relatively to every good I/O pad. Hence, we can associate each cell **I** with a P dimensional cost vector **Vi**, where P is the total number of good I/O pads. The jth component of $\mathbf{V_i}$ is the cost induced by the link of cell **I** to I/O pad **J**.

At the second step the algorithm associates a list **Lj** to each I/O pad **J**. The file **Lj** contains all good cells sorted in an increasing cost order relatively to the I/O pad **J**.

The third step creates groups. A group is formed from each sorted list **Lj**. We take the first four cells in the sorted file. The cost of a group issued from list **Lj**, is the cost of the fourth good cell in this file. The stability of a group is then tested as follows. We make some interchanges between cells in the group and cells out of the group but within the same list **Lj**. One couple of cells is moved at a time. An interchange is accepted if it makes the number of vertical connections decrease within the group without increasing too much the cost function. A simulated annealing approach can be used as well to improve the

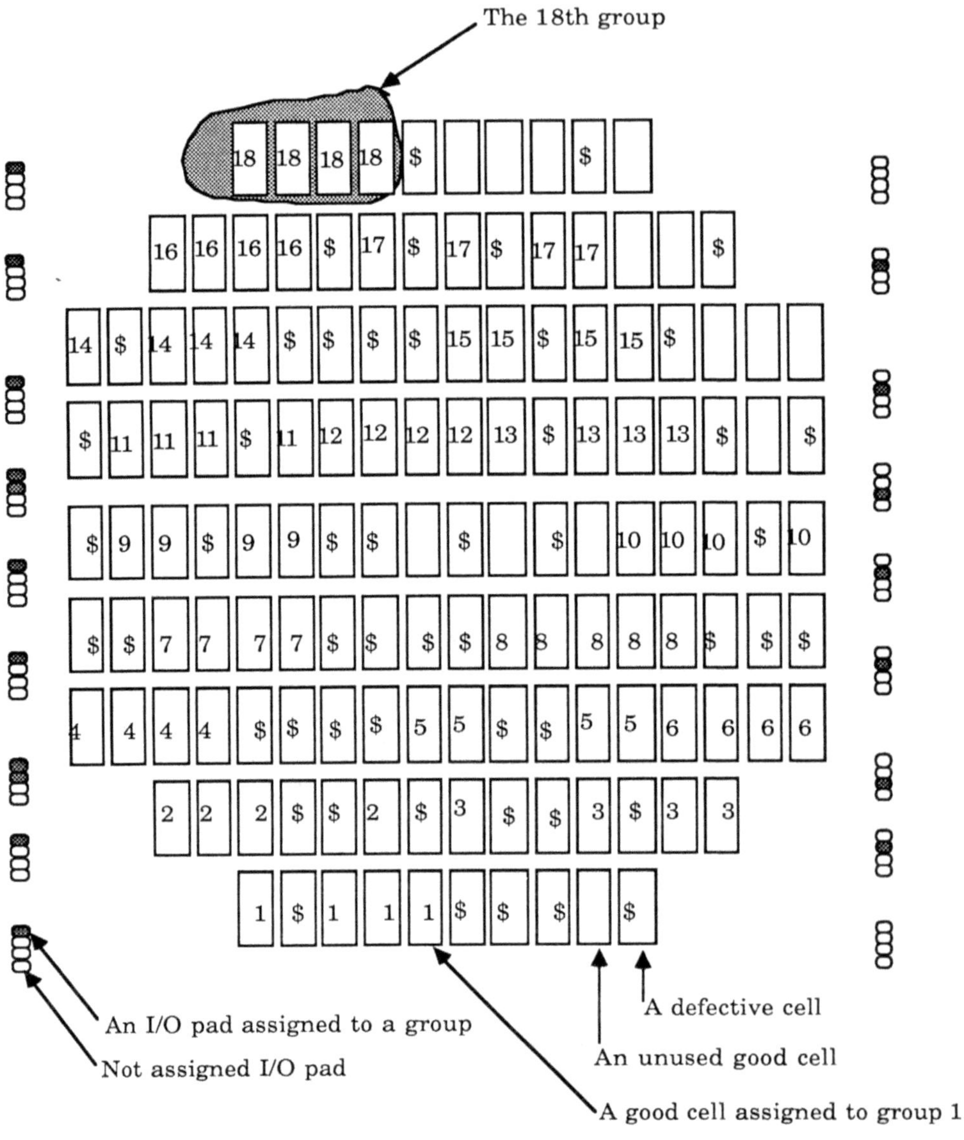

Figure 6: An example of the linear algorithm result from a wafer with a yield equals to 63 %.

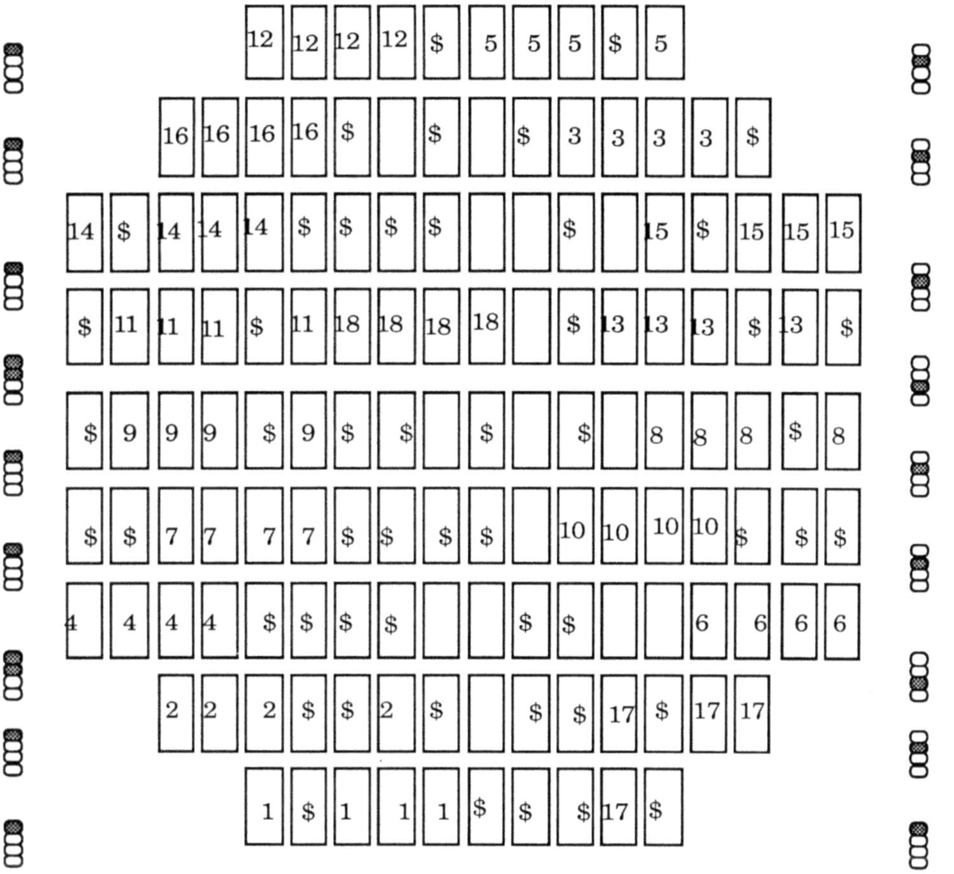

Figure 7: An example of the sorting based algorithm result from the same cartography as presented in figure 8. Group 17 leads to a vertical connection line

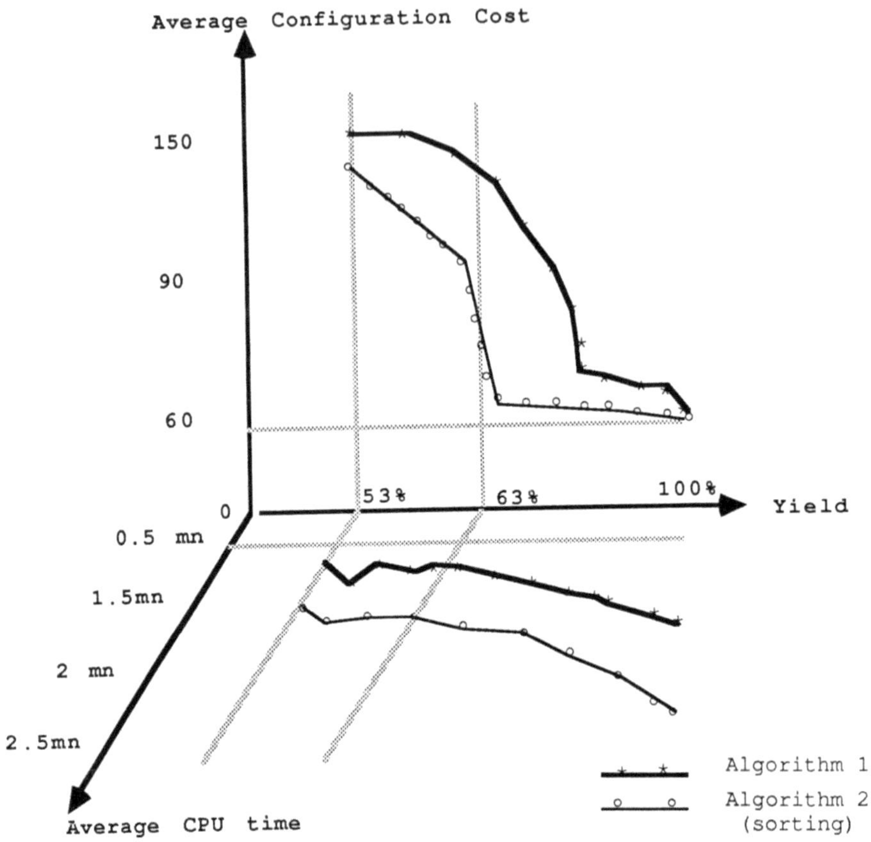

Figure 8: Comparison of the algorithms performances.

grouping of cells. At completion , each list **Lj** receives the cost of the group it has provided.

The fourth step selects the cheapest group among the ones provided by all the lists **Lj**. Let **Lj°** be the elected file. So the group provided by this file will be kept and associated with the I/O pad **J°**. All the cells in this group will be removed from all the lists **Lj**. The elected I/O pad **J°** and its sorted list **Lj°** are then frozen.

The iterations are repeated from step 3 until 18 groups are obtained or the number of remaining resources is less than needed to achieve the target configuration.

This method has been experimented on several examples of simulation with randomly generated wafer cartographies. An exemple of result can be seen in figure 7. Contrary to algorithm 1, groups with vertical connection lines can be formed if necessary. The algorithm copes with high yield values and with low yield values. The proposed method leads to an $O(P^*(N - D)^2\log(N - D))$ algorithm.

These ideas have been embodied in the CRAWL/Memory system, an interactive system which performs **C**onfiguration and **R**econfiguration **A**t **W**afer **L**evel specially dedicated to the 4.5 Mbit SRAM [23].

The performances of the two algorithms are compared on figure 8.

CONCLUSION

The first 4.5 Mbit WSI memories have been sampled to inner customers in 1988. Development is scheduled in 1989 on six inches wafers out of which four inches by four inches will be cut in order to fit in the same package. This new product will be a 8 Mbit WSI memory. A future step will be the replacement of the 64k SRAM by a 1 Mbit SRAM (presently in development in 0.8 µm technology), the memory size achievable, always in the same package, will attain then as much as 64 Mbits. We need to achieve the later objective before 1991 in order to prove a definitive advantage of WSI among classical board and hybrid techniques.

Within the same project there are more sophisticated architectural studies to build a 128 x 128 SIMD array (image processing) and a 16 bits micropocessor tolerant to end of manufacturing defects (building block for microprocessor based system for real time control application). These topics are addressed in other papers.

References

[1] Y. Kitano, S. Khodha, H. Kikuchi and S. Sakai.*IEEE. Trans. Electron Devices, ED. 27-8, Aug. 1980, p. 1631.*

[2] C. Cohen. *Electronics, Jan. 1984, pp. 77.*

[3] Y. Veoka, C. Minagawa, M. Oka and A. Ishimoto. *IEEE J. Solid-State Circuits SC 19-3, june 1984, p. 319.*

[4] R. Aubusson and I. Catt. *IEEE J. Solid-State Circuits, SC 13-3, june 1978, p.339.*

[5] I. Catt. *Wireless World, july 1981, p. 57.*

[6] *Communication at the WSI workshop, Southampton (UK), 1985.*

[7] E. Bennet. *Electron. Des., Dec., 2, 1985.*

[8] R. Hawkins. *Electron. Des., Oct. 1985.*

[9] G. Saucier and J. Trilhe editors, *Wafer Scale Integration, published by North-Holland, 1986.*

[10] *A data sheet of the 64 K SRAM will be available from ST (Carrolton) in 1989.*

[11] M. Ketchen. *IEEE Circuits and Design magazine, jul. 1985.*

[12] C. Staper. *Proc. of the IEEE, Vol. 71, N°4, april 1983.*

[13] K. Hedlund. *PhD thesis dissertation, Purdue university, west Lafayette, nov. 1982.*

[14] NCR Corp., *Semiconductor International, march 1981, p. 87.*

[15] *Ice, ICECAP report, issue 5-5, 1985.*

[16] J. Asaitis, G. Chapman and J. Raffel. *IEEE electron device letters, Vol. EDL 3 N°7, july 1982, p. 184.*

[17] O. Minuto and al., *IEEE Journal of Solid-State Circuits, Vol. SC 17 N°5, oct. 1982, p. 793.*

[18] B. Williams, I. Herman, F. Mitlitsky, R. Hyde and L. Wood. *Appl. Phys. Lett., Vol 43, N°10, nov. 1983, p. 946.*

[19] P. Wyatt, J. Raffel, G. Chapman, B. Mathur, J. Burns and T. Herdon. *Proceedings IEDM 1987, p. 626.*

[20] P. Girard. *Proc. Conference Scanning Microscopy International, St. Louis (Missouri), 1988.*

[21] D. Shaver. *Solid-State Technology, feb. 1984.*

[22] J. Barret. *Proc. IMC, Tokyo (Japan), may 1988.*

[23] B. Nasreddine. *"Conception d'une mémoire reconfigurable intégrée sur tranche". INPG PhD thesis dissertation. Grenoble, July 1988.*

[24] R. Ferrant. *Fast and low power asynchronous state RAM. Centre d'étude et de fabrication - Eurotechnique, March 1985.*

[25] M.B. Ketchen. *Point defect yield model for wafer scale integration. IEEE Circuit and devices magazine, July 1985.*

Appendix 1

Supplier	Size bit	Type : Dynamic Static	Architecture	Techno.	CD µm	Access Time in ns	Size memory point µm2
Mitsubishi	4 M	D	4M x 1 bit 1M x 4 bit	CMOS	0.8	90	10.9
IBM	4 M	D	1M x 4 bit	CMOS	0.8	65	11
Hitachi	4 M	D	4M x 1bit 1M x 4 bit	CMOS	0.8	65	14.7
OKI	4 M	S	512 K x 8 bits	CMOS	1	95	16.8
NTT	16 M	D	2M x 8 bit	CMOS	0.7	70	7.5
Hitachi	1 M	D		BICMOS	1.2	35	36
Matsushita	4 M	D	1M x 4 bit	CMOS	0.8	65	11
Fujitsu	4 M	D	4M x 1 bit 1M x 4 bit	CMOS	0.7	70	7.5
Toshiba	4 M	D	4 M x 1 bit 1 M x 1 bit	CMOS	0.9	60	13.75
Hitachi	256 k	S	32k x 8 bit	CMOS	1	43	3 9.2
IBM	256 k	S	256k x 1 bit	CMOS	0.7	30	30
Motorola	256 k	S	32k x 8 bits	CMOS	1.2	21	
Sony	1 M	S	128k x 8 bit	CMOS	0.8	35	74.2
Hitachi	1 M	S	128k x 8 bit	CMOS	0.8	42	44.7
Mitsubisi	1 M	S	128k x 8 bit	CMOS	0.8	34	44
Toshiba	1 M	S	1 M x 1 bit	CMOS	0.8	25	5

BLOCK ALIGNMENT: A METHOD FOR INCREASING THE YIELD OF MEMORY CHIPS THAT ARE PARTIALLY GOOD

C.H. Stapper

IBM Corporation
General Technology Division
Essex Junction, VT 05452

INTRODUCTION

Redundant word and bit lines have been used successfully for increasing the manufacturing yield of memory chips since 1979 [1, 2, 3]. During early phases of manufacturing, however, there may not be enough redundancy available to repair all the faults on a chip. Such unfixable chips have often been used as partially good chips [4]. The drawback with the use of such chips is the different combinations of failure patterns that can produce the same number of partially good bits. Each pattern requires a different method for connecting the chips. Because of these different requirements, a number of chips with different part numbers result.

To solve the problem of multiple part numbers, Fitzgerald and Thoma introduced data steering with fusible links [5]. Such links are identical to the fuses used for word and bit line redundancy. In this case, however, blown fuses affect entire blocks of data instead of single lines. In this way, blocks of fault free data are always steered to the same output pads. Although this method was originally applied to output data, Anderson *et al.* also used it to steer data away from faulty addresses [6, 7]. Partially good chips that make use of either data or address steering all look the same after the fuses are blown.

The method of data steering can be used to increase the yield of partially good chips. This is possible when the memory arrays are subdivided in a number of sub-arrays or sections. In this approach, blocks of data corresponding to the faulty sections are all aligned in the same address space, or in the same data output bit. This method is therefore referred to here as *block alignment.* It tolerates more faulty sections than is possible with a straight partially good chip scheme.

Block alignment for yield enhancement was first introduced in the IBM 4-Mbit chip that will be produced by the Sematech industrial consortium [8, 9]. This chip, however, is too complex to be adequately described and analyzed in this paper. Instead, we use a hypothetical chip contrived as an example. This example is chosen because it is a simplified version of one eighth (sometimes referred to as an octant) of the IBM 4-Mbit chip. The results, furthermore, amply illustrate the increase in yield through the use of redundant word and bit lines, partially good chips, and block alignment. It also makes it possible to demonstrate the effect of different types of fault clustering on those yields.

AN ILLUSTRATIVE EXAMPLE

The hypothetical chip is assumed to have four redundant word lines or rows, and four redundant bit lines or columns. These redundant lines can replace combinations of defective memory cells, and faulty word or bit lines. The chip is divided into four *islands*, which are known as *quadrants*. Each quadrant contains four sections, where every section has its own word decoders, bit decoders, and sense amplifiers. A conceptual arrangement of the example chip is shown in Fig. 1. Note that the actual layout of a chip of this type could actually be more complex.

The subdivision of memory arrays into sections is typical for high performance dynamic random access memory chips (DRAMS). It results from the short word and bit lines that are needed to achieve the decreased access and cycle times of such chips. Each section consequently is an independent memory array that contains a block of data. Data can therefore be steered towards or around these sections or blocks.

Also shown in Fig. 1 is a group of circuits designated as *supports*. When these circuits are defective, a chip is completely useless. Defects causing support circuit failures are therefore known as *fatal faults*, or as *chip kills*. These faults, by definition, cannot be fixed with redundancy. Neither can chips affected by these faults be used as partially good, or with block alignment.

For the sake of brevity only 3/4 (three-quarter) good schemes are analyzed in this paper. The 3/4 partially good chips are those where all the bits in three of the quadrants are either fault free or else fixable with redundancy. The fourth quadrant, however, is defective to such an extent that there is insufficient redundancy on the chip for fixing it.

It is the purpose of this paper to investigate yields associated with a block alignment scheme and compare these to the yields of partially good chips. For this example, it is assumed that the address steering can be done only within islands. This limits the benefits from the block alignment method. It is, however, a practical limitation carried over from the IBM 4-Mbit chip. Nevertheless, for this hypothetical example chip it still results in 622 combinations of failing sections that can result in block-aligned 3/4 good chips. It is this large number of combinations of block-aligned chips that produces higher yields than those attainable with the simple partially good chips scheme. For the latter there are only four combinations by which three quadrants can be good.

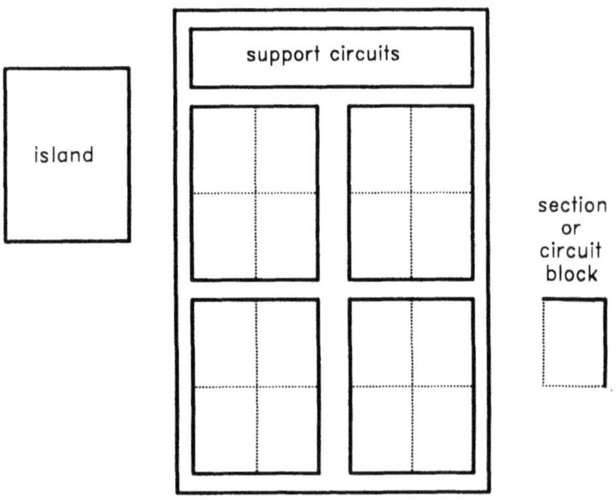

Figure 1. Simplified diagram of a memory chip. This hypothetical chip contains support cicuits and four islands of memory cells known as quadrants. The quadrants are further subdivided into sections. Blocks of data corresponding to these sections can be aligned with each other to increase the yield of partially good chips.

A SIMULATION MODEL

Another purpose of this paper is to demonstrate the effect of defect clustering on yield. Analytical techniques for calculating the perfect chip yields and the fixable chip yields are available in reference [4]. Exact formulas for determining the yield of partially good chips and block-aligned chips with redundancy are possible in principle. Until now, however, none have been available to this author. Simulation models have therefore been used instead. The results from these models are compared to an approximate analytical model.

The simulation program uses pseudo random number generators. These were used to obtain a random number of fixable faults for each section on a chip. The number of fatal faults per chip were generated in a similar fashion. Chips without fatal faults and with 16 fault-free sections were counted as perfect chips. In the same way, the chips without fatal faults and fewer than eight fixable faults were considered repairable and were therefore counted as all-good chips. For simplicity a 100% yield was assumed for the redundant word and bit lines.

Note that in this model no distinction was made between the faults in word lines, bit lines, and single cells. Such differences are accounted for in industrial yield models like the one described in reference [4]. They were not included here in order to keep the computer run times down. This in no way detracts from the didactic value of the results.

The simple 3/4 good scheme is one where three quadrants can be repaired, but where there is insufficient redundancy to fix the fourth quadrant. These were the chips with more than eight fixable faults, but where three quadrants contained eight or fewer faults. The simulated chips that fell in this category and that were free from fatal faults contributed to the 3/4 good yield.

The chips that were 3/4 good with block alignment were determined in a similar way. In this case only the three sections with the lowest number of fixable faults were taken into account in each quadrant. The number of faults in these sections were summed for the four quadrants. Chips for which this sum was less than eight were counted as block-aligned 3/4 good if they did not contain any fatal faults.

YIELDS FOR RANDOM DEFECTS

It has been shown by Glaser and Subak-Sharpe [10] that purely random distinguishable point defects result in a binomial distribution for the number of faults per chip. In the mathematical expression for this distribution, the total number of faults per wafer is a parameter. It has been shown in [10] and [11] that the wafer-to-wafer variation of this number of faults per wafer should be taken into account for a general yield expression. If such variations follow a Poisson distribution, then according to reference [11], the distribution of the number of faults per chip also follows a Poisson distribution, albeit with a different parameter.

The Poisson distribution can readily be simulated with random numbers. A program for doing so was used to randomly generate the number of fixable faults per section. The simulations were done for different averages of the number of fixable faults per chip. The resulting yields are plotted in in Fig. 2 as a function of the average number of fixable faults per chip. The results show clearly the sizable increase in partially good chip yields that result from the use of block alignment. This graph, however, is intentionally misleading. It represents the degree of yield modeling often encountered in the literature on fault tolerant circuitry.

The results in Fig. 2 do not include the effect of faults in the support circuits. This must be taken into account next, and it will be assumed that the average number of fatal faults is equal to the average number of fixable faults in a single section. Also neglected in this plot is the fact that not all the bits are useable in the partially good or block-aligned chips. This is usually accounted for as an *equivalent yield*. For 3/4 good chips the equivalent yield is obtained by multiplying the partially good chip yield and the block-

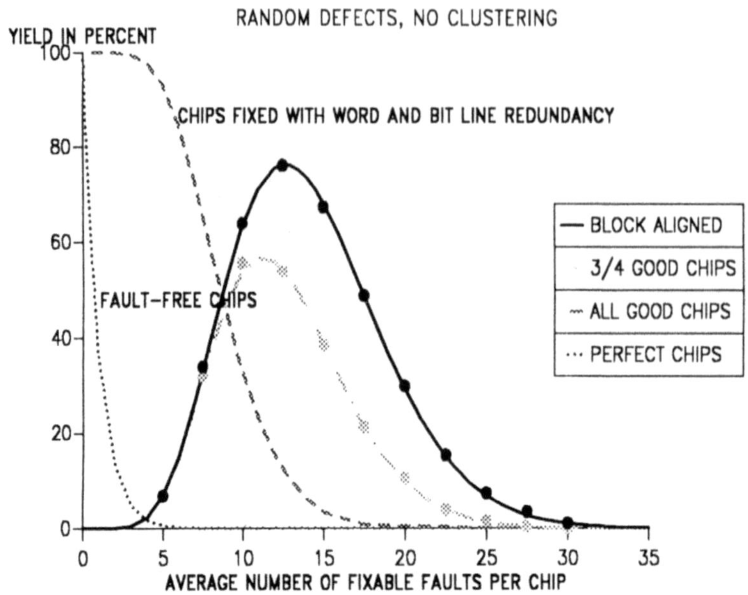

Figure 2. Random defect yields for a memory chip. The redundant word and bit lines can fix up to 8 faults per chip. This redundancy is very effective in increasing the yield of chips with an average number of fixable faults per chip that is less than 10. The partially good chip yields for chips with higher fault averages can be significantly improved with fault alignment. Fatal faults in the support circuits were not included.

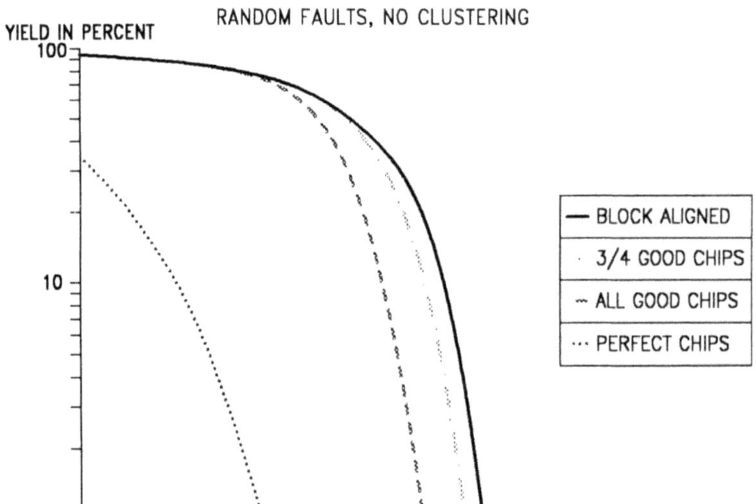

Figure 3. Equivalent yields for a memory chip. The yield of the partially good and block-aligned chips is multiplied by 3/4 and added to the all good chip yield to obtain an equivalent yield. Fatal faults are included in these simulations.

aligned yield by 3/4. Adding this equivalent yield to the all good chip yield results in what is often known as the *all good equivalent yield.*

The perfect chip yield, the all good yield, and the equivalent yields for partially good chips and block-aligned chips are plotted in Fig. 3. The scales are logarithmic. On the vertical scale this is done so that equal multiples of yield improvement show up as equal magnitudes regardless of the level of yield. The horizontal logarithmic scale is used in anticipation of the results of fault clustering. Notice that under these conditions the yield improvements show up much less favorable than they do in Fig. 2.

It is usually necessary at this point to introduce the concept of productivity. This is done to take into account the fact that different fault tolerance schemes require different chip areas, thus affecting the number of bits per wafer. Eventhough fuses were used in the block alignment scheme, their implementation on the IBM 4-Mbit chip required no increase in chip area. It is therefore assumed that for the hypothetical chip in this paper, there also is no increase in size with block alignment. The yield curves obtained here are therefore similar to productivity curves.

FAULT CLUSTERING

A number of papers on fault tolerant integrated circuits still deal solely with random faults as was done in the preceding section. This is unrealistic because it has been known since the 1960s that defect clustering and hence fault clustering does occur in integrated circuits. It has been known since that time that such clustering has a profound influence on the integrated circuit manufacturing yields. As is shown in this paper, it has a similar influence on fault tolerance.

In some papers on fault tolerance Bose-Einstein statistics have been used to account for clustering. This is unrealistic because such statistics have nothing to do with fault clustering. They pertain instead to undistinguishable particles. Integrated circuit defects and faults can be seen, albeit sometimes with difficulty. Such faults are therefore distinguishable and Bose-Einstein statistics does not apply to them.

In practice the frequency distribution of the number of faults per integrated circuit can often be approximated by a negative binomial distribution [4, 12]. Another example is found in the distributions of failing cells, word and bit lines plotted in Figs. 4-7. These data were obtained by H. Boisselle from 420 DRAM chips, each containing 1-Mbit of memory. The data are represented by solid lines in the diagram, while negative binomial distributions are plotted as dotted lines. The agreement is amazingly good. As a result, negative binomial distributions are therefore used in this paper.

The negative binomial distribution can be written in the form

$$P(X = k) = \frac{\Gamma(\alpha + k)}{k!\,\Gamma(\alpha)} \quad \frac{(\lambda/\alpha)^k}{(1 + \lambda/\alpha)^{\alpha+k}}\,, \tag{1}$$

where X is a random variable designating the number of faults per integrated circuit, k an integer equal to 0, 1, 2, etc., α a parameter related to the fault clustering, and λ the average number of faults per circuit. This distribution is key to the cluster simulations of this paper. Three cases are investigated. These deal with fault clusters that are small, intermediate size, and very large.

COMPUTER SIMULATIONS

In the simulation of small clusters complete independence is assumed between the number of faults in adjacent sections. This implies that the clusters are smaller than the sections. This makes it possible to generate the number of faults in the sections with a pseudo random number generator that results in a negative binomial distribution. A value of 0.125 for the cluster parameter α was used in the simulation. This also resulted

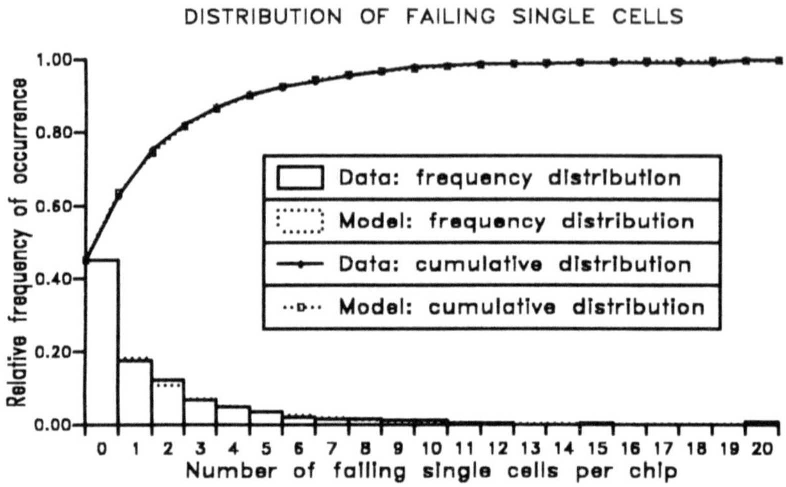

Figure 4. Fault distribution. The frequency distribution of the number of failing single cells can be modeled with a negative binomial probability distribution function. For this example $\lambda = 1.8825$, and $\alpha = 0.5181$.

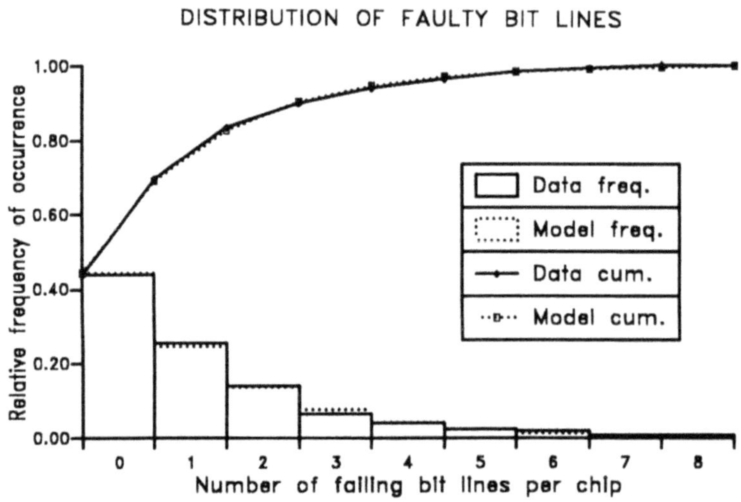

Figure 5. Fault distribution. The frequency distributions of the number of failing bit lines per chip can also be modeled with a negative binomial probability distribution function. In this case $\lambda = 1.2512$, and $\alpha = 0.9955$.

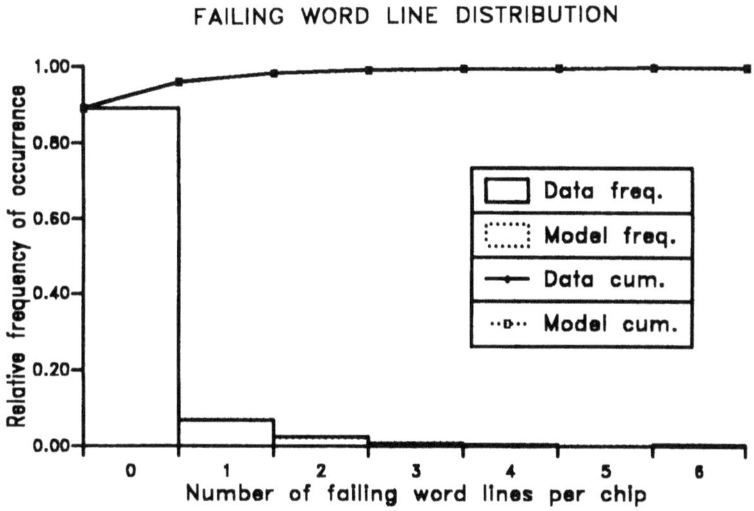

Figure 6. Fault distribution. The negative binomial probability distribution function can also be used to model the frequency distributions of the number of failing word lines per chip. For this case λ =0.1805, and $\alpha = 0.1341$.

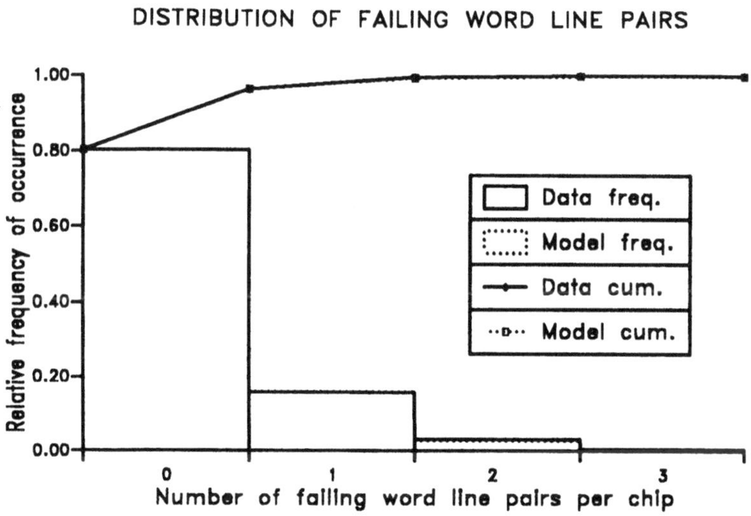

Figure 7. Fault distribution. The negative binomial probability distribution function used to model the frequency distributions of the number of failing word line pairs per chip has $\lambda = 0.2384$, and $\alpha = 1.2565$.

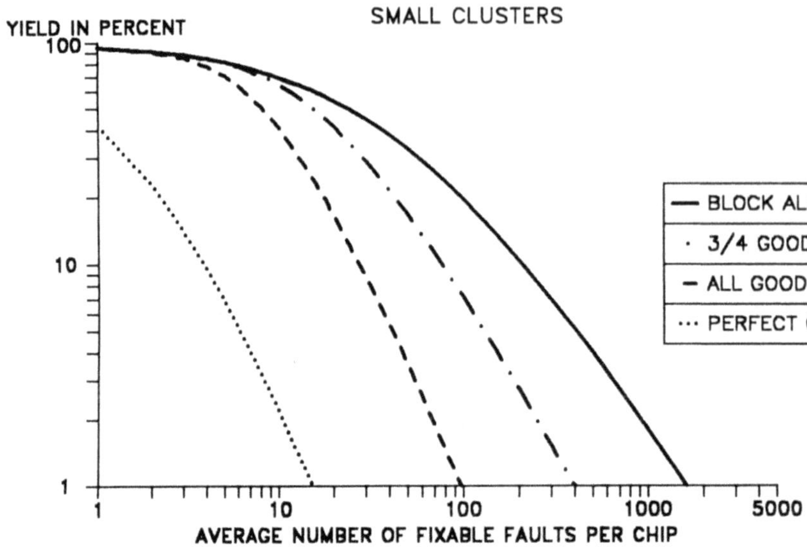

Figure 8. Simulation results for small clusters. The equivalent yields are plotted as a function of the number of fixable faults per chip.

in negative binomial distributions for the frequency distributions of the number of faults per island and per chip. In those cases, however, the average number of faults were 4 and 16 times higher and the cluster parameters were 0.5 and 2 respectively.

Results from the simulation are shown in Fig. 8. They indicate that even with an average of 1000 faults per chip, block alignment makes it possible to get 2% equivalent yield. Block alignment is therefore extremely effective in dealing with fault clustering of this type. This happens because the clusters with a high number of faults are smaller than the circuit blocks. The block alignment scheme can therefore steer around them.

Next let us take a look at fault clusters that are larger than the individual sections. Such clusters can be generated with a technique that randomly generates faults as a function of time. It was shown in Reference [13] that such an approach can result in negative binomial distributions. It does so when, during an incremental time interval Δt, the probability of obtaining a fault in an integrated circuit is linearly related to the number of faults already in that circuit. A Poisson distribution results when this probability is constant. We use this technique here with a unique modification.

In the simulation program a random number generator is used to determine whether a fault is to be added to a section during a time interval Δt. The probability of adding a fault to a section is assumed to be linearly dependent not only on the number of faults in a section, but also on the number of faults in its four nearest neighbors.

The program starts with no faults in any of 160,000 sections and 40,000 additional circuits. It slowly builds up a pattern of faults by stepping through consecutive time increments. The process stops when the desired average number of faults is reached. A typical result of fault patterns obtained with this program is shown in Fig. 9. In this diagram the section boundaries are indicated by dotted lines, and the chip boundaries by solid lines. The number of faults in each section is represented by randomly placed dots.

Fault patterns generated with this program result in negative binomial distributions for the number of faults in the sections, islands and chips. The cluster parameters depended on the average number of faults, and for the section they were found to vary between 0.6 and 0.85, while those for the islands between 0.8 and 1.1. The value of α was fixed at two for the complete chip for all averages of the number of faults per chip. In this way the perfect and fixable chip yields corresponded to those obtained for the small clusters.

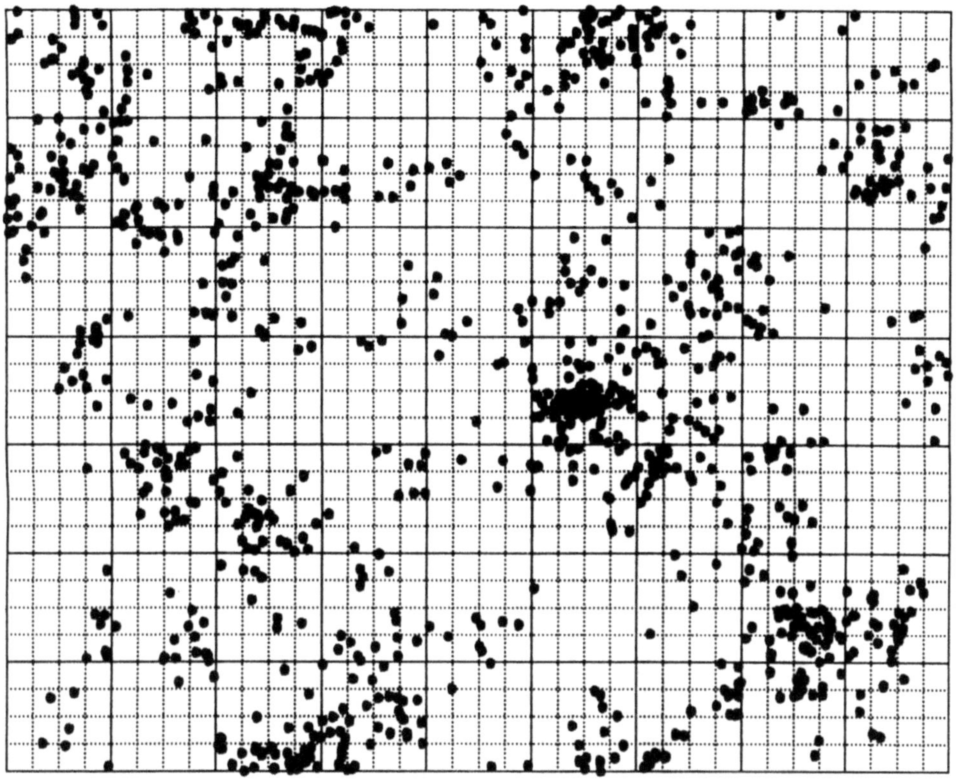

Figure 9. Clustered faults. This map was generated for intermediate size clusters. The chips are indicated by solid boundaries, the islands by dotted boundaries. The average number of faults per section is slightly more than one.

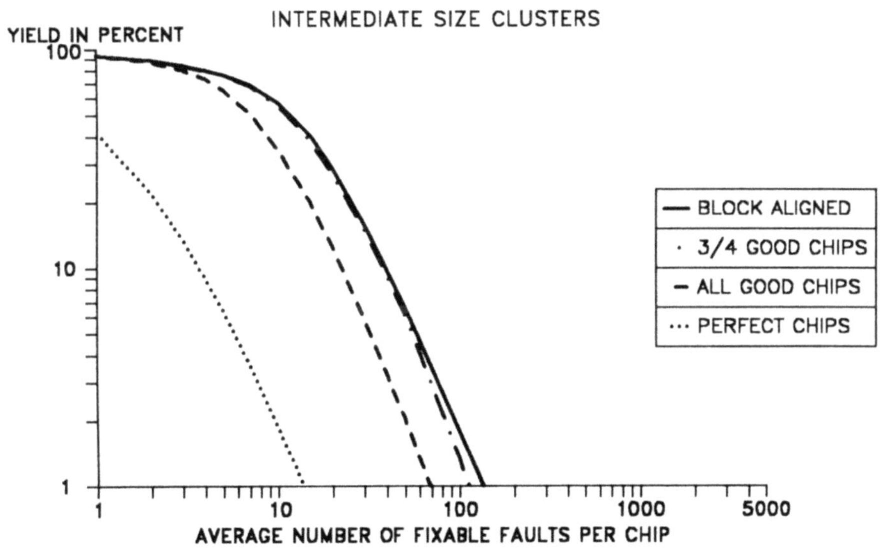

Figure 10. Simulation results for intermediate size clusters. The resulting yields are plotted as a function of the average number of fixable faults per chip.

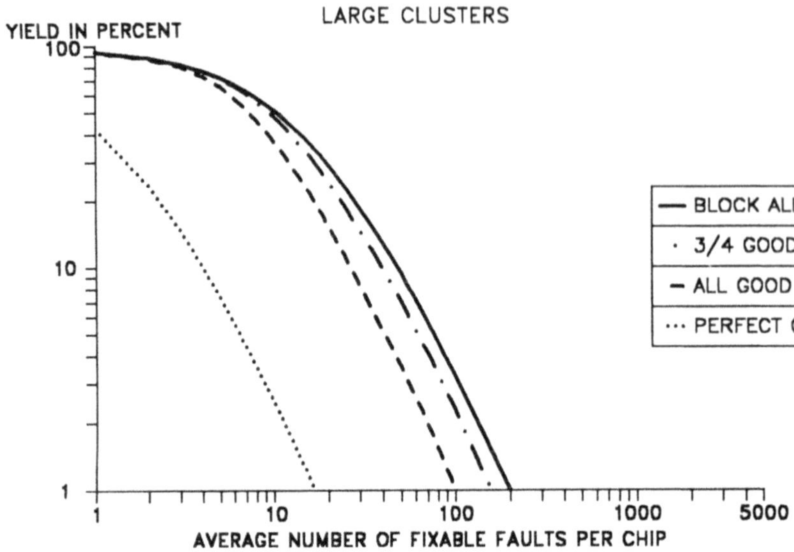

Figure 11. Simulation results for large area clusters. This type of clustering is somewhat more forgiving than the intermediate size clustering.

The resulting yields for partially good chips and block-aligned chips are shown in Fig. 10. Note that in this case these schemes for fault tolerance produce much less yield benefit than did the preceding examples. This, however, stands to reason, because of the correlation between the number of faults in adjacent sections. When a section fails, the adjacent sections are likely to fail also. Block steering within an island cannot fix such fault clusters and its effectiveness is therefore limited. As can be seen in Fig. 9 the fault clusters generated by this program extend beyond the chip boundaries. As a result, even yields of the regular 3/4 partially good chips are reduced.

In the third case of interest the fault clusters are assumed to be very much larger than the chips. The faults within such clusters are assumed to be randomly distributed. This condition is simulated here by randomly generating an average number of faults for each chip from a gamma distribution. This average is then used with a Poisson distribution to randomly generate the faults for each section on a chip. This also results in negative binomial distributions for the number of faults per section, island, and chip. In this case, the cluster parameter α is the same for all three. For the simulations the value of this parameter was fixed at two.

The results of the large area cluster simulation are shown in Fig. 11. They do not differ much from those obtained for intermediate size clusters. It is this similarity that makes it possible to approximate both large and small clusters with an analytical model. Such a model is extremely useful because the computer simulations are expensive and very time consuming.

ANALYTICAL APPROACH

The yield of partially good chips, and block-aligned chips without redundancy can be calculated analytically for all clustering schemes. This is best done by defining a yield function Y_{nS} for n sections in the form

$$Y_{nS} = \{1 + (\lambda_{CK} + n\lambda_S)/\alpha_{nS}\}^{-\alpha_{nS}}, \tag{2}$$

where λ_{CK} is the average number of fatal, or chip kill faults per chip, λ_S the average number of faults per section, and α_{nS} a cluster parameter that varies with the number of sections n. Knowledge of this dependency is crucial for the correct calculations of yields for the partially good and block-aligned chips. When the clusters are smaller than the sections, this parameter is directly proportional to the number of sections. In the case of very large clusters it has a contant value. In most practical applications α_{nS} has a non-zero value at $n = 0$ and increases in value when the number of sections increases.

Use of expression (2) for determining the yield of 3/4 good chips results in the formula

$$Y_{34} = 4(Y_{12S} - Y_{16S}), \tag{3}$$

where the subscripts 12S and 16S indicate that 12 and 16 sections are involved. Similarly for the block-aligned chips the yield is given by

$$Y_{BA} = 256Y_{12S} - 896Y_{13S} + 1246Y_{14S} - 812Y_{15S} + 206Y_{16S}. \tag{4}$$

In this case the number of sections involved are 12, 13, ... , 16. The formulas in expressions (3) and (4) are exact for small and large clustering. No approximations had to be made to obtain them. They are approximate for the intermediate size fault clusters. Similar formulas can be derived for other partially good chip schemes and other schemes for block alignment. Furthermore, the technique of expanding the yield expression, as is done here, can be applied to the yield calculation for any method used for fault tolerance.

The use of redundancy complicates the yield calculations somewhat. The all-good chip yield is given by

$$Y_{AG} = \sum_{k=0}^{8} \frac{\Gamma(\alpha_{16S} + k)}{k!\,\Gamma(\alpha_{16S})} \; \frac{(16\lambda_S/\alpha_{16S})^k}{\{1 - (\lambda_{CK} - 16\lambda_S)/\alpha_{16S}\}^{\alpha_{16S}+k}}, \tag{5}$$

where the cluster parameter α_{16S} pertains to an entire chip, including the support circuits. This yield expression is also exact, but it is now used to find an approximation for the average number of faults that are left in a section after fixing the chip with redundancy. This quantity is designated by λ_{SAF}.

Although the actual distribution of the number of unrepaired faults on a chip is quite complex, we approximate this distribution here to be negative binomial. This distribution is assumed to have an average number of faults $\lambda_{CK} + 16\lambda_{SAF}$, and a cluster parameter α', where $\alpha' \leq \alpha_{16S}$. As a result the all-good chip yield can be expressed as

$$Y_{AG} = \{1 + (\lambda_{CK} + 16\lambda_{SAF})/\alpha'\}^{-\alpha'}. \tag{6}$$

Solving this equation for λ_{SAF} produces

$$\lambda_{SAF} = \frac{1}{16}\{\alpha'(\sqrt[\alpha']{1/Y_{AG}} - 1) - \lambda_{CK}\}. \tag{7}$$

This approach may be crude, but it can now be used to estimate yields with the formula

$$Y_{nS} = \{1 + (\lambda_{CK} - n\lambda_{SAF})/\alpha'\}^{-\alpha'}. \tag{8}$$

These yields can then be used in formulas (3) and (4) to determine the yield of partially good and block-aligned chips that have redundancy. Results of such calculations with $\alpha_{nS} = 2$ for all values of n, and $\alpha' = 2$ are shown in Fig. 12.

DISCUSSION

The effect of fault clustering on partially good chips and block alignment depends completely on the nature of the clusters. Block alignment is particularly effective in

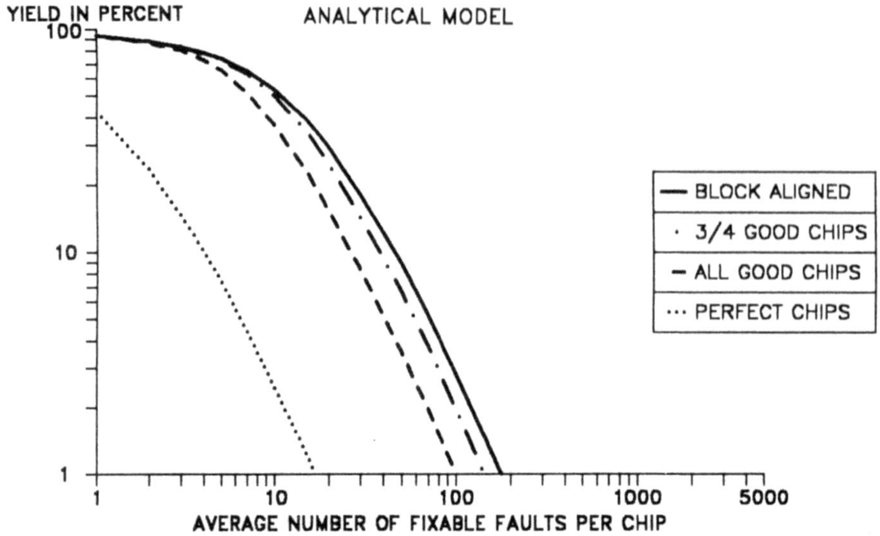

Figure 12. Results from the analytical model. The equivalent yields resulting from analytical calculations are shown as a function of the average number of fixable faults per chip.

dealing with small fault clusters. These type of clusters are often encountered during early phases of manufacture. They can result from scratches, large photolithographic defects, and residual spots left over from chemical processes. Block alignment makes it possible to use chips that are affected by such defects, thus allowing for early delivery of chips for the construction of advanced engineering hardware. Such chips also can find their way into memory expansion of personal computers and systems.

The example of block alignment in this paper has been used to show how to deal with fault clustering in yield calculations for fault tolerant chips. Such fault clustering is often neglected in the literature. However, as has been shown here, it can have a profound effect on the yield. The approach can readily be extended to other schemes for fault tolerance. Yields for 1/2 good and 1/4 good chips with block alignment have been calculated with these techniques for the IBM 4-Mbit chip.

The models that have been described in this paper were simplified for didactic reasons. The industrial yield model used for 4 Mbit chip deals with 45 different defect types such as photolithographic defects, oxide pinholes, and junction leakage defects. That model also distinguishes between 13 different failure types, such as failing cells, word lines, bit lines, sections, islands, and chips. Nevertheless, the principles discussed in this paper apply to that model as well.

REFERENCES

[1] R. R. DeSimone, N. M. Donofrio, B. L. Flur, R. H. Kruggel, H. L. Leung, and R. Schnadt, "Dynamic memories," in *1979 IEEE Int. Solid-State Circuits Conf., Dig. Tech. Papers*, pp. 154-155, Feb. 1979.

[2] R. P. Cenker, D. Clemens, W. R. Huber, J. B. Petrizzi, F. J. Procyk, and G. M Trout, "A fault tolerant 64k dynamic RAM," in *1979 IEEE Int. Solid-State Circuits Conf., Dig. Tech. Papers*, pp. 150-151, Feb. 1979.

[3] R. P. Cenker, D. Clemens, W. R. Huber, J. B. Petrizzi, F. J. Procyk, and G. M Trout, "A fault tolerant 64k dynamic random access memory," *IEEE Trans. Electron Devices*, vol. ED-26, pp 853-860, June 1979.

[4] C. H. Stapper, A. N. McLaren, and M. Dreckmann, "Yield model for productivity optimization of VLSI memory chips with redundancy and partially good product," *IBM J. Res. Develop.*, vol. 24, pp 398-409, May 1980.

[5] B.F. Fitzgerald, and E. P. Thoma, "A 288K-bit dynamic RAM," in *1979 IEEE Int. Solid-State Circuits Conf., Dig. Tech. Papers*, pp. 68-69, Feb. 1982.

[6] K. L. Anderson, A. R. Ellenberger, W. F. Ellis, and J. P. Streck, "Partially good chip part number reduction by dynamic relocation," *IBM Technical Disclosure Bulletin*, vol. 27, no. 11, pp. 6393-6394, April 1985.

[7] K. L. Anderson, A. R. Ellenberger, W. F. Ellis, and J. P. Streck, "Dynamic relocation circuit," *IBM Technical Disclosure Bulletin*, vol. 30, no. 1, pp. 462-463, June 1987.

[8] R. A. Parent, D. G. Morency, C. A. Kilmer, D. K. Tewarson, R. E. Newhart, J. S. Kosson, M. P. Clinton, T. A. Bronson, D. M. Plouffe, M. H. Bus, J. R. Morrish, E. P. Thoma, R. E. Busch, T. M. Redman, "A 4Mb DRAM with double buffer static column architecture," *1987 IEEE Int. Solid-State Circuits Conf., Dig. Tech. Papers*, pp. 13-14, Feb. 1987.

[9] C. A. Kilmer, T. A. Bronson, M. P. Clinton, J. S. Kosson, D. G. Morency, J. R. Morrish, R. E. Newhart, R. A. Parent, D. M. Plouffe, D. K. Tewarson, "4Mb DRAM Circuit Features," *Thirteenth European Solid-State Circuits Conference, Digest of Papers*, 1987, pp. 61-64.

[10] A. B. Glaser, and G. E. Subak-Sharpe, *Integrated Circuit Engineering - Design, Fabrication, and Applications*, Reading, MA: Addison-Wesley Publishing Co., 1977, Ch. 16, pp. 773-774.

[11] C. H. Stapper, "The effect of wafer to wafer defect density variations on integrated circuit defect and fault distributions," *IBM J. Res. Develop.*, vol. 29, pp 87-97, Jan. 1985.

[12] R. Glang, "Measurement and distribution of faults on defect test site chips," *Proceedings of the IEEE International Workshop on Defect and Fault Tolerance in VLSI Systems*, University of Massachusetts, Amherst, pp.3.1-1 - 3.1-12, Oct. 1988.

[13] C. H. Stapper, F. M. Armstrong, and K. Saji, "Integrated circuit yield statistics," *Proc. IEEE*, vol. 71, pp 453-470, April 1983.

FAULT TOLERANT INTEGRATED MEMORY DESIGN

C.A. Njinda, C.G. Guy*, and W.R. Moore

Oxford University
Department of Engineering Science
Oxford OX1 3PJ, UK

INTRODUCTION

With today's levels of integration, it is increasingly common to have complete systems such as digital controllers and other application specific processors implemented as single chips. As the number of devices in such chips increases the reliability of the system becomes an important issue. Memory has been singled out as the source of the majority of failures in such systems. Consequently, there is a perceived need to make these memory devices highly reliable.

Memory faults can be permanent (hard) or transient (soft). Transient faults have been found to account for nearly 90% of all memory failures. These faults are mainly caused by electro-migration, voltage variation, alpha particle radiation, temperature changes, vibrations and to a lesser extend improper design[1,2,3]. To be able to test for faults in memories, techniques of off-line testing have been extensively developed. Such techniques, however, are inadequate for detecting transient faults and usually offer no possibility of correcting the resulting errors. Therefore, on-line or "Concurrent Error Detecting" (CED) techniques have become very important when designing memory circuits for high reliability applications. They usually involve some form of coding and decoding and hence have penalties associated with their use in terms of area overhead, power dissipation and access time.

Several manufacturers already have standard integrated circuits for memory Error Detection and Correction (EDAC). These are external to the memory array and act as an interface between the central processing unit and its storage[4]. Such circuits are often large and complex and should only be used with large memory arrays if their presence is not to degrade the overall reliability of the system. Also available on the market are memories with integrated EDAC[5,6]. So far, these devices have been 256K*1 bit RAMs, and hence are mainly suitable for large memory systems unless they are treated as serial storage devices. However, many application specific processors also have a high reliability requirement over considerable periods of time. When such systems require only small amounts of read/write memory (say less than 4K bytes), it is quite common to find that the available solutions are

*Dept. of Electronic Eng., UCL, Torrington Place, London WC1E JE, UK.

too expensive and actually turn out to make the overall system less reliable[7]. Therefore in this paper, we examine how to design a Fault Tolerant Integrated Memory (FTIM). In our case the EDAC circuits are to be implemented on the same chip as the rest of the memory circuits.

The first section of the paper deals with the theoretical requirements that need to be satisfied by the device. In the next section, we present a summary of the coding techniques required for the array and the supporting circuits (error detecting and correction circuits). We also present results showing the variation of Reliability and Mean Time to Failure (MTTF) of the various coding techniques. In order to illustrate the procedure required in the design of a highly reliable memory device, we consider a case study of the design of a 1K byte FTIM in the next section. Here we show how it might be possible to use the EDAC circuits to aid in testing the device. Finally a brief summary of the paper is given.

THEORETICAL REQUIREMENTS FOR A FTIM

The criteria that need to be satisfied by any FTIM depend on the application for which the device is intended. Obviously, the reliability improvement is the prime criteria. In applications like digital controllers for nuclear reactors, medical or aerospace computers, access time may be the most important factor, while in other cases such as in fully integrated memory and processor, the area taken up by the design may be more important. The parameters that are usually considered are summarised as follows:

Minimal redundancy. It is usually essential to keep the overhead due to coding as small as possible. A large die area generally results in more manufacturing defects. This provides a larger target area for alpha particle radiations resulting in increased failure rates with a degrading effect on the reliability.

User independent. No special software should be required. The device should simply replace a less reliable device which was designed to the same specification in a circuit. Or it could be used as a normal device with no error detecting or correction capabilities if required.

Actual reliability enhancement. It is obviously important that when all the additional circuits have been included, the reliability of the device should not be less than that of a non coded device.

Ease of implementation. There are various elaborate mathematical schemes available by which errors could be detected and corrected in any device[8,9]. To use any of these schemes in memories depends on the ease of implementation. A coding scheme which is difficult to implement will in most cases lead to greater redundancy and increases in the access times.

Minimum access time penalty. Any coding scheme used should not seriously compromise the access times. It would be unreasonable to have a highly reliable device with very long access times.

In most of the applications for which an FTIM is suitable, static memory devices are required, hence this is used as one extra criteria.

CODING

In order to apply coding as a means of EDAC in memory devices, it is necessary to divide the structure into two separate blocks. These are the Memory Array and the Supporting Circuitry. The Memory Array includes the memory cells, the pre-charging circuits, the array pass transistors and finally the sense amplifier circuits. The supporting circuits consist of the EDAC circuits, the addressing, input/output and the chip enable circuits.

The Memory Array

A detailed study of the failures in this region of the device[10,11], suggests that the following assumptions can be made:

i. Logical determinate (s-a-1, s-a-0) faults occur in the memory array and lead to either single, double or multiple errors at the output of the array.
ii. All logical errors at the output of the memory array can be classified as the result of failures in the memory cells in the appropriate columns.
iii. If more than a certain number of cells in a column experience the same fault effect, it is assumed the fault is as a result of failures in the bit lines for that column.
iv. If all the corresponding columns in different storage arrays experience a similar fault effect, it is assumed the fault is due to failures in the sense amplifier for that column.

Also, from previous works[12], it has been observed that the majority of failures in the Memory Array are actually due to Single Bit Errors (SBE) occurring in a memory word. In order to improve the reliability of these devices, codes are chosen for their ability to detect this class of errors. Following a detailed study[10] of several error detecting and correction schemes, coding techniques such as the BCH and Arithmetic codes were ruled out as being inappropriate. This was due to the much higher levels of redundancy required by them and difficulties in implementing them coupled with the probability of a significant increase in the access times. The most appropriate codes satisfying the theoretical requirements in section 1 were:

i. Product (Block) Coding,
ii. Triple Modular Redundancy (TMR),
iii. The Hamming Single Error Correcting and Double Error Detecting (SEC-DED) code.

In order to apply Product codes to a memory array, the array is divided into smaller blocks[14,15]. As data is made available to the data lines, an on-chip encoder computes the row parity of the word. When this data is stored in the array, the vertical parities of the block where the data has been stored is computed. On reading the data from the memory array, the horizontal parity is re-computed and compared with the stored value. If there is a discrepancy, then a error has occurred. The vertical parities are re-computed to locate the faulty data/parity bit which is then complemented.

TMR relies on triplication of the memory array. That is, the same data bit is stored in three different location[16]. When data is read from the memory array, a self-checking comparator circuit compares these three data bits. The correct output will be produced providing no more than one of the bits have changed state. Typically, any multiplicity of bit errors in a word are corrected. As in previous works[13,14], we make the assumption that the probability of two corresponding bits failing simultaneously is infinitessimal.

Hamming coding is a technique whereby several check or parity bits are computed for various combinations of the data bits in a word[18]. On reading a word from the storage array, a new set of check bits are computed and compared with the stored values. The information resulting from this computation, the syndrome data, is used for error detection and correction.

The Supporting Circuitry

In order to make the reliability of small (less than 4K bytes) memories less dependent on the reliability of the supporting circuitry, it was necessary to include in these circuits some form of error detection and (possibly) correction capability. To achieve this, the use of self-checking or fault tolerant circuits is highly desirable. To design such circuits, we first need to set out the list of faults most likely to occur. These were assumed to be stuck-at-1 (s-a-1), stuck-at-0 (s-a-0), and stuck-open faults. These faults affect different circuits in different ways but in the majority of cases lead to unidirectional errors at the output of the circuit. Any coding scheme to be employed must therefore have the ability of detecting such errors.

Following a detailed study of various coding techniques, the following results were obtained.

(i) If only concurrent error detection is required in the supporting circuitry, a basic duplication of the circuit, with the second copy performing the inverse function will suffice[10]. This produces a k-out-of-2k code word at the circuit output. The output can hence be monitored for possible error indication by a self-checking checker. This is the case where random logic is used. If a PLA is to be used for the design of these circuits, several error detecting schemes are available for this class of circuits[17]. These approaches use coding techniques such as Berger, modified-Berger, t-unidirectional error detecting and the general m-out-of-n codes. It was observed that if all unidirectional errors are to be detected, the Berger, modified-Berger coding or m-out-of-n codes are best. In cases where only a fixed number of unidirectional errors are to be detected, the t-unidirectional codes presented by Bose and Der Jer Lin[18] were more appropriate. In this case the coded PLA was less redundant.

The use of a PLA in the design of EDAC circuits will however depend on the design conditions such as access time, area overhead and power consumption. In general PLAs are slower than random logic circuits and might dissipate higher levels of power depending on the technology used for the implementation. For example, a pseudo-NMOS PLA would dissipate more power than a fully static CMOS random logic circuit. If a fully static CMOS PLA circuit is used, the area penalty will be higher.

In the above self-checking circuits, we made the assumption that the in-coming data is always correct or has been separately checked. Any error at the output of the circuit must therefore be a result of a fault in the circuit itself. Also all the circuits satisfy the code-disjoint property, in which case, it maps code inputs to code outputs and non code inputs to non code outputs. This is the basis of the "Principle of Error Propagation" used in the placement of self- checking checkers for these circuits[10,19].

(ii) In the second approach, some form of error correction is required in the supporting circuitry. These are in applications where extremely highly reliable circuits are warranted and there is a very remote possibility of manual interface. Such cases like aerospace and nuclear applications often demand correct functionality for considerable periods of time and a significant improvement in the MTTF. In these cases, we used a triple modular redundancy approach in the design of the circuits. Here the basic circuit is triplicated with self-checking comparator circuits used to monitor the outputs. Again the principle of error propagation is used to determine suitable positions for the placement of self-checking checkers.

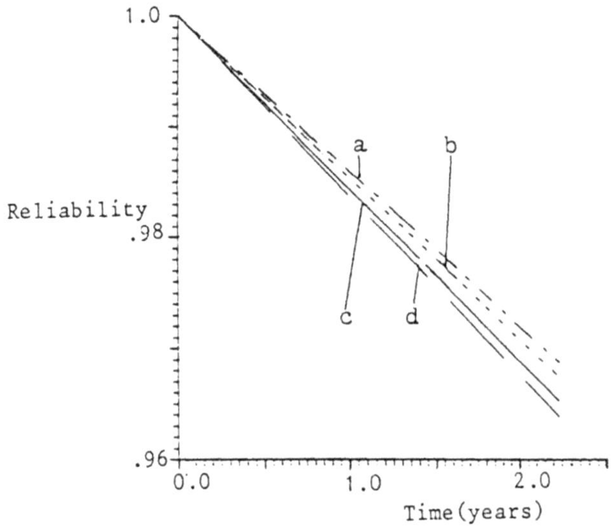

a - Hamming coded array with TMR supporting circuitry
b - Hamming coded array with no error detection in the supporting circuitry
c - TMR on entire device
d - Hamming coded array with self-checking supporting circuitry

Fig. 1. Reliability of coded memory devices

Faults in the address circuits will in general lead to one of the following:

i. The addressed cells may not be accessed but instead non-addressed cells will be accessed
ii. More or fewer cells may be accessed than the number addressed.

In case (i), the cells which would have been accessed are assumed to be s-a-0. In the case of the multiple access condition (case (ii)), the addressed circuit fault can be visualised as coupling faults in the selected cells. In general therefore, faults in the addressing circuits can be modelled as array failures. However, if it is essential for the controller to know where the fault actually occurred, a concurrent error detecting PLA will be suitable for the implementation of the address circuits.

Results from Reliability and MTTF Computations

In order to compare the various methods of coding, the Reliability and MTTF values were used. To compute the reliability of a FTIM it is necessary to divide the device again into the two blocks as before (that is the Memory Array and the Supporting Circuitry). We use the approach presented by Mikhail et al.[20]. In this approach the device is translated into an abstract logical device with physical faults map unto logical faults. Using the model to compute the reliability of various memory configurations, it was observed that if Hamming coding is used in the Memory array and TMR in the Supporting Circuitry, the most reliable device is obtained (Figure 1). Also this device provided the best MTTF values (Figure 2). It optimises the trade off between reliability, die area and access time. The approach is suitable for the design of strongly reliable devices. When self-checking is used in the design of the

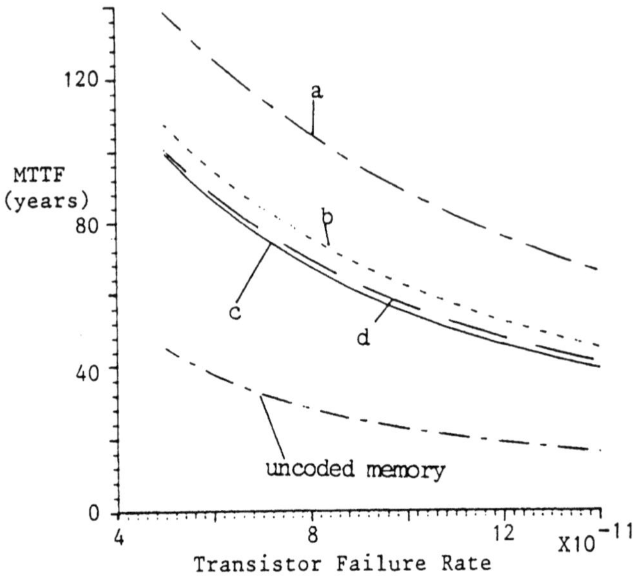

Fig. 2. MTTF versus transistor failure rate

Supporting Circuitry, faults are detected but not corrected, therefore the increase in the reliability figure is less. If no coding is applied in the supporting circuitry, the reliability of the device is better than the case where self-checking is employed. This is expected since by coding the Supporting Circuitry, the number of transistors is almost doubled. This increases the complexity of the circuit with the consequence of increased failure rates. The advantage in the self-checking approach is that the source of the fault can be identified.

In the case where TMR is used on the entire device, a moderately reliability figure was obtained over the usable period of two years. The worst MTTF values were obtained when compared to the other coded devices. Also a very high penalty in terms of die area is paid for the improved reliability. On the plus side, the best access times were obtained since the signal is only delayed through the self-checking comparator circuits.

As pointed out earlier, if the supporting circuitry reliability is improved, the device reliability will also be significantly improved. In Figure 3, it is shown that the memory reliabilities strongly depend on these component reliabilities. Thus, for extra high memory reliability, it is necessary to apply error correcting techniques at this component level. As seen in Figure 1, the Hamming coded devices all have the same array reliability but because of the extra error correction in the supporting circuit an improved reliability figure is obtained for the device with Hamming coding in the Memory Array and TMR in the supporting circuits.

CASE STUDY

To illustrate the design strategy used in obtaining a reliable memory device for applications such as in digital controllers, we consider the design of a 1K byte FTIM. The following user design conditions were suggested:

i. Reliability of the device must not be less than 96% over a two year period.

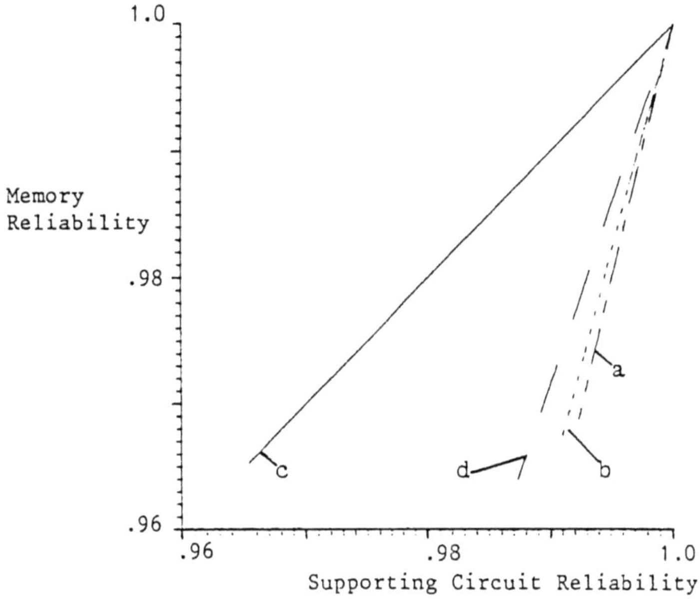

Fig. 3. Memory reliability versus supporting circuit reliability

ii. The Mean Time to Failure of the device must be at least twice that of the non coded device.
iii. No need for error correction in the supporting circuits since the area is limited, but errors must be detected concurrently so that if the supporting circuitry is the faulty section, it could be disabled and the rest of the device used as a normal memory.
iv. Allowable access time increase of no more than 30ns.
v. Fast turn around time in the design phase.
vi. There should be a possibility of using the supporting circuitry to aid in testing the device for initial acceptance.

Design

Because of the area limitation, a TMR approach on the entire memory is not feasible. The only approach which does not place a serious constraint on the area and provides concurrent error detection in the supporting circuitry is the Hamming coded array with self-checking supporting circuits. As can be seen from Figure 1, the reliability of of such a device does not fall below the required 96% value over a two year period. In Figure 2, the MTTF is about 2.5 times that of the non coded device.

To ensure that coding does not significantly degrade the access times, the approach adopted was to design a totally asynchronous device. This might result in race hazards but the speed advantages outweigh this. In such a device, the chip enable line controls all the other processes[21]. When the chip enable line is active, a write or read operation can be performed by the device. A block diagram of the 1K FTIM is displayed in Figure 4.

To satisfy the fast turn around time condition, the EDAC circuits were implemented

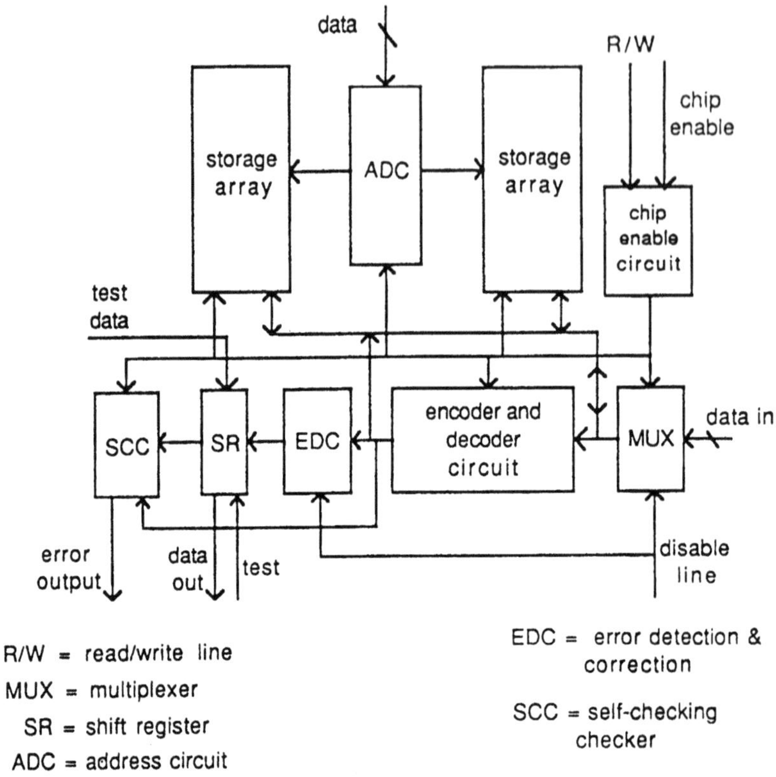

Fig. 4. Block diagram of a prototype device for a Fault Tolerant Integrated Memory

using PLA circuits which are far easier to design than random logic circuits. Since all modelled faults (stuck-at, stuck-open, shorts, missing and extra device faults) in PLAs result in unidirectional errors[22], it is therefore certain that all faults occurring in the supporting circuits in normal operation will be detected. We used the Berger coding technique in designing the self-checking PLAs.

In the memory array, to prevent faults in a cell affecting other cells in the same word, a technique called "Bit-Scattering" is used[23]. This is an organisation procedure where two bits of the same word are never located adjacent to each other. A coupling fault in a cell can only affect a different word. This prevents the alignment of faults in a word resulting in uncorrectable errors.

Device Operation. In the normal mode of operation of the device, both the disable and the test-pin lines are in the inactive low state. During the write cycle, the first stage of this operation is for a valid data byte to appear at the input lines. The encoder circuits computes the check bits and after a period of 15ns (from simulations), by which time the check bits have been computed, the chip enable line becomes active high. Five nano seconds later, a valid address appears at the address input lines. It takes a further 25ns for the data and check bits to be written into the storage array. This gives a total write cycle of 45ns. In the read cycle the chips enable line first becomes active. Five nanoseconds later, a valid address is placed on the address input lines. From simulations, it takes a further 57ns for the data to be read, decoded and corrected. This makes a total read cycle of 62ns.

Initial Testing

Testing of the FTIM device is divided into two processes, first testing of the supporting circuits, and then testing of the memory array.

Testing of the Supporting Circuits. Since the EDAC circuits are monitored by the self-checking checker circuits, we first start by testing these circuits. In the block diagram of Figure 4, the self-checking checker circuit, monitor to see if the encoder/decoder circuit produces a 4-out-of-8 output. This checker is tested by eight test vectors each containing four ones and four zeros. These patterns once loaded into the shift register latches can be used for testing the circuits. In the fault free condition, the error output pins will be either 01 or 10. In the case of a faulty circuit, both outputs will be 1 or 0.

Next is the testing of the encoder, decoder and error correction circuits. To test these circuits, we use all the 256 input vectors (2^8 for a byte wide memory device). The error output pins of the self-checking checker are monitored to determine the state of the EDAC circuits.

Testing the Memory Array. Once the supporting circuits have been tested, they can be used to aid in testing of the Memory Array. At each memory location, an all zero and an all 1 data pattern is written. The data is read and the output pins monitored for error (single or double bit) indication by data error pins. Also an input pattern such as [01001000] is used to test for stuck-at-0 faults at the check bit location. The syndrome bit can also be made available at the output to give the precise location of the cell fault.

In normal operation if the EDAC circuits are known to be faulty, then the operation being performed by the system should be stopped. These circuits are tested off-line to determine the type of fault. If permanent, and if we require use of the rest of the device, the disable line is set active high. In this case, the memory functions as a normal device with no EDAC facilities.

We have implemented a prototype device satisfying most of the above features. Extra facilities were provided to insert errors in the memory array. When tested, all single bit errors were corrected and double bit errors detected. The state of the data read was portrayed by two data output pins. Faults in the supporting circuits were also detected. The measured access time of the device was 65ns. It was also noted that the access time penalty due to coding was no more than 25ns. The power dissipated while the device was been accessed was 1mW.

CONCLUSIONS

In this paper, we have examined the design approaches suitable for a small fault tolerant integrated memory. If fast access is required and area penalty is of little significance, then a TMR approach to the device design is best. In cases where the area overhead is required to be smaller and when a very high reliable device with long Mean Time to Failure values is required, a Hamming coded array with TMR supporting circuits will be more suitable. When only concurrent error detection is required in the supporting circuits, a PLA with a well defined fault set is better for these circuits.

A prototype device has been design using Hamming coding with the supporting circuits made self-checking. In the device all single bit errors are corrected and double bit errors detected. The measured access time of the device is 65ns which includes the time required for error detection and correction. We have also presented an approach whereby the EDAC circuits could be used to aid in testing of the device for initial acceptance.

We have shown that it is feasible to design a small fault tolerant memory circuit provided the error detection and correction circuits are implemented on the same chip as the memory array. The area penalty in all the cases though high is acceptable when the size of the memory is small (less than 4K bytes). Furthermore, if Hamming coding is used for error detection and correction in a memory device, the yield is significantly improved.

REFERENCES

1. X. Castillo, S. R. McConnel and D. P. Siewiorek, "Derivation and calibration of transient error reliability model", IEEE Trans. on Comput., Vol. C-31, No. 7, pp. 658-671, (1982).
2. Y. W. Yak, T. S. Dillon and K. E. Forward, " Bounded-set approach to evaluation of the relibility of fault tolerant systems. Part 1: methodology, powered spares", Proc. IEEE, Vol. 132, No. 4, pp. 217-223, (1985).
3. W. R. Moore, "A review of fault-tolerant techniques for the enhancement of integrated circuit yield", Proc. IEEE, Vol. 74, No. 5, pp. 684-698, (1986).
4. A Texas Instrument Appl. Report, "Error detection and correction using SN54 / LS630 or SN54 / 74LS631, Bulleting CA-201.
5. Micron Technology, inc., Shortform DRAM, (1986).
6. J. Yamada et al., "Submicron 1 Mbit dynamic RAM with 4-bit-at-a-time built-in ECC circuit", IEEE J. of Solid State Circuits, Vol. SC-29, No. 8, pp. 627-633, (1984).
7. C. G. Guy, "Fault tolerant random access memories for small idgital controllers", University of Liverpool Microprocessor Workshop, 31st March - 1st April 1980.
8. W. W. Peterson and E. J. Weldon, "Error correcting codes", MIT Press, Cambridge, Mas., (1972).
9. C. L. Chen and M. Y. Hsiao, "Error correcting codes for memory array applications: a state-of-art review", IBM J. Res. and Develop., Vol. 28, pp. 314-322, (1984).
10. C. A. Njinda, "Techniques for enhancing the reliability of small integrated random access memory arrays", Ph.D. Dissertation, (1987).
11. C. A. Njinda and C. G. Guy, "Fault tolerant integrated memory", IEE Colloquium on Fault Tolerant ICs/Wafer Scale Integration, Digest No. 1986/23, (1986).
12. A. Khan, "Fast RAM correct errors on chip", Electronics pp. 126-130, (1983).
13. Levine and W. Mayers, "Semiconductor memory reliability with error detection and correcting codes", IEEE Trans. on Comput., pp. 310-315, (1976).
14. D. B. Sarrazin and M. Malek, "Fault tolerant semiconductor memories", IEEE Compt., pp. 49-56, (1984).
15. R. M. Tanner, "Fault tolerant 256K memory designs", IEEE Trans. on Comput., Vol. C-33, No. 4, pp. 124-134, (1984).

16. P. K. Lala, "Fault tolerant and fault testable hardware designs", Prentice-Hall International Inc., London, England, (1985).

17. C. G. Guy, C. A. Njinda, K. J. Steptoe and G. Santoro, "An automatic concurrent detecting tool for PLA generation", Microelectron. Relia., Vol. 28, No. 3, pp. 395-405, (1988).

18. B. Bose and D. J. Lin, "Systematic unidirectional error-detecting codes", IEEE Trans. on Comput., Vol. C-34, No. 11, pp. 1026-1031, (1982).

19. J. E. Smith and P. Lam, "A theory of totally self-checking system design", IEEE Trans.on Comput., Vol. C-32, No. 9, pp. 831-844, (1983).

20. W. F. Michail, R. W. Bartoldus and R. A. Rutledge, "The reliability of a memory with single-error correction", IEEE Trans. on Comput., Vol. C-31, No. 6, pp. 560-564, (1982).

21. S. E. Shuster et al., "A 15ns CMOS 64K RAM", IEEE J. of Solid-State Circuits, Vol. SC-21, No. 5, pp. 704-712, (1986).

22. G. M. Mak, J. A. Abraham and E. S. Davidson, "The design of PLAs with concurrent error detection", Proc. 12th Annual Symp. on FTC, pp. 303-310, (1982).

23. F. J. Aichelmann Jr., "Fault-tolerant design techniques for semiconductor memory applications", IBM Res. and Devel., Vol. 28, No. 2, pp. 177-183, (1984).

9 RECONFIGURABLE ARRAYS

PROBABILISTIC ANALYSIS OF YIELD AND AREA UTILIZATION OF RECONFIGURABLE RECTANGULAR PROCESSOR ARRAYS

Ajai Jain and **Janusz Rajski**

McGill University
Department of Electrical Engineering
Montreal, Quebec H3A 2A7, Canada

INTRODUCTION

The increase in the number of components on a single chip has forced the designers and manufacturers to consider carefully, the issues related to the reliability, yield and effective silicon area utilization. Several reconfiguration schemes have been suggested in the literature to make a chip fault-tolerant by adding redundancy at various logical and physical levels.

To analyze the reconfiguration schemes, some authors use Markov models[1,2], while others use combinatorial approach[3,4,5,6,7]. In some cases, where it is difficult to obtain analytical solutions, simulation[8] is used to determine the yield and area utilization. We use combinatorial approach to develop the analytical solutions for various classes of reconfiguration schemes for rectangular arrays of identical processors. Two approaches are used for analyzing the schemes. In the *continuous* approach, the yield of an array is analyzed as a function of the probability of a PE being faulty. In the *discrete* approach, the yield is analyzed as a function of the number of faults in the array. The reconfiguration schemes have been classified according to the way the spares are used in an array. A method is developed to calculate the area utilization of the representative schemes in each class, so that the analysis can be extended to other schemes in the same class.

If there are no spare PEs in an array, then a single faulty PE in the array will make it unusable. In fault-tolerant arrays, spare PEs are added to the target array to replace faulty PEs. The complexity of switches and amount of additional interconnect area depend on the reconfiguration scheme and amount of redundancy introduced. The number of spares and additional interconnect area contribute to the total area of a fault-tolerant chip and affect the overall area utilization.

Faults in a processor array can occur both in the processing elements (PEs) and in the interconnection network. The actual area utilization is a product of two components. The first component depends on the logical reconfigurability of the scheme, when the faults are only in the PEs, while the second component depends on the interconnect area overhead and on the probability of having any fault in the "kill area" of array which leads to the failure of array.

A commonly used assumption in the literature, is that the faults in an array are limited to the PEs only. In the following discussion, faults are assumed to be randomly distributed in the complete physical array (including spare PEs) according to a binomial distribution. The effect of faults in the interconnection network is also analyzed. In the following discussion m, n, p, Y, A denote, respectively, the number of rows, number of columns in target array, probability of a PE being fault - free, yield and area utilization of a scheme.

YIELD AND AREA UTILIZATION BASED ON THE LOGICAL RECONFIGURABILITY

In this section, the analysis is done for the first component of yield and area utilization, which depend only on the logical reconfigurability of an array in the presence of faulty PEs. The second component of yield and area utilization, which depends on the probability of having any failure in the interconnection network and the increase in interconnect area, is analyzed separately in the next section. In this section, Y_l and A_l denote the first components of yield and area utilization respectively. All the results shown are for a target array of 10×10 and probability range of 0.9 to 1.00, though the discussion is not limited to this target size and probability range. All the figures for area utilization have the probability of a PE being fault-free on the x-axis and the percentage area utilization on the y-axis.

Global Spares

This is an ideal case, as it assumes that a spare PE can replace a faulty PE anywhere in an array. The array can be reconfigured if the number of faults in the array does not exceed the number of spares[9]. For s spare PEs,

Continuous approach. Let $P(i)$ be the probability of having exactly i faults in the array. Then,

$$P(i) = \binom{m\times n+s}{i} p^{m\times n+s-i}(1-p)^i$$

The yield, in this case, is the probability that there are at most s faults in the array. Therefore,

$$Y_l = \sum_{i=0}^{s} \binom{m\times n+s}{i} p^{m\times n+s-i}(1-p)^i \quad \text{and} \quad A_l = Y_l \times ((m\times n)/(m\times n+s))$$

Discrete approach. Let f be the number of faults in the array. Then,

$$Y_l = \begin{cases} 1, & \text{if } f \le s; \\ 0, & \text{otherwise.} \end{cases} \quad \text{and} \quad A_l = Y_l \times ((m\times n)/(m\times n+s))$$

Fig. 1 shows the area utilization with global spares. The yield in this probability range is almost 100% with 20 or more spare PEs. However, the maximum area utilization with 20 spare PEs is only about 83% since the spare PEs contribute to about 17% of the total area of the array. Using more spares leads to a further reduction in the area utilization since the additional spares do not contribute to any significant increase in the yield. However, an increase in the redundancy leads to an improvement in area utilization in lower probability range due to the increased yield. The analysis of other results shows that for a given amount of redundancy, at lower probabilities the area utilization decreases with an increase in array size.

As mentioned earlier, having global spares is an ideal case from the point of view of the logical reconfigurability of an array. The analysis gives an upper bound on the area utilization of any reconfiguration scheme using the same number of spare PEs. However, due to the complexity of switches and large interconnect area, the schemes with global spares have very limited practical value.

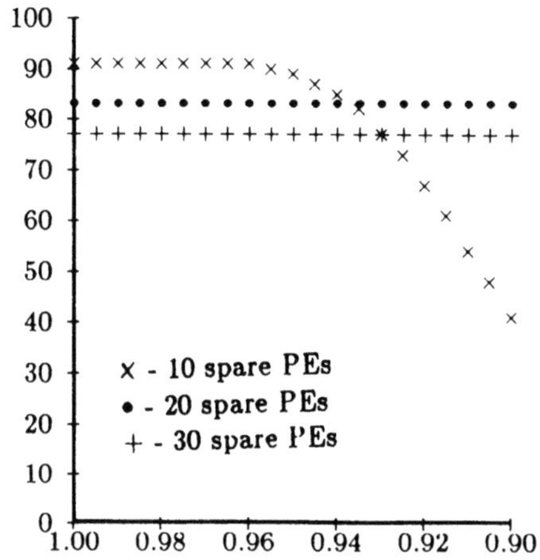

Fig. 1. Area utilization with global spares

Spare Row/Column

In this class of schemes a complete row (column) is bypassed in case of a faulty PE with in the row (column). One such scheme with spare columns is MPP designed by Goodyear Aerospace[10]. For s spare columns,

Continuous approach. Let P_c be the probability that a column has at least one faulty PE. Then,

$$P_c = 1 - p^m$$

The yield, in this case, is the probability that at most s columns are faulty. Therefore,

$$Y_l = \sum_{i=0}^{s} \binom{n+s}{i} P_c^{\,i} (1 - P_c)^{n+s-i} = \sum_{i=0}^{s} \binom{n+s}{i} p^{m \times (n+s-i)} (1 - p^m)^{i},$$

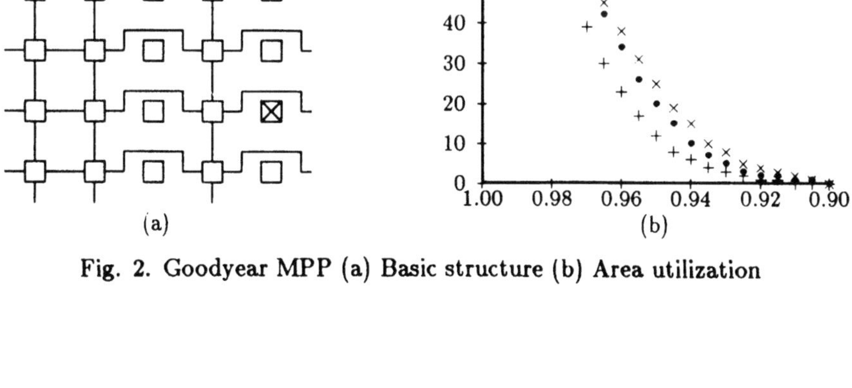

Fig. 2. Goodyear MPP (a) Basic structure (b) Area utilization

$A_l = Y_l \times (n/(n+s))$

Discrete approach. Let f be the number of faults in the array, let c_i be the number of combinations in which the faults are located in exactly i columns and C_i be the number of combinations in which the faults are limited to i columns. Then,

$$c_1 = \binom{m}{f}, \quad C_i = \binom{i \times m}{f} \quad \text{and} \quad c_i = C_i - \sum_{j=1}^{i-1} \binom{i}{j} c_j$$

The yield is calculated as the ratio of the number of combinations in which the faults are limited to any s columns of the array to the total number of possible combinations of faults in the array,

$$Y_l = \left(\sum_{i=1}^{s} \binom{n+s}{i} c_i\right) / \binom{(n+s)m}{f} \quad \text{and} \quad A_l = Y_l \times (n/(n+s))$$

Fig. 2(b) shows the area utilization for the Goodyear MPP. For a given amount of redundancy, the area utilization increases with increase in the probability until a saturation occurs at high probabilities, i.e., almost 100% yield is achieved. The probability at which this saturation occurs, decreases with an increase in redundancy. Similarly, for a given probability, the area utilization increases at lower probabilities with an increase in redundancy but decreases at higher probabilities. The analysis of other results show that more redundancy is required to get the same level of area utilization for larger arrays.

Since a single faulty PE within a row (column) leads to the bypassing of the entire row (column), the spare utilization for the schemes in this class is poor. The scheme needs very simple switches and have small interconnect area overhead. The kill area of the array is the interconnect area between adjacent pairs of rows (columns). It is possible to tolerate some faults in the interconnect with in a row (column) by bypassing the faulty row (column).

Spares Local to a Row or a Column

In this class of reconfiguration[11,12], each row (column) has some spare PEs which can be used to replace the faulty PEs only within the row (column). The yield of any scheme in this class depends on the number of PEs which can be used to replace a faulty PE. The yield and area utilization of such schemes can be calculated by finding the probability that all rows (columns) can be successfully reconfigured. An upper bound on the yield for such schemes can be calculated assuming a global utilization of spares

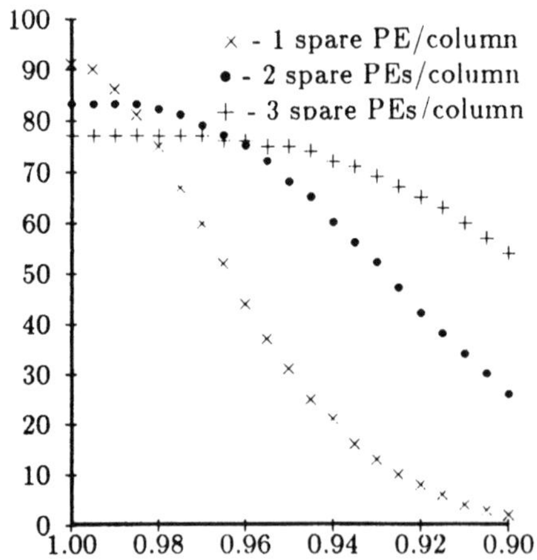

Fig. 3. Area utilization with spares local to columns (upper bound)

within a row (column). Let s be the number of spare PEs per column in an array and P be the probability that a column has at most s faults. Then,

$$P = \sum_{i=0}^{s} \binom{m+s}{i} p^{m+s-i}(1-p)^i$$

The upper bound on the yield (Y_{ub}) is the probability that each column has at most s faults. Thus,

$$Y_{ub} = P^n = \left(\sum_{i=0}^{s} \binom{m+s}{i} p^{m+s-i}(1-p)^i\right)^n \quad \text{and} \quad A_{ub} = Y_{ub} \times (m/(m+s))$$

Fig. 3 shows the upper bound on the area utilization for 1, 2 and 3 spare PEs per column. The amount of redundancy required to obtain the optimal area utilization decreases with an increase in the redundancy.

One example of a scheme with spares local to a column is the McGill Reconfigurable Massively Parallel Processor (RMPP)[11] (Fig. 4(a)). In this scheme, each PE can be connected to one of the two neighboring PEs in each direction (N,S,E,W). Furthermore, the array has a toroid structure in N-S direction. For one spare PE per column, the array can be reconfigured if no column has more than one faulty PE.

Continuous approach. For one spare PE per column, the yield and area utilization can be obtained by putting s=1 in the expression for the upper bound. Therefore,

$$Y_l = \left(p^{m+1} + (m+1)(1-p)p^m\right)^n \quad \text{and} \quad A_l = Y_l \times m/(m+1)$$

Discrete approach. The array can be reconfigured if and only if each fault is in a different column. Let f be the total number of faults in the array, C_t be the total number of possible combinations of these faults in the array and C_p be the number of permissible combinations of these faults in the array. Then,

$$C_t = \binom{n \times (m+1)}{f}, \quad C_p = \binom{n}{f} \times (m+1)^f$$

The yield is calculated as the ratio of the number of permissible combinations of the faults to the total number of possible combinations of the faults,

$$Y_l = C_p/C_t = \left(\binom{n}{f}(m+1)^f\right) / \binom{n \times (m+1)}{f} \quad \text{and} \quad A_l = Y_l \times (m/(m+1))$$

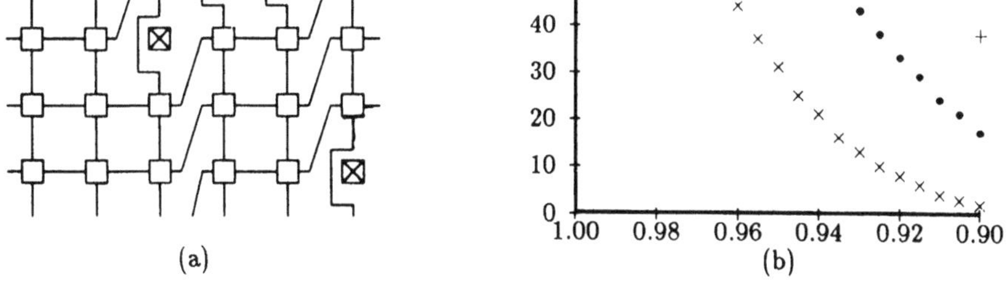

Fig. 4. Mcgill RMPP (a) Basic structure (b) Area utilization

If there are s spare PEs ($s > 1$) in each column, then due to the limited connectivity of interconnection network, some fault-free PEs may not be accessible for reconfiguration due to the "shadowing effect" of neighboring faulty PEs. The behavior of an array, in the presence of faults, depends on the specific fault pattern. It may not be possible to have a general closed form solution for all array sizes, which takes into account the shadowing effect of all possible fault patterns. However, the results have been obtained for yield and area utilization using Monte Carlo simulations.

Fig. 4(b) shows the area utilization for the RMPP for 1, 2 and 3 spare PEs per column. The results show that increase in redundancy leads to an improvement in the area utilization at lower probabilities, but in the higher probability ranges extra redundancy leads to a reduction in the area utilization. We can determine the effect of shadowing on the area utilization using the analytical results for the upper bound (Fig. 3) and the simulation results (Fig. 4(b)). The comparison of the figures show that the shadowing has more effect at lower probabilities. The analysis of other results show that the shadowing effect increases with an increase in the redundancy at lower probabilities but decreases at higher probabilities.

Since the spares can be shared within a row or column and faults need not be limited to a given number of rows or columns, the schemes in this class provide more efficient utilization of spares than the schemes of previous class. However, the switches are more complex than simple row/column bypass switches and interconnect area is also larger. Most of the interconnect area contributes to the kill area of the array.

<u>Spares Local to a Group of Rows and Columns</u>

In this class of reconfiguration schemes, each PE can be used to replace a faulty PE in a neighboring row or column. One example of schemes in this class is the direct reconfiguration scheme[13]. In this scheme, both spare rows and columns are used to provide fault-tolerance (Fig. 5(a)). For r spare rows and c spare columns, the reconfiguration is done as follows. The first r faults in each column are declared as vertical faults (spare rows are used for the reconfiguration) and the remaining faults as horizontal faults (spare columns are used for the reconfiguration). A fatal failure condition occurs when a row has more than c horizontal faults.

To analyze the scheme, we find recursively (starting from the bottom most row), the probability $P_r(j)$ that a fatal failure condition occurs in the first j rows. A fatal failure will occur in the first j rows if either a fatal failure occurs in the first $j-1$ rows or if the first fatal failure occurs in j^{th} row. Let $P_d(i)$ be the probability that a row has exactly i faults, $P_c(j)$ be the probability that a column has at least r faults in the first $j-1$ rows and $P_k(i,j)$ be the probability that at least $c+1$ out of i columns have r faults or more in the first $j-1$ rows of the array. Then,

$$P_d(i) = \binom{n+c}{i} p^{n+c-i}(1-p)^i, \qquad P_c(j) = \sum_{k=r}^{j-1} \binom{j-1}{k} (1-p)^k p^{j-1-k}$$

$$P_k(i,j) = \sum_{l=c+1}^{i} \binom{i}{l} (P_c(j))^l (1-P_c(j))^{i-l} \quad \text{and}$$

$$P_r(j) = \begin{cases} 0, & \text{if } j \leq r; \\ P_r(j-1) + \{1 - P_r(j-1)\} \times \sum_{i=c+1}^{n+c} (P_d(i) P_k(i,j)), & \text{otherwise.} \end{cases}$$

The yield, in this case, is the probability that no fatal failure occurs in any of the $(m+r)$ rows. Therefore,

$$Y_l = 1 - P_r(m+r) \quad \text{and} \quad A_l = Y_l \times ((m \times n)/((m+r) \times (n+c)))$$

Fig. 5(b) shows the area utilization for 1 spare row and 0, 1 and 2 spare columns. As the probability increases, fewer spare columns are needed for better area utilization. Similar results are obtained for more spare rows. The analytical results are used to find the number of spare rows and columns needed for optimal area utilization. The results show that redundancy may be needed in different forms for optimal utilization at different probabilities. For example, for a target array of 20 × 20, 2 spare rows, 2 spare columns and 1 spare row, 3 spare columns give optimal utilization at p= 0.95 and 0.96 respectively. The results also show that larger arrays give better area utilization in this probability range.

In the schemes belonging to this class, a PE can be used to replace a faulty PE both in a row or column. This leads to a more efficient utilization of spares than the schemes from the two previous classes. However, the switches are relatively more complex and interconnect area is larger. As in case of previous scheme, most of the interconnect area contributes to the kill area of array.

Hierarchical Schemes

In this class of schemes, the physical array is divided into several logical levels of modules and the redundancy is introduced at each level. The yield and area utilization for such schemes can be calculated by finding the logical reconfigurability at each level starting from the lowest level. Two schemes from this class are analyzed here.

In the first scheme[14], each 'logical' row of the array consists of $(2n+2)$ PEs and n switches (Fig. 6(a)). Each switch can be connected to any of the 4 neighboring PEs according to the priority A,B,C,D. Spare rows are provided to add second level redundancy. A row can be successfully configured if each switch can be assigned an unique fault-free PE, otherwise the row is bypassed. Let P_i^0 be the probability that in a row, switches 1 to $(i-1)$ have been assigned fault-free PEs (starting from the leftmost switch) but no PE can be assigned to switch i from column i of PEs, P_i^1 be the probability that the switches 1 to $(i-1)$ have been assigned fault-free PEs and a fault-free PE from column i is available to the switch i and P_{row} be the probability that a row can be successfully reconfigured. Then P_n^0 and P_n^1 can be calculated by using the following relations recursively,

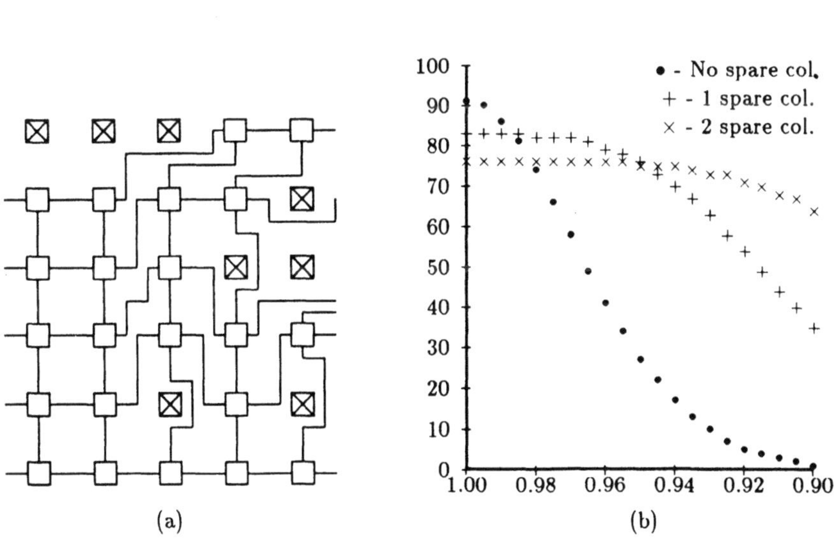

Fig. 5. Direct Reconfiguration (a) Basic structure (b) Area utilization

$$P_1^0 = (1-p)^2,\ P_1^1 = 1-(1-p)^2$$

$$P_i^0 = (1-p)^2 P_{i-1}^1 + 2p(1-p)P_{i-1}^0, \quad P_i^1 = \left(p^2 + 2p(1-p)\right) P_{i-1}^1 + p^2 P_{i-1}^0$$

Now, a row can be reconfigured if the switches 1 to $n-1$ have been assigned fault-free PEs and the switch n can be assigned a free PE either from column n or from column $n+1$. Therefore,

$$P_{row} = P_n^1 + \left(1-(1-p)^2\right) P_n^0$$

The yield, in this case, is the probability that at most k rows are faulty. Therefore,

$$Y_l = \sum_{j=0}^{k} \binom{m+k}{j} (1-P_{row})^j {P_{row}}^{m+k-j}$$

and

$$A_l = Y_l \times (n/(2n+2)) \times (m/(m+k))$$

Fig. 6(b) shows the area utilization for 0, 1 and 2 spare rows. Even though the yield of the scheme is very good, (almost 100% with 3 spare rows at probability as low as 0.7), the area utilization is always less than 50% due to an inherent redundancy of more than 100%. Thus, the scheme gives competitive area utilization only at lower probabilities. The redundancy required for the optimal area utilization increases with a decrease in the probability or an increase in the array size.

Since we need $(2n+2)$ PEs to configure one logical row of n PEs, the scheme provides very inefficient utilization of spares. The switches need 4 to 1 multiplexing within a logical row and row bypass switches are needed to bypass faulty rows. The interconnect area can be significantly large.

In the second scheme from this class [15], the fault-tolerant array consists of a number of rows of modules (Fig. 7(a)). Each module is a subarray of PEs. The redundancy is introduced as spare rows in a module and as spare modules in a row of modules. If there is any faulty PE in a row, the row is declared faulty. If the number of the faulty rows in a module is more than the number of spare rows, the module is declared faulty. The array fails if the number of faulty modules in a row of modules are more than the number of spare modules.

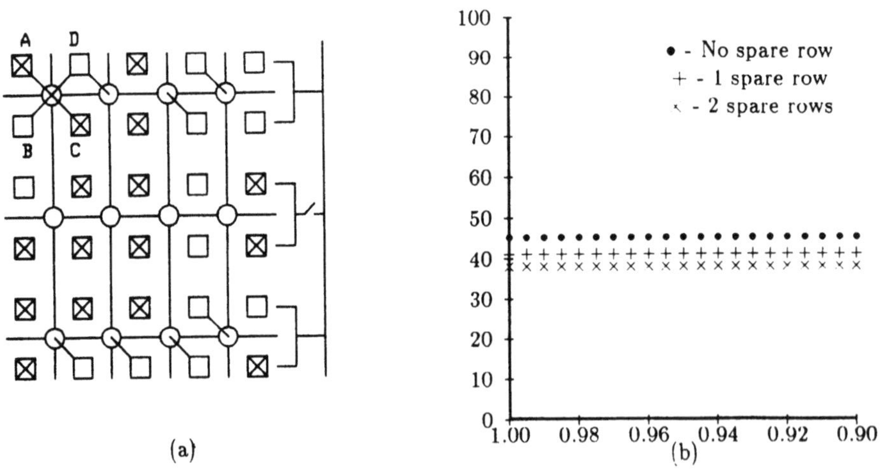

Fig. 6. Hierarchical scheme (J & B) (a) Basic structure (b) Area utilization

Let n_r, n_p, s_r, s_m be the number of rows in a module, number of PEs in a row of module, number of spare rows in a module and number of spare modules in a row of modules respectively. Then, for a target array with m rows and n columns of PEs, we need $m/(n_r - s_r)$ rows of $(s_m + n/n_p)$ modules. Let p_r be the probability that a row of PEs within a module is fault-free, p_m be the probability that a module is working, i.e., at most s_r rows are faulty in the module and p_{row} be the probability that a row of modules can be reconfigured. Then,

$$p_r = p^{n_p}, \quad p_m = \sum_{i=0}^{s_r} \binom{n_r}{i} {p_r}^{n_r - i}(1-p_r)^i = \sum_{i=0}^{s_r} \binom{n_r}{i} p^{(n_r-i)\times n_p}(1-p^{n_p})^i$$

and

$$p_{row} = \sum_{i=0}^{s_m} \binom{s_m + n/n_p}{i} {p_m}^{s_m + n/n_p - i}(1-p_m)^i$$

The yield, in this case, is the probability that all rows can be reconfigured. Therefore,

$$Y_l = {p_{row}}^{m/(n_r - s_r)} \quad \text{and} \quad A_l = Y_l \times ((n_r - s_r)/n_r) \times ((n/n_p)/(s_m + n/n_p))$$

Fig. 7(b) shows the area utilization for a module size 3×2 (one spare row of PEs in each module). The results show that adding spare modules in a row of modules results in an improvement of area utilization in lower probability range but leads to a decrease in the area utilization at higher probabilities. Similar results are observed for spare rows within a module. Further analysis of results show that for small arrays, increase in the array size improves the optimal area utilization in high probability range. The analysis can be used to determine the array sizes which give best area utilization at different probabilities.

Since the spares can be shared only within a module at each level of redundancy, spare utilization is not very efficient. Switches required within a module are simple bypass switches and more complex switches are needed for bypassing the modules and connecting the rows of modules. Since routing is required at several logical levels, the interconnect area is also large.

One significant advantage of hierarchical schemes is their capability of tolerating the faults in part of the interconnect area. If there is any fault in the interconnect of

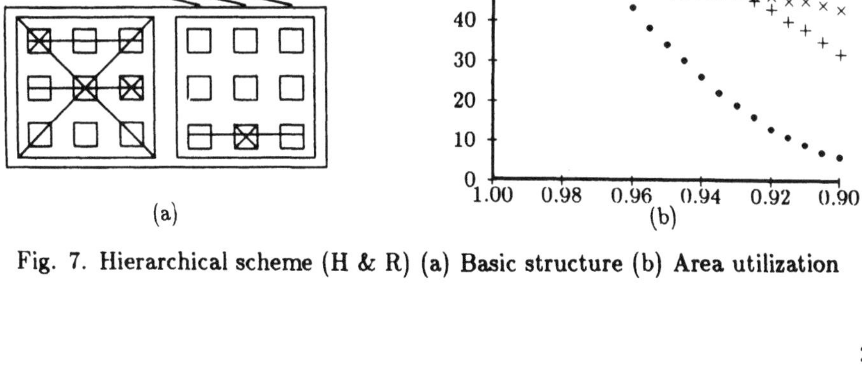

Fig. 7. Hierarchical scheme (H & R) (a) Basic structure (b) Area utilization

a lower level module, the module can be bypassed without killing the complete array. Thus the kill area of array in these schemes is only a part of total interconnect area.

Comparison of Schemes on the basis of logical reconfigurability

In order to compare different schemes, we calculate their optimal area utilization as a function of various parameters like module size and amount of redundancy. Fig. 8 shows the optimal area utilization achievable for various schemes for a target array of 10×10 and probability range 0.9 to 1.0. The results show that if we consider only logical reconfigurability, then as expected, global spares give best results for all probabilities. Simple row/column bypass schemes, e.g., the Goodyear MPP give good performance only at very high probabilities (> 0.99). Schemes with spares local to row/column, e.g., the RMPP give better results than the Goodyear MPP for almost complete probability range. Schemes with spares local to a group of rows and columns, e.g., the direct reconfiguration scheme are slightly better than the RMPP in this probability range. The hierarchical schemes provide very good yield at low probabilities, but the area utilization for these schemes is lower than other schemes in higher probability range since the spares can be shared only within a module and thus the spare utilization, in general, is poor for these schemes. For example, the hierarchical scheme by Jesshope and Bentley [14] gives comparative area utilization only at low probabilities (0.7 or less) even though it provides very good yield. The modular scheme by Hwang and Raghavendra[15] shows slightly lower area utilization than the direct reconfiguration scheme but provides better performance than the RMPP and the Goodyear MPP.

THE EFFECT OF THE INTERCONNECT OVERHEAD ON AREA UTILIZATION.

In the previous section, we analyzed the yield and area utilization for various classes of reconfiguration schemes based on the logical reconfigurability provided by the schemes. In this section, we discuss the second component of the yield and area utilization. The increase in the interconnect area affects the area utilization in two ways:

1. Direct reduction in the fractional area of the chip used by the PEs. This can be calculated by finding the percentage increase in interconnect area due to redundancy.

2. Reduction in area utilization due to the loss of yield because of increased probability of a fault in kill area.

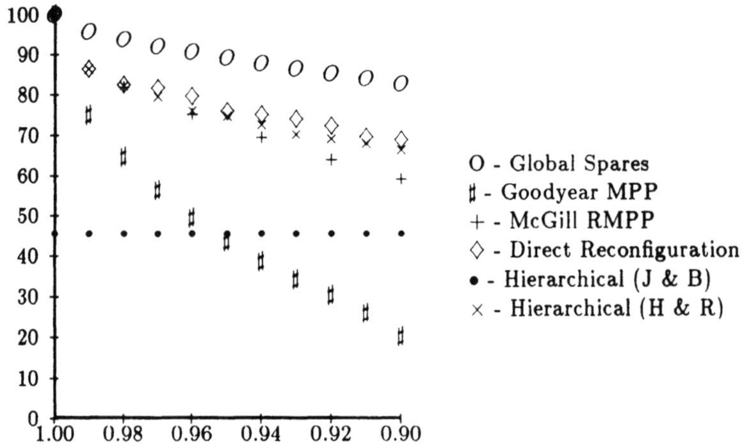

Fig. 8. Optimal area utilization

For calculating the area utilization for hierarchical schemes, let a_i be the kill area of ith level module, a_p be the area of a PE and p_k be the probability that there is no fault in the kill area of module. Then,

$$p_k = p^{a_i/a_p}$$

If the logical reconfigurability of module at level i (as calculated in previous section) is p_r, then the effective reconfigurability of module

$$p_{eff} = p_r \times p_k$$

P_{eff} should be used to calculate the reconfigurability of module at next level. For single level redundancy schemes (except for row/ column bypass schemes), most of the interconnect area of an array is the kill area and therefore, the second component of the yield and area utilization can be directly obtained by calculating p_k for the complete array.

The interconnect overhead of a particular implementation of a reconfiguration scheme can be calculated from the complexity of switches and the total interconnect area. As an example of the effect of interconnect overhead, Fig. 9 shows the second component of area utilization of an array with 120 PEs (e.g., a 10 $\times$ 10 array with 20 spare PEs) for the interconnect overhead of 5%, 10% and 15% and a probability range of 0.9 to 1.0. The area utilization decreases very sharply with the increase in the interconnect overhead. For example, if two single level redundancy schemes A and B have the interconnect overheads of 5% and 15% respectively, then the probability of having no faults in the kill area for the schemes A and B are approximately 0.5 and 0.12 respectively at p=0.90. Thus, even if the scheme B has slightly better logical reconfigurability, the actual area utilization for the scheme B will be about one fourth of the scheme A. The decrease in area utilization with the increase in the interconnect overhead is much sharper for larger arrays. These result show the advantage of using schemes with simple switches (e.g. MPP) and small interconnect area (e.g. RMPP) over the schemes with complex switches and large interconnect area (e.g. global spares and direct reconfiguration scheme). This also indicates that the hierarchical schemes will give better area utilization than the single level schemes since the kill area of the hierarchical schemes is only a fraction of the total interconnect area.

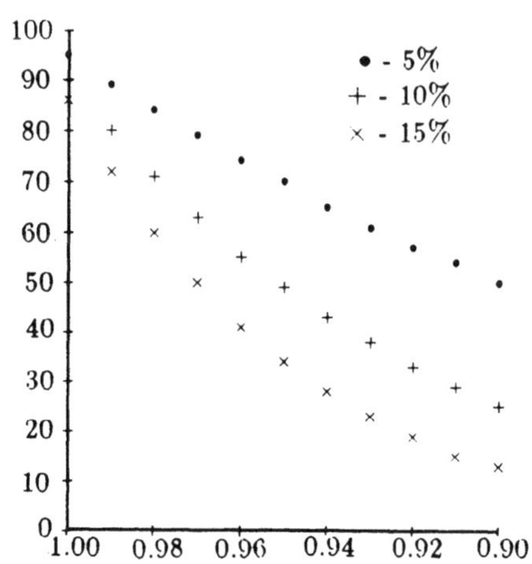

Fig. 9. Effect of interconnect overhead on area utilization

CONCLUSION

A methodology has been developed to analyze various classes of reconfigurable rectangular processor arrays for their yield and area utilization. It was demonstrated that in order to analyze the schemes for their yield and area utilization, both factors related to the logical reconfigurability and interconnect overhead should be taken into account. The relative importance of these two factors depend on the target array size, the minimum interconnect area needed for the non-redundant array of same size and the quality of the manufacturing process. The quality of the manufacturing process can be characterized either by the average number of faults in the array or by the probability of a PE being faulty. It was observed that the hierarchical schemes can give a good area utilization, since for these schemes, it is possible to tolerate some faults in the interconnect area also. The analysis of yield and area utilization can be used to select an optimal scheme for a given target array size and the quality of the manufacturing process. In the actual implementation of the reconfiguration schemes, the analysis can be used to determine the various parameters like array size, module size and the amount of redundancy for obtaining optimal area utilization.

REFERENCES

1. I.Koren and M.A.Breuer, "On area and yield considerations for fault - tolerant VLSI processor arrays", IEEE Trans. Comp. , Vol. C-33, pp. 21-27, Jan. 1984.
2. I.Koren and D.K.Pradhan, "Introducing redundancy into VLSI designs for yield and performance enhancement", Proc. International Symposium on Fault - Tolerant Computing , pp. 330-335, June 1985.
3. J.M.Daughton and B.K.Koenemann, "Cost tradeoffs in wafer-scale integration", Proc. IEEE International Conference on Computer Design , pp. 88-93, Oct. 1984.
4. K.Hedlund and L.Snyder, "Systolic architecture - A wafer scale approach", Proc. IEEE International Conference on Computer Design , pp. 604-610, Oct. 1984.
5. T.E.Mangir and A.Avizienis, "Fault-tolerant design for VLSI : Effect of interconnect requirements on yield improvement of VLSI design", IEEE Trans. Comp. , Vol. C-31, pp. 609-615, July 1982.
6. T.E.Mangir, "Sources of failure and yield improvement for VLSI restructurable interconnects for RVLSI and WSI: Part I- Sources of failure and yield improvement for VLSI", Proc. IEEE , Vol. 72, pp. 690-708, June 1984.
7. S.C.Seth and V.D.Agarwal, "Characterizing the LSI yield equation from the chip test data", Proc. IEEE ICCC-82 , pp. 556-559, Sept. 1982.
8. P.Frauzon,"Interconnect strategies for fault - tolerant 2D VLSI arrays", Proc. Int. Conf. on Computer Design , pp 230-233, Oct. 1986.
9. A.L.Rosenberg, "The Diogenes approach to testable fault - Tolerant arrays of processors", IEEE Trans. Comp. , Vol. C-32, pp. 902-910, Oct. 1983.
10. K.E.Batcher, "Design of a Massively Parallel Processor", IEEE Trans. Comp. , Vol. C-29, pp. 836-840, Sept. 1980.
11. H.Cox, S.Gaiotti, A.Jain, R.Tio, K.Faddallah, M.Malowany, B.Mandava, J.Rajski and N.C.Rumin, "A processing element for a Reconfigurable Massively Parallel Processor", Proc. Canadian Conference on VLSI , pp. 241-246, Oct. 1987.
12. S.Pateras and J.Rajski, "Self-reconfigurable interconnection network for a fault - tolerant mesh - connected array of processors", Electronics Letters , Vol. 24, pp. 600-602, 12th May 1988.
13. M.G.Sami and R.Stefanelli, "Reconfigurable architectures for VLSI processing arrays", Proc. IEEE , Vol. 74, pp. 712-722, May 1986.
14. C.Jesshope and L.Bentley, "Techniques for implementing two - dimensional wafer - scale processor arrays", Proc. IEE , Vol. 134, pp. 87-92, March 1987.
15. J.H.Hwang and C.S. Raghavendra, "VLSI implementation of fault - tolerant systolic arrays", Proc. IEEE International Conference on Computer Design , pp. 110-113, Oct. 1986.

FABRICATION-TIME AND RUN-TIME FAULT-TOLERANT ARRAY PROCESSORS USING SINGLE-TRACK SWITCHES

S.Y. Kung*, **S.N. Jean**, and **C.W. Chang**

***Princeton University**
Department of Electrical Engineering
Princeton, NJ 08544

University of Southern California
Department of Electrical Engineering
Los Angeles, CA 90089

Abstract

This paper addresses the important fault-tolerance issue for arrays of large number of processors. An array grid model based on single-track switches is adopted. Single track requires less hardware overhead and suffers less from possible faults on switches. More significantly, we are able to establish a very useful necessary and sufficient condition for the *reconfigurability* of the array. This is indeed the theoretical footing for two reconfiguration algorithms: one adopts global control for the (fabrication-time) yield enhancement and the other is a distributed scheme for the (run-time) reliability improvement. For the fabrication time reconfiguration algorithm, the task can be reformulated as a *maximum independent set* problem. An existing algorithm in graph theory is adopted to effectively solve this problem. The simulations conducted indicate that the algorithm is computationally very efficient; therefore, it is also very suitable for the compile-time fault-tolerance. In contrast, for the real time reconfiguration algorithm, it is more suitable to adopt a distributive method for (asynchronous) array processors. The algorithm has several important features: (1) it is *distributively* executed by the processor elements (PEs); (2) no global information is required by the individual PEs; (3) the time overhead for reconfiguration is independent of the array size; (4) *transient faults* are handled by retries or by *deactivating/reactivating* the temporarily failed PE. Based on simulations, the performance of the algorithms and the tradeoffs between *fault-tolerance capability* and *hardware complexity* for various kinds of spare PE distributions are evaluated.

1 Introduction

Two popular array processors are systolic and wavefront arrays. They feature the important properties of modularity, regularity, local interconnection, and a high degree of pipelining. They are very suitable for most signal and image processing algorithms. According to Kung and Leiserson [6], "A systolic system is a network of processors which rhythmically compute and pass data through the system". The wavefront array does not employ global synchronization; instead, each PE has its own local clock (self-timed) and exchanges data with neighboring PEs by asynchronous handshaking. Thus the requirement on correct *timing* in the systolic array is now substituted by the correct *sequencing* in the wavefront array, opening up new flexibilities in fault tolerant designs [9].

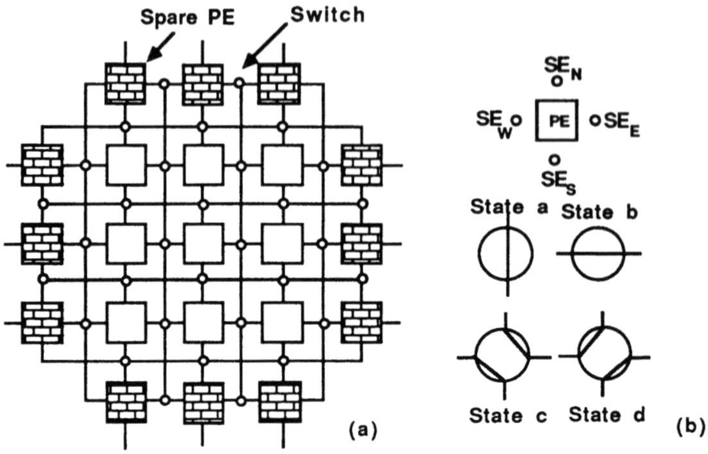

Figure 1: (a) The physical array with double-row-column spare PEs, and (b) the switch functions.

When implemented in a wafer-scale-integration (WSI) or an ultra-large-scale-integration (ULSI), the array with a very large number of PEs are squeezed into a wafer or chip. Manufacturing defects on one or more PEs in such wafers or chips are almost inevitable, so the yield is often very poor. Therefore, some fault-tolerance techniques may be employed at the fabrication stage to enhance the *yield*. On the other hand, in run time, it is almost impossible to guarantee such an array to have all the PEs running correctly over the time span of a mission. To combat this problem, run-time fault-tolerance techniques must be incorporated to improve the *reliability*.

In Section 2, a single-track switch model is proposed for fault-tolerance. Necessary and sufficient conditions for *reconfigurability* are established. This provides the theoretical footing for the new reconfiguration algorithms proposed in the subsequent sections. In Section 3, the task of finding a feasible reconfiguration using *global control* is formulated as a *maximum independent set* problem. An existing algorithm in graph theory is then adopted to solve this problem. The simulations conducted support the efficiency of the algorithm. In Section 4, we consider the *reconfiguration* of a wavefront array in real-time environment. The algorithm can be *distributively* executed by all PEs without the broadcasting of global information and the time overhead of the algorithm is independent of the array size.

2 Reconfiguration Theory for Single-Track Switch Model

To develop a theoretical analysis of the reconfiguration algorithm, we introduce two kinds of arrays: *physical array* and *logical array*. The *physical array* is the array of PEs which may be contaminated by manufacturing defects (or by operational faults). The *logical array* on the other hand represents the desired array structure, as specified by the intended application. The objective of the fault tolerance design is to maximize the probability that the desired logical array may be mapped to the healthy PEs in the physical array.

2.1 Single-Track Array Grid Model

As depicted in Figure 1, the basic switch grid model used in our discussion consists of PEs, double-row-column spares, and switches, with the following assumptions.

1. PEs, including faulty PEs, can be converted into connecting elements. This assumption is adopted in many designs [4,7,12].
2. The switching elements and the interconnection wirings are fault-free.

These assumptions are justifiable by claiming that the switches/wires use much less hardware as compared to the PEs so they are less vulnerable to faults. In [3] [8], our algorithm is extended to cover the situations of switch/wire/connection failures.

Our scheme is different from previous works in that only single-track switches are used. Here single-track means *only one communication path along each horizontal/vertical channel.* (The communication path includes both the data and control signals between two adjacent PEs.) Note that the connecting function of PEs actually provides one more track *within PEs*, which helps enhance the reconfiguration capability. The switching functions provided by a single-track switch are shown in Figure 1(b).

Previous works on this problem tend to use multi-track switches (or multiplexers) (e.g., 6-track in [11] [12]) and centralized global reconfiguration scheme to improve the yield and/or array reliability. The simplicity of the single-track switch not only saves hardware but also reduce the probability of faulty switches and interconnection wiring. Note that the major area saved is not for the control part of the switches but for the communication channels. For example, the one-track switch chip implemented with 3.5μm NMOS technology in [2]. In the total area of 600μm $\times$ 822 μm, over half of it is used for the 8-bit data "track".

2.2 Reconfiguration Theory

Throughout this paper, the logical array is indexed from [1, 1] to [N, M] and the physical array is from [0, 0] to [N + 1, M + 1] (without the four corner PEs), cf. Figure 1. Reconfiguration in array processors comprises two tasks (1) *placement* and (2) *routing.*

1. *Placement*: A *placement* is a one-to-one function $P(\cdot,\cdot)$ which maps all the PEs on the logical array to (good) PEs on the physical array. A mapping $P(i,j) = (x,y)$ means that the logical PE (i,j) is placed at the physical location (x,y). Here the vertical axis is indexed with x (i) and the horizontal axis with y (j).
2. *Routing*: On the physical array, a path linking $P(i,j)$ and $P(i+1,j)$ is called a *vertical path* and a path linking $P(i,j)$ and $P(i,j+1)$ is a *horizontal path.* A valid *routing* means the establishment of *all* the horizontal paths and vertical paths (for $i = 1,\ldots,M-1, j = 1,\ldots,N-1$) on the physical array.

Definition For a non-spare faulty PE, (x, y), the *compensation path* is a sequence of physical PEs, (PE_0, PE_1, PE_2, ..., PE_k), where PE_0 =(x, y) and PE_{i+1} = $P(PE_i)$, for $i = 0,1,2,\ldots,k-1$.

If there is no compensation path passing a PE, then the logical index for that PE is the same as the physical index, i.e. P(i,j) = (i,j). Note that a placement is uniquely determined by a set of compensation paths. It will become clear later that, regarding reconfiguration, only *straight* compensation paths are of interest. For each non-spare faulty PE with physical location (x, y), there are four possible *straight* compensation paths. The South, North, East, and West compensation paths are denoted as [x^+, y], [x^-, y], [x, y^+], and [x, y^-], respectively. For example, the *east* compensation path, [x, y^+], connects PEs (x, y), (x, $y+1$),..., and (x, $M+1$), and the *west* compensation path, [x, y^-], connects PEs (x, y), (x, $y-1$),..., and (x, 0).

Definition of "Near-Miss" Two horizontal compensation paths, [x_1, y_1^+] and [x_2, y_2^-], are declared "near-miss" if $|x_1 - x_2| = 1$ and $y_1 < y_2$. Similar definition goes for two opposite vertical compensation paths. All the other situations are not considered "near-miss."

Reconfigurability Theorem *Given an $(N+2)\times(M+2)$ physical array, it is reconfigurable into an $N \times M$ logical array using one-track routing if and only if (1) there exists a set of continuous and straight compensation paths covering all the faulty PEs and (2) there is neither intersection nor "near-miss" among the compensation paths.*

Table 1: The definition of the vertical routing state (VRS).

0	not on a vertical compensation path
1	not faulty and on an upward compensation path
2	faulty and on an upward compensation path
3	faulty and on a downward compensation path
4	not faulty and on a downward compensation path

Table 2: (a) The switch element state SW2 is a function of VRS_W and VRS_E. (b) The corresponding function, where "x" stands for "don't care".

$HRS_W \backslash HRS_E$	0	1	2	3	4
0	b	c	c	d	d
1	d	b	d	x	x
2	d	c	x	a	x
3	c	x	a	x	d
4	c	x	x	c	b

Proof: The sufficient part can be proved in a constructive way (see [8]). This means that, once the compensation paths are determined, the corresponding routing scheme can be constructed. The proof for the necessary part of the theorem can also be found in [8].

The construction procedure provides a basis for the two reconfiguration algorithms, one for fabrication-time (see Sec. 3) and the other for real-time (see Sec. 4).

2.3 Routing State

The routing is determined by the switching states in the *switching elements* (SE), which are solely determined by the compensation paths. There are two kinds of switches: The first kind, denoted as SW1, is located between two vertical neighboring PEs (i.e., SE_N and SE_S). The other kind, denoted as SW2, is located between two horizontal neighboring PEs (i.e., SE_E and SE_W).

Let us look into the switch control design for SW2 type switches. Such a switch control is determined by the so-called *routing states* of its two horizontal neighboring PEs. This state, as defined in Table 1, will be called the *vertical routing state* (VRS). The state of the SW2 switch is a simple function of VRS_W and VRS_E as shown in Table 2(b). Similarly, SW1 type switches may be controlled by the horizontal routing state (HRS) of its two vertical neighboring PEs.

3 Fabrication-Time Reconfiguration for Yield Enhancement

Given a defect distribution, we want to determine the directions of the compensation paths for all the non-spare faulty PEs. Let us denote the number of non-spare faulty PEs as F. Since there are four alternative compensation paths for each non-spare faulty PE, the total number of possible choices is 4^F. For example, if $F = 10$, there are more than one million possible choices. However, only a small fraction of them will be valid, i.e., not contradicting the four conditions described below.

3.1 Contradiction Graph

Definition of Contradiction Graph: Given a fault pattern, a ***contradiction graph*** is an undirected graph $G(V, E)$ where V denotes the set of all vertices, each representing a candidate compensation path; and E denotes the set of (contradiction) edges. An edge exists between u and v, where $u, v \in V$, if and only if u and v can not coexist. Note that u and v can not coexist if (1) Both u and v are for the same faulty PE (since the compensation path for each non-spare faulty PE is unique); or (2) The coexistence of both u and v violates the the "non-intersecting" and "non-near-miss" conditions.

Here we use a simple example to illustrate the contradiction graph. A fault pattern with three faults, two for non-spare PEs and one for spare PE, is shown in Figure 2(a). The corresponding contradiction graph has two "kites" as shown in Figure 2(b). In general, the contradiction graph is an F-kite graph, where F is the number of non-spare faulty PEs. Since some compensation paths will be proclaimed as non-candidate in the first place, their corresponding vertices may be "crossed" out. In Figure 2(b), 2S is "crossed" out because there is another faulty PE on the compensation path. The graph is thus reducible to Figure 2(c).

Graph Terminologies Given an undirected graph G, an *independent (vertex) set* is a set of vertices of G so that no two vertices of the set are adjacent. An independent set is *maximal* when there is no other independent set that contains it. The set is called *maximal independent set.* If Q is the family of maximal independent sets, then the number $\max_{S \in Q} |S|$ is called the *independence number* of the graph and the set S^* from which it is derived is called *maximum independent set.*

Corollary *Given a fault pattern in the $(N+2) \times (M+2)$ one-track physical array, there exists a valid routing scheme (to support an embedded $N \times M$ logical array) if and only if the independence number of the contradiction graph is F.*

Based on the Corollary, the problem of finding a placement which satisfies the theorem is equivalent to finding a *maximum* independent set of the contradiction graph with independence number equal to F.

3.2 Reconfiguration Algorithm

Reconfiguration includes placement and routing. *The placement algorithm consists of two stages: (1) Construct a contradiction graph for a given distribution of faulty PEs; and (2) Based the contradiction graph, find a maximum independent set with size equal to F.* The routing algorithm is simply based by Tables 1 and 2.

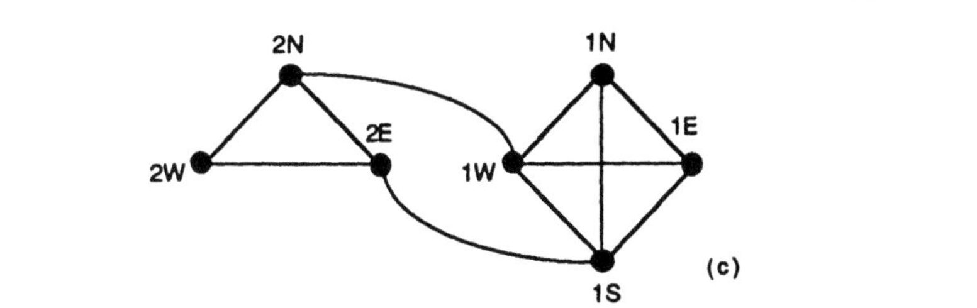

Figure 2: Illustration of contradiction graph.

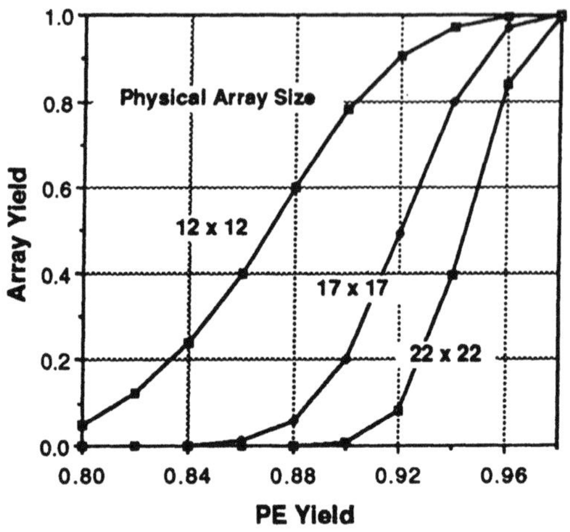

Figure 3: Array yield simulation results.

An effective method for finding *all the maximal* independent set is proposed by Bron and Kerbosch in [1], which is essentially a tree search method. The algorithm consists of *forward step* and *backtracking step*. In a forward step, say at the stage k, an independent vertex set S_k is augmented by another proper vertex to produce a new independent set S_{k+1}. In smooth condition, the forward step continues until the set becomes a *maximal* independent set. In some other conditions, backtracking step has to be used to achieve the objective.

For our application, we are interested in finding *one maximum* independent set with the special interest in checking whether the size of the maximal independent set is F. The testing condition of the Bron and Kerbosch's algorithm can be modified according to our special need. The detailed procedure can be found in [3], [8].

Screening It is computationally advantageous to further reduce the number of vertices in the contradiction graph. The following "screening" procedure is very useful for this purpose. Note that a vertex should be adopted if (1) all the other three vertices in the same kite are already crossed out; or (2) it is not connected to any vertices in another kite. Once a vertex is adopted, all its adjacent vertices should also be crossed-out (Otherwise, contradiction could occur). Such a "screening" process may be recursively executed to cross out a large number of vertices in the contradiction graph.

The screening may also be applied before the contradiction graph has been constructed [3]. A specific compensation path is chosen for a fault if (1) the other three compensation paths are blocked; or (2) the compensation path does not conflict with the compensation paths of other faults. The choice of a specific compensation path for a fault (and therefore the "crossing out" of the other three compensation paths for the fault) may (1) block some more compensation paths and (2) remove conflicts from some compensation paths. Thus the screening process can be recursively executed to "solve" a large number of faults without using the contradiction graph.

Simulations With the proposed algorithm which uses tree search, a feasible placement can always be found if it exists. Thus the algorithm always leads to the best yield enhancement for the single-track grid model. To estimate the yield of the proposed model and algorithm, Monte-Carlo simulations are performed for 12 × 12, 17 × 17, and 22 × 22 physical arrays. (Simulation results on clustered faults can be found in [8].) Here all the PEs are assumed to have independent failures and the array yields are summarized in Figure 3. In all the simulations, the reconfiguration for each fault pattern consumes in average less than 100 msec in a SUN/4-280 workstation (with screening). Note that the algorithm provides very good enhancement for smaller array sizes. The results also suggest that the partitioning scheme (cf., Sec. 3.4) may be desirable for larger arrays.

3.3 Compile-Time Fault-Tolerance

Two kinds of run-time environments need to be distinguished: *compile-time* versus *real-time*. It is considered a *compile-time* environment if the detection and location of the operational faults and the reconfiguration can be done off-line. Otherwise, it becomes a *real-time* environment, which will be addressed in the next section. The simulations conducted indicate that the algorithm originally proposed for the fabrication-time fault-tolerance is computationally very efficient. Therefore, the same reconfiguration algorithm may also be very suitable for the compile-time reconfiguration. To adapt our algorithm to the compile-time environment, we note that the array may have some compensation paths pre-selected as the result of reconfiguration taken place in the fabrication time. This information can be reflected in the initial contradiction graph (for the compile time) which comprises all the selected compensation paths during the fabrication time. The new "kites" will be added to the initial contradiction graph in the run-time, and the same algorithm for maximum independent set may be applied.

3.4 Spare PE Distributions and Partitioning Scheme

Since the spare PEs are used to handle defects, the number of spare PEs used and their distribution can greatly influence the yield. The kinds of spare PE distributions include (1) *Double-Row-Column*, (2) *Single-Row-Column*, (3) *Double-Column*, and (4) *Single-Column*. Tradeoffs between the array reliability and the hardware/software complexity may be made.

Partitioning Scheme With single-track switches, the reconfiguration capability is limited and the yield enhancement may not be satisfactory for large arrays. Our approach is to partition a large array (eg., 32 × 32) into many smaller subarrays (e.g., 8 × 8), each equipped with spare PEs (see Figure 4). Note that there is only a single spare column between two neighboring subarrays and a spare PE may be used by either (but not both) of the two subarrays. Therefore the reconfiguration processes for two neighboring subarrays become coupled. This means that some extra edges have to be added to the contradiction graph in the algorithm.

Simulations for Partitioning Scheme To estimate the yield when extra spares are incorporated with the partitioning scheme, Monte-Carlo simulations are adopted. Here all the PEs are assumed to have independent failures. The simulation results are summarized in Figure 5. The simulations indicate that very good yield enhancement may be achieved with a moderate increase of *redundancy overhead*, which is defined as the number of spares over the number of *logical* PEs.

Figure 4: A partitioned array.

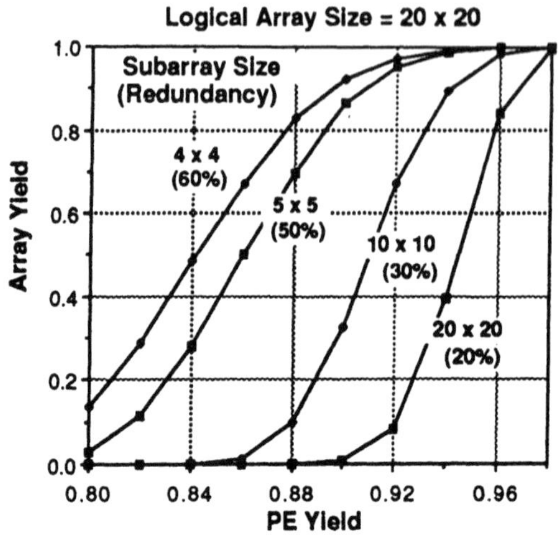

Figure 5: The array yield with partitioning.

4 Real-Time Reconfiguration for Reliability Improvement

In this section, we consider the *reconfiguration* of a wavefront array in real-time environment. A new reconfiguration algorithm, based on the reconfigurability theorem, will be developed. For run-time fault-tolerance design, the *physical array* is considered to be an array which has been successfully tested and delivered by the manufacturer for real mission operations. Operational faults during the mission time are expected, although the operational faults have a much lower probability of occurrence as compared to production defects [5]. The *logical array* again represents the desired array structure corresponding to the application. When operational faults in the physical array occur, the run-time reconfiguration techniques will be applied so that the logical array can be successfully mapped to the *working* PEs in the physical array.

Wavefront Arrays versus Systolic Arrays For fabrication-time or compile-time fault-tolerance, there are no apparent reasons to favor either a systolic array or a wavefront array. However, for the run-time fault-tolerance, the wavefront array is superior to the systolic array, since it can cope with on-line reconfigurations due to its asynchronous processing capability.

To illustrate the asynchronous reconfiguration process, let us first look into a linear array consisting of 6 PEs with the rightmost PE as a spare one. The PEs are indexed, from left to right, 1 to 6. Assume that there is no global control line for this array. Once a fault occurs, say at PE(2), the information about the failure of PE(2) is transmitted to PE(1) and PE(3). The output of PE(2) is discarded and the unprocessed inputs are transferred to PE(3) to be executed at the next clock period. PE(2) is then converted into a connecting element. At the next clock period, PE(3) will notify PE(4) and the job originally assigned to PE(3) will be reassigned to PE(4). In this fashion, the spare PE will eventually become active.

4.1 Distributed Reconfiguration Algorithm

The basic assumptions on the array grid model in Section 2 are still valid in this section. Here we make an additional assumption that the (built-in) fault detection, the reconfiguration states, and the communication among PEs, are all fault free.

To achieve real-time reconfiguration, distributive processing is necessary. Therefore, it is important that each individual PE keeps updating the information concerning the reconfigurations incurred in its neighborhood. For distributive routing, the *routing state* discussed in Section 2 can still be used. For distributive placement, we need to define a ***placement state*** within each PE to specify the available options of compensation paths to choose from.

Placement State and Its Updating To prevent compensation paths from intersecting and overlapping, every PE keeps a 4-bit *placement state.* Each bit indicates the availability of one compensation path direction.

In our reconfiguration scheme, both the deactivating and the reactivating are considered. The *deactivating* program will be initiated when a PE is declared faulty, while the *reactivating* program will be executed when the failed PE is recovered. The updating of placement state depends on which program is currently executed.

1. *Deactivating:* Each compensation path for a faulty PE may "block" some future compensation paths. The location and the direction of the newly generated compensation path are propagated to all the other PEs. Based on these data, all PEs update their own placement state. The updating is based on the region it locates with respect to the newly generated compensation path.
2. *Reactivating:* Unlike permanent faults, transient faults may be recovered. (The occurrence of transient faults, mainly due to temporary environmental changes, is reportedly about 10 times more frequent than that of permanent faults [13,10].) When a faulty PE is recovered, its corresponding compensation path will be canceled. The cancellation of a compensation path will influence the placement states of the other PEs since some originally "blocked" compensation paths become permissible. To indicate whether a compensation path is allowed, there are four counters used in the scheme for each PE, one for each possible compensation path direction. For example, the upward-counter records the number of compensation paths which prevents the upward compensation path for the PE. These include the horizontal compensation paths which pass directly below the PE and the downward compensation paths which overlap with the upward compensation path for the PE. The counter is increased (decreased) by the creation (cancellation) of such a special compensation path. When a counter is decreased to zero, the corresponding compensation path becomes allowed and the placement state of the PE should be upgraded.

Distributed Reconfiguration Algorithm Assume that the *fault detection* is concurrently performed in every PE by means of some *on-line self-testing* circuits, e.g., via duplication of arithmetic and logic units with matching circuits. To avoid unnecessary reconfigurations, a PE failure will not immediately activate the reconfiguration operation. Instead, the *retry* is repeatly performed by the faulty PE until either the fault disappears or the number of retry exceeds a bound.

1. If the fault disappears, it is viewed as a transient fault and the PE is never considered as faulty. Because of the data-driven property of the wavefront array, the activities of all the neighbor PEs are suspended during the retry period. Once the transient fault recovers, the PE and all the neighbor PEs can continue as normal.
2. If the fault persists beyond a certain number of retries, the PE is declared faulty and the *deactivation* and *reactivation* processes are initiated.
   - **Deactivation of a faulty PE:** When a fault is declared, the faulty PE issues an interrupt to its neighbor PEs and chooses either a vertical or a horizontal compensation path. The choice is based on the current placement state of the faulty PE. The information about the physical location of the faulty PE and the type of the compensation path is propagated out not only along the compensation path, but also across almost the entire array similar to an electromagnetic wavefront propagation (see Figure 6).

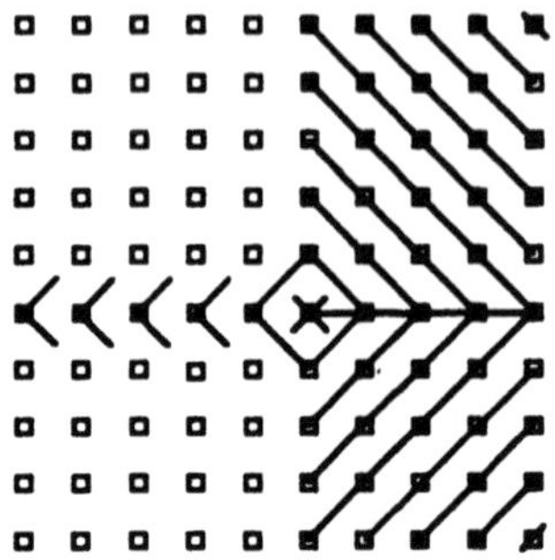

Figure 6: Wavefronts for the propagation of information.

(a) The PEs not on the compensation path will (1) update their placement states according to the fault position and the type of compensation path and (2) function like a wave media, i.e., interrupt the neighbor PEs and pass information to them.

(b) For PEs on the compensation path, in addition to the above tasks, each has the following activities: (1) checking the placement state to see if there are other compensation path passing through the PE. If yes, the array is declared failed. Otherwise, (2) and (3) are executed. (2) updating its VRS and HRS states (cf. Table 1) and thus changes the state of its neighboring switches (cf. Table 2), and (3) reassigning the job originally belonging to it to the next PE on the compensation path.

- **Reactivation of a recovered PE:**

  Once a fault is declared, the faulty PE becomes a connecting element and enters a dormant state. In the dormant state, the PE tests itself to check its status repeatedly. If the fault is transient and the transient fault is removed, the dormant PE initiates an interrupt to its neighboring PE, takes back the job originally assigned to it, changes its own states (including placement states, HRS, and VRS), and then continues its unfinished work.

4.2 Example: Single-Row-Column Spare Distribution

For illustration, we apply the above general scheme to the simpler special case of the single-row-column spare distribution (see Figure 7).

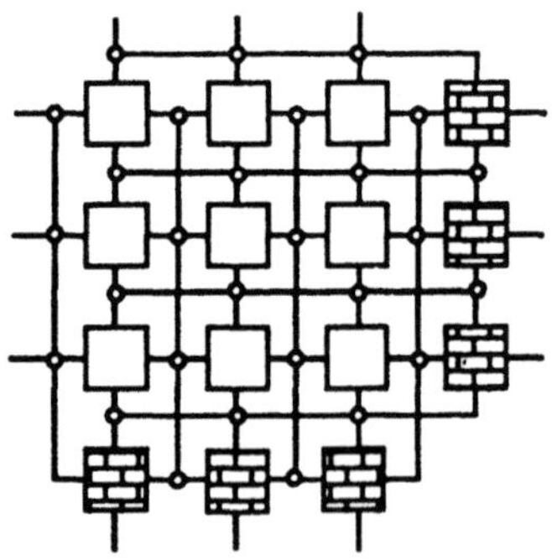

Figure 7: The physical array with one row and one column of spare PEs.

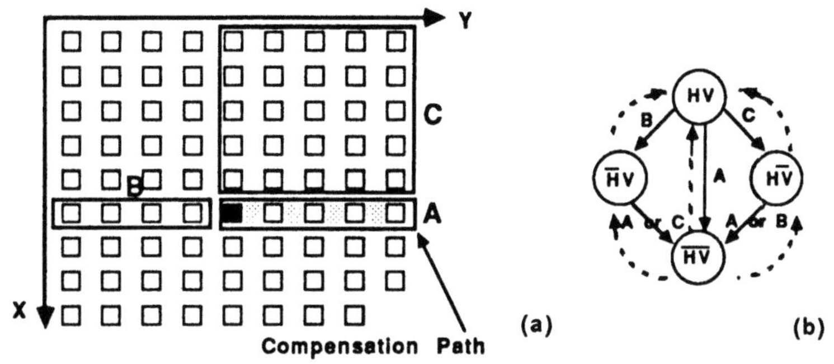

Figure 8: (a) Placement state updating regions. Similar regions can be defined for a vertical compensation path. (b) State transition diagram.

Reconfigurability Theorem for Single-Row-Column Spares: *Given an $(N+1) \times (M+1)$ one-track physical array, it is reconfigurable into an $N \times M$ logical array if and only if there exists a set of continuous and straight compensation paths covering all the faulty PEs and they are non-intersecting.*

Since this is a special case of the reconfigurability theorem in Section 2, the proof is self-evident.

Placement State Every PE keeps a 2-bit *placement state*, which provides sufficient information to prevent the compensation paths from intersecting.

1. **HV**: Both horizontal and vertical compensation paths are allowed to pass through this PE. When a PE in this state fails, the shorter compensation path will be chosen to decrease the possibility of future intersection.
2. $\mathbf{H\overline{V}}$: Only horizontal compensation path is allowed to pass through this PE.
3. $\mathbf{\overline{H}V}$: Only vertical compensation path is allowed to pass through this PE.
4. $\mathbf{\overline{HV}}$: No more compensation path is allowed to pass through this PE.

All PEs have the same initial placement state, **HV**. The updating is based on the region it locates with respect to the newly generated compensation path (see Figure 8(a)). The resulting state *transition diagram* is shown in Figure 8(b) where the dashed lines indicate the state transitions due to fault recoveries. If any PE in the state $\mathbf{\overline{HV}}$ is declared faulty, the system fails.

Routing State The routing state for single-row-column spares is a special case of what discussed in Section 2. Only three VRS states, i.e., 0, 1, and, 2 (see Table 1) are required since there is no downward compensation path. The corresponding switch control function can also be reduced to the one within the bold-faced square in Table 2(b)).

Monte-Carlo Simulation To get some ideas about a real-time fault-tolerance system, let us consider a system with the following system parameters:

- The PE computation clock rate is 10 MHz and the system communication clock rate is 100 KHz.
- The failure rate of permanent faults, λ, is 10^{-6} per second. If disregard the transient faults, this means the PE reliability is 0.99 in a mission time (t) of 10^4 seconds. Here the PE reliability is the probability that a PE works during the mission time given that the PE works initially and is equal to $e^{-\lambda t}$.

Table 3: Simulation results of the system reliability.

Logical Array Size	Without Fault-Tolerance	Deactivation Only	Retry (0.5 Leakage)	Retry (No Leakage)
Equivalent r	0.896	0.896	0.942	0.99
4 × 4	1.726×10^{-1}	8.289×10^{-1}	9.457×10^{-1}	9.984×10^{-1}
8 × 8	8.866×10^{-4}	2.670×10^{-1}	6.556×10^{-1}	9.886×10^{-1}
12 × 12	1.356×10^{-7}	1.901×10^{-2}	2.720×10^{-1}	9.608×10^{-1}
16 × 16	6.178×10^{-13}	1.734×10^{-4}	5.165×10^{-2}	9.208×10^{-1}

- The failure rate of transient faults is 10 times that of permanent faults. If the transient faults are treated the same as permanent faults in the reconfiguration scheme, the corresponding PE reliability is 0.896 in a mission time of 10^4 seconds.
- The average duration of transient faults is 10 system communication clocks. That is the recovery rate for transient faults is 10^4 per second.

The following two observations are useful in simplifying our simulation for the system reliability.

- For transient faults the recovery rate (10^4 per second) is much higher than the failure rate (10^{-5} per second). Thus the system reliability should be approximately the same if we disregard the influence of transient faults which are recovered later. Since in our scheme the recovery of transient faults is by retrying the faulty task, a leakage may be defined as the percentage of transient faults which can not be recovered by a finite number of retries (and thus is handled as a permanent fault). In our scheme, the number of retries can be infinite. The leakage should be very small. As an example, if the leakage is 0.5, the equivalent PE reliability is

$$r = e^{-(10^{-6}+0.5\times 10^{-5})10^4} = 0.942$$

- The failure rate of permanent faults is 10^{-6} per second, or equivalently, 10^{-11} per system communication clock period. Thus for an array with a moderate size, e.g., 20 × 20, the probability that a permanent fault occurs during the propagation of the compensation path for another permanent fault is very low.

Assume that all the PEs have independent failures and the same PE reliability, r, then the array reliability for a physical array with size $N \times M$ is

$$\sum_{i=0}^{K} \binom{NM}{i} C_i \, r^{NM-i} \, (1-r)^i \tag{1}$$

where C_i is the probability that a faulty pattern with i faulty PEs does not cause array failure and K is the number of spare PES. Based on a Monte-Carlo simulation, we estimate C_i and compute the array reliability. Since the effect of the *reactivating* process can be incorporated as leakages, only the *deactivating* process is simulated. The simulation results are summarized in Table 3.

4.3 Partitioning and Stealing

With single-track switches, the reconfiguration capability is limited and the array reliability becomes not satisfactory for large arrays. The partitioning solution as mentioned in the previous section can still be applied.

Table 4: The probabilities of array failure for a 32 × 32 logical array ($r = 0.99$).

Number of Spare PEs	Row-Column (No Stealing)	Column (No Stealing)	Row-Column (Stealing)	Column (Stealing)
1024	0.0665	0.0973	0.0428	0.0055
512	0.1096	0.1415	0.0305	0.0516
256	0.1132	0.2220	0.0967	0.0809
128	0.2596	0.3563	0.2226	0.2070
64	0.5006	0.5474	0.5006	0.4832

Note that when each subarray has single-row-column (resp., single-column) spares, the reconfiguration algorithm for double-row-column (resp., double-column) spares may be adopted by stealing[12] spare PEs from neighboring subarrays.

Array Reliability Simulations For the single-column-spare scheme, a PE row works only if the number of faulty PEs on the (physical) row is less than 2 and an array works only if all the PE rows work. Thus the array reliability for a physical array with size $N \times M$ is

$$\left[r^{M} + M \times r^{M-1}(1-r)\right]^{N}$$

For the other schemes, especially for those with stealing, Monte-Carlo simulations are performed to estimate C_i (see Eq. 1). Table 4 summarizes the probabilities of array failure for different schemes by assuming a 32 × 32 logical array. We observe that the row-column-spare scheme performs better than the column-spare scheme. This is at the expenses of extra control and doubling the number of switches. As expected, stealing improves the array reliability at the expense of extra control.

5 Conclusion

An array grid model based on single-track switches is shown to offer very good performance in the reconfigurability while keeping the hardware overhead minimum. The necessary and sufficient condition for the *reconfigurability* leads to reconfiguration algorithms: one adopts global control for the (fabrication-time) yield enhancement and the other is a distributed scheme for the (run-time) reliability improvement. To enhance the yield during fabrication-time, the proposed reconfiguration algorithm can systematically enumerate all the placement possibilities. The algorithm is efficient enough to be adopted for compile-time applications. In real-time fault-tolerance designs, reconfiguration time overhead and handling of transient faults are two important factors to be taken into account. Wavefront arrays offer the asynchronous data-driven computing feature which facilitates fault-tolerant designs for real time environment. This leads to a distributed reconfiguration scheme with the reconfiguration time overhead independent of the array size.

Acknowledgment

This research was supported in part by the National Science Foundation under Grant MIP-87-14689, and by the Innovative Science and Technology Office of the Strategic Defense Initiative Organization, administered through the Office of Naval Research under Contract No. N00014-88-K-0515 and N00014-85-K-0599.

References

[1] C. Bron and J. Kerbosch. Algorithm 457–Finding all cliques of an undirected graph. *Comm.* of ACM, 16, p. 575, 1973.

[2] J. R. Hwang. The reconfiguration and switch design of array processors. Master's Thesis, The Institute of Electronics, Chiao Tung University, Taiwan, 1987.

[3] S. N. Jean and S. Y. Kung. Necessary and Sufficient Conditions for Reconfigurability in Single-Track Switch WSI Arrays. in *Proc. of International Conference on Wafer Scale Integration*, January 1989.

[4] I. Koren. A reconfigurable and fault-tolerant multiprocessor array. In *8th Annual Symposium on Computer Architectures*, pages 425–442, 1981.

[5] I. Koren and D. K. Pradhan. Modeling the effect of redundancy on yield and performance of VLSI systems. *IEEE Transactions on Computers*, 344–355, March 1987.

[6] H. T. Kung and C. E. Leiserson. Systolic Arrays (for VLSI). in *Sparse Matrix Symposium*, pp. 256–282, SIAM, 1978.

[7] H. T. Kung and M. S. Lam. Wafer-scale integration and two-level pipelined implementations of systolic arrays. *J. Parallel and Distributed Computing*, 32–63, 1984.

[8] S. N. Jean, *Mapping/Matching Algorithms to Reconfigurable Mesh Arrays*, PhD thesis, University of Southern California, Los Angeles, California, 1988.

[9] S. Y. Kung, S. C. Lo, S. N. Jean, and J. N. Hwang. Wavefront array processors: from concept to implementation. *IEEE Computer Magazine*, 18–33, July 1987.

[10] S. R. McConnel, D. P. Siewiorek, and M. M. Tsao. The measurement and analysis of transient errors in digital computer systems. In *Proc. IEEE Fault-Tolerant Comput. Symp.*, pages 67–70, 1979.

[11] M. Sami and R. Stefanelli. Fault-tolerance and functional reconfiguration in VLSI arrays. In *Proc. Int. Conf. Circuits and Systems 1986*, pages 643–648, 1986.

[12] M. Sami and R. Stefanelli. Reconfigurable architectures for VLSI processing arrays. *Proc. IEEE*, 712–722, May 1986.

[13] D. P. Siewiorek and R. S. Swarz. *The Theory and Practice of Reliable System Design*. Digital Press, 1982.

AN EFFICIENT RESTRUCTURING APPROACH FOR WAFER SCALE PROCESSOR ARRAYS

Adit D. Singh and **Hee Yong Youn***

University of Massachusetts
Department of Electrical and Computer Engineering
Amherst, MA 01003

***University of North Texas**
Department of Computer Science
Denton, TX 76703

INTRODUCTION

Parallel processing using processor arrays is being widely investigated to overcome the performance limitations of traditional uniprocessor computer systems. Some inherent problems with the board level implementation of highly parallel systems are separate component packaging costs, assembly cost on printed circuit boards, and low reliability due to the complex pin to pin interconnections on the boards. Especially significant is the large signal *propagation delay* in MOS VLSI technology required to drive signals off chip. Wafer Scale Integration (WSI)[1] promises a solution to this problem by integrating the entire processor array and the interconnection structure on a single packaged wafer. Thus WSI will make it possible to eliminate the off chip signal drivers required within the processor chips, and the complex board level interconnections among the processors. As a result, signal delays can be substantially reduced allowing faster operation. In addition, system reliability may also be improved because of elimination of the mechanical and electrical failures frequently observed at the pins and interconnections in traditional designs.

Although WSI has many attractive features, the low yield problem due to the large chip area[2] must be overcome before such circuits can become practical. In conventional VLSI designs, the entire circuit is discarded if it contains even a single defect that is capable of causing a logical fault. For large area circuits, which have a high likelihood of containing at least one defect, this leads to extremely low yield. To overcome this problem, an on-chip fault tolerance scheme, employing redundant components and a reconfigurable interconnection structure is required. Using electrical[14] or laser[13] programming or discretionary wiring, such a scheme can allow proper operation even in the presence of defects. This will increase the yield of 'good' circuits at the manufacturing stage, and can perhaps also be used to increase the reliability of the system in the operating stage.

This work was supported in part by the National Science Foundation under grant MIP-8808325.

A number of fault tolerance schemes for VLSI processor arrays[3-9] have been proposed. The objective is to reconfigure the failure free processors in the physical array into a desired specific logical computational topology to best match a given parallel algorithm. Proposed fault tolerance schemes employ redundant buses and switches to route signals around faulty processors. It is important to recognize that this redundant circuitry, introduced to support reconfiguration, can significantly add to the overall array area unless the processors themselves are large as compared to the switches and interconnections, i.e. the reconfiguration is carried out at a relatively large granularity. However, large processors are more susceptible to manufacturing defects. A significant number of them can, therefore, be expected to fail due to manufacturing defects. Unfortunately, most proposed fault tolerance schemes for processor arrays are effective against only a few failures and are not very efficient in their use of silicon area for realistic processor yields in the range of 50-90%. In this paper, we present a general fault tolerance approach that enables the efficient restructuring of regular array structures by mapping the reconfiguration problem to a matching problem on bipartite graphs. This approach was first presented in 15,16 for rectangular arrays. In this paper we show that several important computational topologies - the rectangular and hexagonal array, and complete binary tree architecture, can be efficiently reconfigured using this scheme even when a significant number of processors are faulty.

The rest of the paper is organized as follows. In Section 2, we present our basic reconfiguration as applied to rectangular arrays. The proposed scheme is compared with other designs in Section 3. Sections 4 and 5 show how the complete binary tree architecture and the hexagonal array can be reconfigured using the strategy proposed for the rectangular array. Section 6 concludes the paper.

2. RECTANGULAR ARRAYS

The two dimensional rectangular array has received much attention because it is a useful planar array structure for parallel computation, and many parallel algorithms for such arrays have been suggested. In this section, we present a Processor Switch Array (PSA) structure which can be used to efficiently reconfigure the rectangular array in the presence of defects.

PSA Structure

The PSA structure consists of a regular array of connected switches and processors as shown in Figure 1. The switches are interconnected in the *desired rectangular topology*. Also, each switch has connections to four neighboring processors such that any one of these four processors can be assigned to the switch. Thus, if a failure free processor is assigned to each switch, we obtain the desired rectangular processor array. Following fabrication of a wafer, we propose to first test all processors. Then by assigning a failure free processor to each switch (assumed to be failure free - we address the problem of switch and inter-switch connection failures later), we achieve the desired final failure free rectangular processor array. Because any one of the four neighboring processors can be assigned to a switch (and each processor can be assigned to any one of its four neighboring switches) there is considerable flexibility in assigning a processor to a switch. This greatly increases the chances that a failure

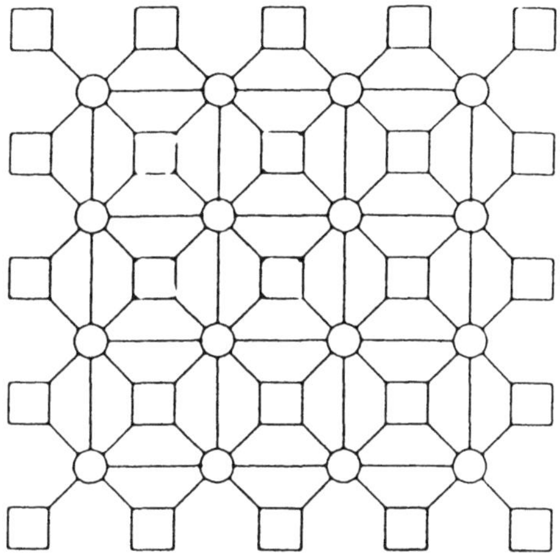

Figure 1: Structure of Processor Switch Array (PSA).

free computational array can be configured, even in the presence of multiple failures. This flexibility in matching the processors and switches is the primary unique characteristic of the PSA approach. It makes the structure highly fault tolerant because, depending on the status of neighboring processors, up to three faulty processors out of four processors can be tolerated at each switch site. The switches are assumed to be electrically or laser programmable links and are programmed (open or closed) according to the pattern of processor assignments.

Note that while we explicitly discuss a switch based implementation here, our approach can also be easily adapted to descretionary wiring technology. In such an implementation, the switches have only a logical, but not physical meaning. Assigning a processor to a logical switch is equivalent to assigning a logical array index to the good processors which are then connected by discretionary wiring.

An n x n PSA array contains $2n + 1$ extra processors; therefore the inherent redundancy is $2n+1$. It is easily seen that as a percentage of the number of processors in the computational array, the inherent redundancy of the 8 x 8 and 16 x 16 processor switch array is 27% and 13%, respectively. Next we show how to assign failure free processors to the switches so as to obtain the desired rectangular array in the presence of failures.

Reconfiguration Algorithm . The desired rectangular processor array can be obtained by matching a unique failure free processor to each switch which is already interconnected in the rectangular array structure. The matching between the switches and the processors is a *bipartite matching* because there is no direct connection between the processors. The failure free elements of the PSA structure can be divided into two sets, the switch set and the processor set, connected by processor-switch interconnections. A rectangular processor array of maximum size can be obtained by finding the maximum matching between these two sets. The time complexity of the best known matching algorithm[10] is $O(|V|^{1/2} \bullet |E|)$ where $|V|$ and $|E|$ are the number of the nodes (processors and switches) and edges (processor-switch interconnections), respectively, in the bipartite graph. Thus the processor assignment can be made in small polynomial time.

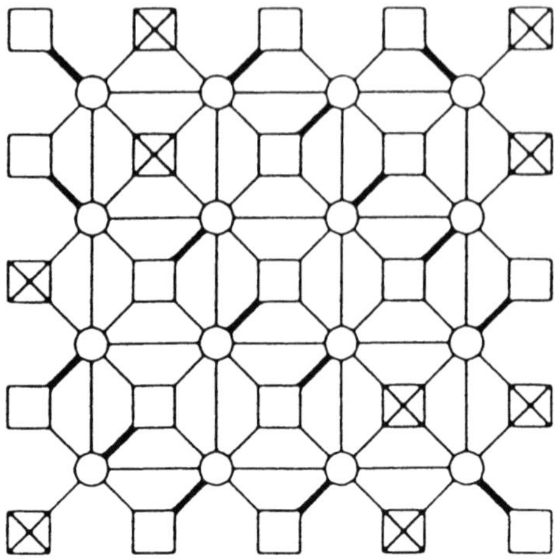

Figure 2: Reconfiguration of a 4x4 array using bipartite matching.

When we assume that only the processors can be faulty, up to $2n+1$ faulty processors can be tolerated in an n x n switch array. For example, the 4 x 4 switch array has 25 processors, and up to 9 faulty processors can be tolerated as shown in Figure 2. In this example, all the failure free processors were utilized to construct a rectangular array.

Of course the proposed PSA structure cannot always utilize all the faulty processors for every fault distribution. The above reconfiguration strategy will be unable to configure a failure free array in the following cases.

1. The number of faulty processors is greater than the number of the inherent redundant processors $(2n + 1)$.
2. Even though the number of faulty processors is less than $2n + 1$, all four processors connected to a switch are faulty, or the faulty processors are clustered in such a way that all switches cannot be assigned.
3. A switch or an inter-switch connection is faulty.

For the first two cases we adopt the row and column exclusion strategy[3] to exclude the rows and columns of switches and thereby reduce the size of the active switch array to a level where all the active switches can be matched by failure free processors. The non-faulty switches in the excluded rows and columns are converted into simple connecting elements. The same strategy can be used to exclude switches and inter-switch connections that are faulty from the active switch array, covering the last case.

Whenever a row and column is excluded, the likelihood of extracting a fully matched rectangular processor array with a size smaller than the original switch array is increased because the number of switches to be matched decreases while the number of failure free processors stays unchanged. The row and column selected to be excluded are those with the largest number of unmatched switches. We cannot be sure that this selection is optimum for the extraction of the largest possible array, but we propose this suboptimal strategy because the problem of finding an optimal selection is NP complete.

The row and column exclusion process is repeated until we find a fully matched rectangular array. The following theorem predicts the lower bound of the number of exclusions.

Theorem 2.1: For n x n switch array, let us assume that there are more than $(2n+1)$ faulty processors and let the number be f. Then at least $d = f - (2n+1)$ switches must be removed before a unique processor can be matched to each switch. If we let the minimum number of row and column exclusions be k, then k should satisfy the following inequality.

$$n^2 - (n-k+1)^2 \leq d \leq n^2 - (n-k)^2$$

Proof:

(right hand side inequality)

$(n+1)^2$ is the number of the processors in the original array. $(n-k)^2$ is the number of the processors actually used in the extracted array. The difference between these two numbers should be greater than the number of faulty processors. Therefore $f \leq (n+1)^2 - (n-k)^2$ and $(2n+1) \leq n^2 - (n-k)^2$.

(left hand side inequality)

$(n-k+1)^2$ is the number of the processors used in the array when we exclude k - 1 rows and columns. Then the difference between $(n+1)^2$ and $(n-k+1)^2$ should be less than f. Therefore $(n+1)^2 - (n-k+1)^2$ and $n^2 - (n-k+1)^2 \leq f - (2n+1)$. □

Solving the inequality equations in Theorem 2.1 we get the following lemma.

Lemma 2.1: For the same condition as Theorem 2.1

$$k = \lceil n - \sqrt{n^2 - d} \rceil$$

Because of the flexibility of the association between the processors and switches allowed in the PSA structure, we can generally expect that eliminating k rows and columns (with k defined as above) will yield a complete match between the failure free switches and processors for most fault distributions, i.e. we will generally need only a minimum number of exclusions. This means that we will often extract the largest array obtainable from the available non-faulty processors, leading to high area efficiency. Additional rows and columns may need to be excluded only for very unique distributions of processor faults that are unlikely to occur in practice except in large array with low processor yield. The PSA structure works better at relatively high processor yields because the more the faulty processors, the greater the possiblity of fault distributions with clustered faulty processors that need additional extra row and column exclusions. The array yield can be expected to be relatively poor for low processor yields. The computer simulations presented in the next section confirm this.

Bypassing Scheme

We can solve the low array yield problem for relatively low individual processor yield by modifying the simple exclusion scheme to include a switch bypass capability. This

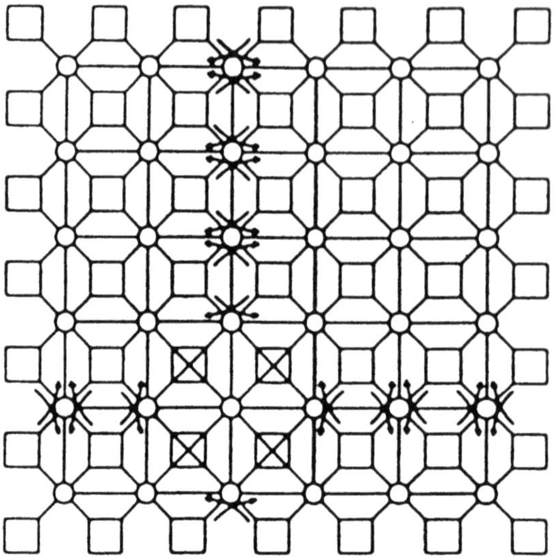

Figure 3: Bypassing Scheme (Scheme 2).

bypassing scheme, needs a somewhat more complex switch structure as compared with that of the simple exclusion scheme. In the bypassing scheme, instead of the excluded switches just acting as connecting elements, they have a further bypassing capability that allows processors connected to them to be connected (assigned) to adjacent switches as shown in Figure 3. In Figure 3, the third column of processors can be matched to the fourth column of switches (by bypassing the third switch column) and the fourth column of processors can be matched to the second column of switches. Consequently, the switches next to the excluded switches have two more processors with which they can be matched. This increased connectivity can save additional exclusions, hence increasing area efficiency for low processor yield. The relative efficiency of the simple exclusion scheme and the bypassing scheme will be manifested by the computer simulation in the next section. From now on, we will refer to the simple exclusion scheme as *Scheme 1* and the bypassing scheme as *Scheme 2*. We next evaluate the PSA scheme on area efficiency and compare it with other designs proposed in the literature.

3. EVALUATION AND COMPARISON OF THE PSA SCHEME

In modeling the PSA structure for computer simulation, we assume that the fault distribution is positionally uniform. In other words, the processor yield (the probability of each processor being failure free) is the same for all the processors on the wafer. This assumption is not totally valid in practice. Several researchers[2,11,12] have reported skews in the spatial distributions of defects on a wafer, with clustering of defects and more observed defects on the edges than near the center of the wafer. But this variation is small and difficult to model. We have assumed a uniform defect distribution to simplify the simulation. We also assume that the switch and interconnections are failure free.

Simulation Model. First, we select the faulty processors randomly in proportion to the processor yield, with a uniform distribution across the whole wafer. After assigning faults to processors in the original array structure, we attempt to obtain a rectangular array using the bipartite matching algorithm, using the row and/or column exclusion scheme if neeeded. We evaluate our structure for the two different reconfiguration schemes for 8 x 8 processor switch arrays, while varying the processor yield (P) from 0.9 to 0.4 in steps of 0.1.

10,000 simulation runs were executed and the results then averaged for each value of P. To measure the efficiency of the reconfiguration scheme, we consider the Silicon Area Utilization Factor[9] (SAUF), defined by

$$\text{SAUF} = \frac{\text{number of processors in the computational array} \times \text{yield}}{\text{number of processors in the original array} \times \text{P}}$$

This factor measures the fraction of failure free processors that are utilized (on the average) in implementing an array on silicon. Clearly a design that maximizes this fraction will maximize chip area utilization. We compare SAUF data for our scheme with other two fault tolerant designs discussed in the literature - the Hierarchical approach due to Hedlund[4], the Row Major redundancy scheme in Greene[5] and Interstitial Redundancy[9]. While the results of the last approach were directly obtained from Singh[9], we have estimated SAUF for the other two schemes as follows.

Hierarchical Redundancy Scheme

This scheme uses a hierarchical method for extracting the desired rectangular array by using redundant submodules. The objective is to ensure, through redundancy, a very high probability that a failure free 2 x 2 submodule can be configured at each submodule site, so that row and column exclusion, employed at a higher level to protect against such failures is very rarely needed. Here we need to find the optimum size for the larger redundant submodule that guarantees a 2 x 2 failure free submodule array (processors in the submodule are assumed fully connected). The yield and SAUF can be obtained by

$$\text{yield} = (\sum_{x+4}^{M} \binom{M}{x} P^x (1-P)^{M-x})^{nsub} \tag{1}$$

$$\text{SAUF} = \frac{4 \times \text{yield}}{M \times \text{P}} \tag{2}$$

Here M is the number of processors in the original submodule which might contain some faulty processors (proportional in number to the processor yield) and *nsub* is the number of submodules implementing the desired array. From the above equations we find a submodule size to obtain the maximum SAUF for each processor yield.

Row Major Redundancy Scheme

An n x $m(n < m)$ array is used for the extraction of an n x n rectangular array in this scheme. By extracting a size n failure free linear array from each row and then connecting the n failure free linear arrays vertically (assuming failure free interconnections), the desired $n \times n$ computational array can be realized. Again, the yield and SAUF can be obtained by

$$\text{yield} = (\sum_{m}^{x=n} \binom{m}{x} P^x (1-p)^{m-x})^n \tag{3}$$

$$\text{SAUF} = \frac{n \times \text{yield}}{m \times P} \tag{4}$$

Here n is the number of rows in the desired rectangular array and m is the number of columns of the original array. Again we get the maximum SAUF value by varying m.

Comparisons. Figure 4 shows the comparisons of SAUF for the different schemes. Also plotted in the Figure is a theoretically optimum scheme which always ensures proper reconfiguration if enough good processors exist. (No such scheme has yet been developed). As we can see from the Figure, the PSA Scheme 2 is the most efficient design for 8 x 8 array. Note that the SAUF data of Row Major Redundancy scheme was obtained without the restriction on the number of tracks or interconnection lengths. With such restrictions its SAUF degrades significantly. The Interstitial Redundancy Scheme displays the shortest interconnection lengths in the restructured array.

4. THE BINARY TREE ARCHITECTURE

A Processor Switch Array (PSA)structure for the fault tolerant embedding of a four level tree is shown in Figure 5. This consists of a regular array of interconnected switches and processors. A 5 x 5 array containing 25 processors is used for embedding a four level tree which requires 15 failure free processors. The processor redundancy is therefore 67%. The switches in Figure 5 are interconnected in the *desired computational tree topology* (shown in bold lines).

Higher Level Fault Tolerant Tree Embedding

Larger trees with more than four levels can be embedded using this four level fault tolerant tree module in a hierarchical pattern. First, a five level tree can be obtained by just connecting together two basic four level fault tolerant modules. For larger tree size, a H-tree pattern connection using the fault tolerant five level tree modules

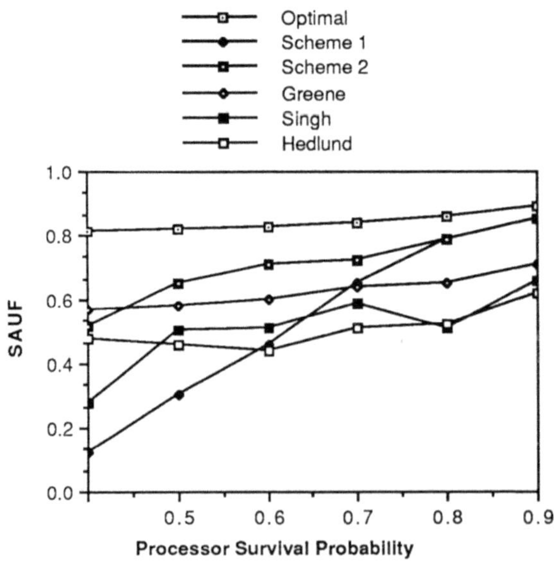

Figure 4: SAUF comparisions for reconfiguring an 8x8 array.

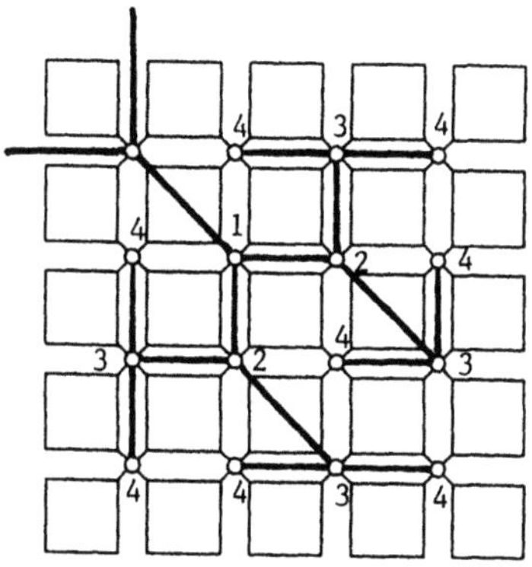

Figure 5: PSA structure for reconfiguring binary tree architecture.

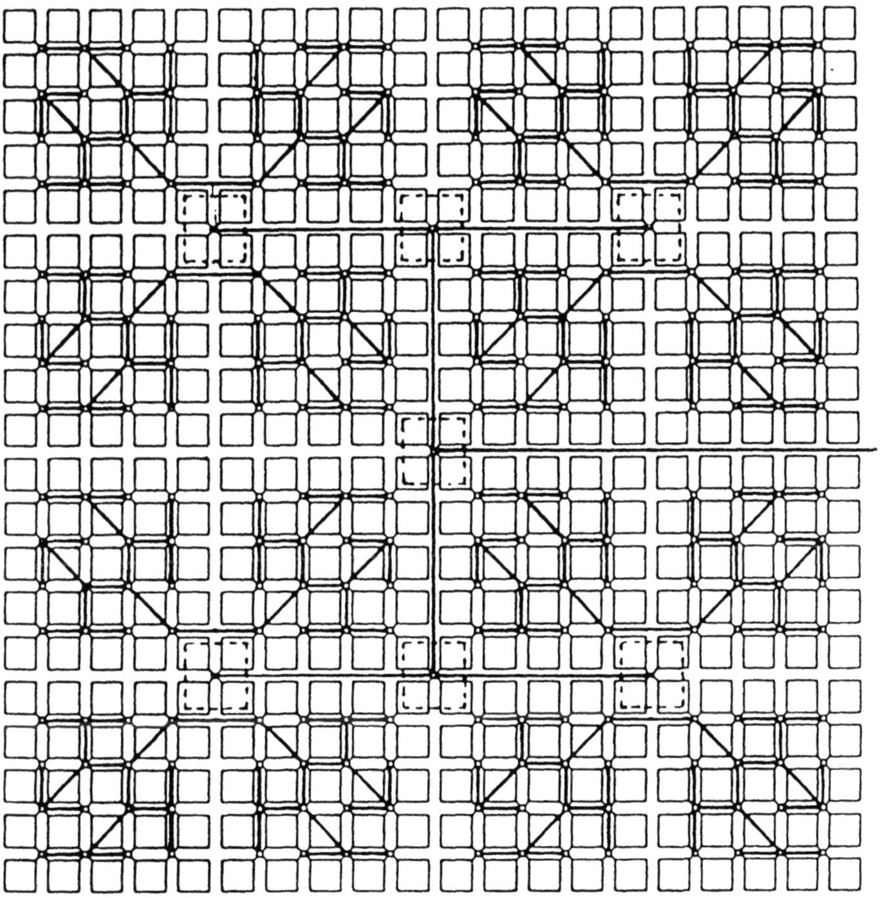

Figure 6: An eight level tree embedding using PSA scheme.

is used as shown in Figure 6 basic (four level) modules are used in the Figure to construct a eight level tree. Note that one of the four corner processors of four basic modules at the point where they come together is used as the node at a higher level in the tree than the four level contained in the basic module. All higher level trees can be embedded by the same way. Therefore, no extra chip area is needed between the basic modules for the accomodation of nodes at a higher level in the tree as was needed in the case of the failure free design.

Tree Yield

The yield data for the basic four level tree modules was obtained by computer simulations for processor yield (P) over the range 0.7 to 1.0 in steps of 0.05. Upper and lower bound yield estimates for larger trees were obtained from these estimates. For tree sizes of practical interest, up to 8 levels (255 processors), the yield is over 0.6 when P is greater than 0.75. However, the yield estimates (probability that the complete binary tree can be configured) obtained above do not provide a good measure of the efficiency of a fault tolerance scheme because they do not account for the extra area required by the redundancy employed. To better evaluate the area efficiency, we consider the silicon area utilization factor (SAUF) which was defined in the previous section.

Using the yield estimates obtained earlier, the SAUF values can be calculated easily. Figure 7 depicts the relation between SAUF and P for trees of level 4, 5 and 8. The SAUF is over 0.6 for trees of level 4 and 5 when P is greater than 0.7. Note that the upper bound and lower bound SAUF values (obtained from the upper and lower bound yield estimates) for tree level 8 are very close when P is over 0.8. Even for smaller P, there is generally less than 10% difference between the two bounds. As expected, the SAUF for trees of the smaller size is better than that for larger ones.

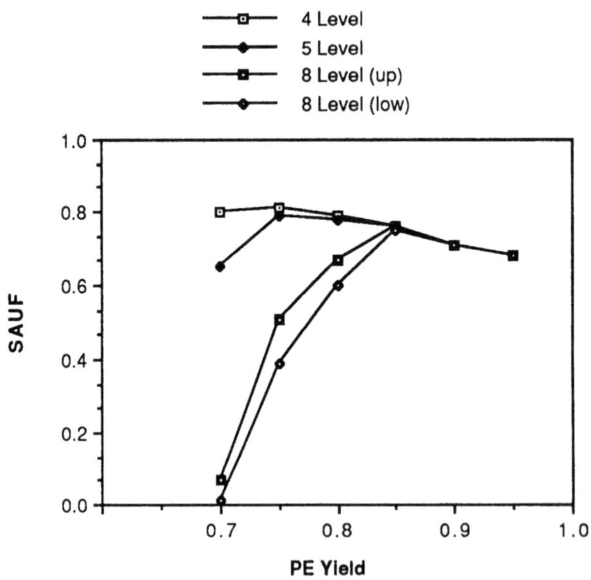

Figure 7: SAUF data for trees of level 4, 5 and 8.

5. HEXAGONAL ARRAYS

The PSA structure employed to obtain a 4 x 4 failure free hexagonal processor array is shown in Figure 8. The switches are interconnected in the desired hexagonal topology, while each processor (except processors on the boundary) is connected to three neighboring switches. Again a failure free hexagonal processor array can be obtained from the PSA structure by assigning a failure free processor to each switch. Note that a PSA with an n x n switch array contains $2n^2 + 4n$ processors. Thus the structure contains the $n^2 + 4n$ redundant processors. The reconfiguration algorithm is same as that for the other two topologies. Figure 9 shows that a row and column exclusion due to a clustered faults.

6. CONCLUSION

We have presented an approach for efficiently reconfiguring desired array architectures on a wafer in the presence of failures caused by defects. Our approach is very

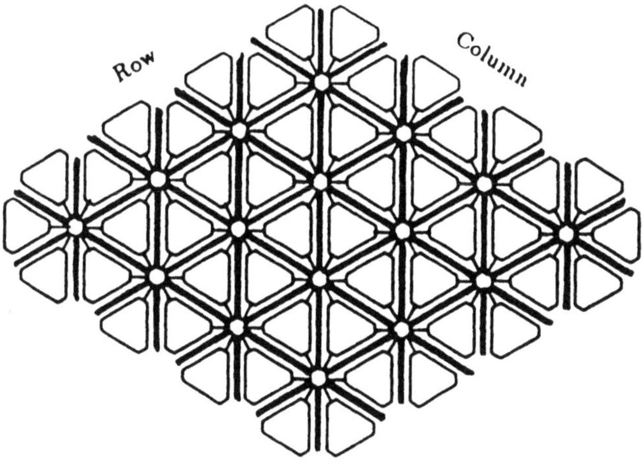

Figure 8: PSA scheme for hexagonal array.

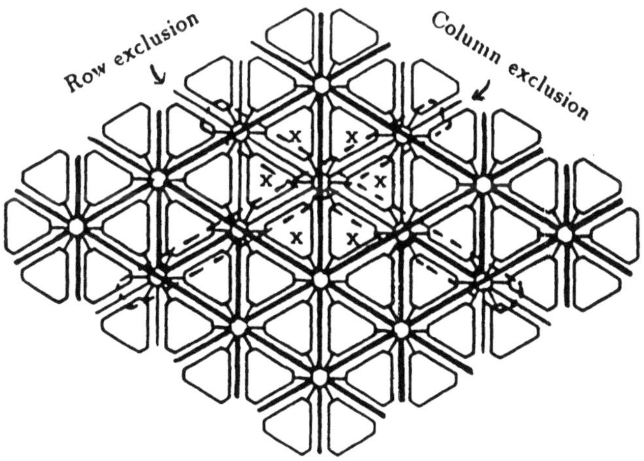

Figure 9: A row and column exclusion due to a clustered fault.

general and can be applied to many different computational topologies. Furthermore, our designs can take advantage of different restructuring technologies.

Several problems remain to be addressed. Most important is the efficient design of switch and interconnection structures to support the restructuring schemes presented in this paper. Complex switch structures connected to wide buses are clearly unacceptable because of the large area and yield penalties associated with them. Also, multistage switches can very substantially degrade performance. At this time our research indicates that a descretionary wiring approach, employing two or three tracks between the processors, is viable and attractive. We will report on it more completely in future publications.

REFERENCES

[1] J.F. McDonald, "The Trials of Wafer Scale Integration", *IEEE Spectrum*, Oct. 1984, pp. 32-39.

[2] C. Stapper, "On Yield, Fault Distributions, and Clustering of Particles", *IBM Journal of Res. and Devel.*, Vol. 30, No. 3, pp. 326-338, May 1986.

[3] D. Fussel and P. Varman, "Fault-Tolerant Wafer-Scale Architectures for VLSI", *Proc. of the 9th Annual Symp. on Comp. Arch.*, May 1982.

[4] K.S. Hedlund, "WASP - A Wafer Scale Systolic Processor", *Proceedings 1985 International Conference on Computer Design*, October 1985.

[5] J.W. Greene and A. El Gamal, "Configuration of VLSI Arrays in the Presence of Defects", *JACM*, Vol. 31, No. 4, October 1984, pp. 694-717.

[6] M.C. Howells and V.K. Agarwal, "Yield and Reliability Enhancement of Large Area Binary Tree Architectures", *Proceedings of the 15th Annual Symposium on Fault Tolerant Computing*, pp. 290-295, June 1987.

[7] T. Leighton and C.E. Leiserson, "Wafer-Scale Integration of Systolic Arrays", *IEEE Transactions on Computers*, Vol. C-34, pp. 448-461, 1985.

[8] A.D. Singh, "A Reconfigurable Modular Fault Tolerant Binary Tree Architecture", *Proceedings of the 15th Annual Symposium on Fault Tolerant Computing*, pp. 298-303, June 1987.

[9] A.D. Singh, "Interstital Redundancy: An Area Efficient Fault Tolerance Scheme for Large Area VLSI Processor Arrays", *IEEE Transactions on Computers*, Nov. 1988.

[10] C.H. Papadimitriou and K. Steiglitz, "Combinational Optimization: Algorithms and Complexity", Prentice Hall, NJ, 1982.

[11] C.H. Stapper, A.N. McLaren and M. Dreckman, "Yield Model for Productivity Optimization of VLSI Memory Chips with Redundancy and Partially Good Product", *IBM J. Res., Dev.*, Vol. 24, No. 3, pp. 398-409, May 1980.

[12] Mark B. Ketchen, "Point Defect Yield Model for Wafer Scale Integration", *IEEE Trans. on Computers*, p. 24, 1985.

[13] J.I. Raffel et al., "Wafer Scale Digital Integrator Using Restructable VLSI", *IEEE J. of Solid State Circuits*, Vol. SC-20, No. 1, Feb. 1985.

[14] D.C. Shaver, "Electron-Beam Customization, Repair and Testing of Wafer-Scale Circuits", *Solid State Technology*, Feb. 1984, pp. 135-139.

[15] A.D. Singh, "A Fault Tolerant Processor Switch Array Structure for Wafer Scale Integration", *Proceedings 22nd Annual Allerton Conference*, 1984, pp. 854-863.

[16] A.D. Singh, "An Area Efficient Redundancy Scheme for Wafer Scale Processor Arrays", *Proceedings International Conference on Computer Design*, October 1985, pp. 505-509.

ORTHOGONAL MAPPING: A RECONFIGURATION STRATEGY FOR FAULT TOLERANT VLSI/WSI 2-D ARRAYS[+]

Leonardo Jervis*, **Fabrizio Lombardi**, and **Donatella Sciuto**

***Politecnico di Milano**
Department of Electronics
Milano, Italy

Texas A&M University
Department of Computer Science
College Station, TX

1. INTRODUCTION

Fault tolerance techniques have been adopted historically in discrete-component computing systems as a consequence of particular applications which demanded high reliability and availability.

The requirements of high reliability and long lifetime of complex VLSI/WSI devices [1,2] also demand new techniques. So-called modular redundancy and fault avoidance approaches are unrealistic due to large overhead [3]: rather, it is necessary to use design techniques that allow to overcome faults through structural redundancy and reconfiguration techniques both at production-time and run-time. Homogeneous VLSI/WSI architectures such as arrays of processing cells, are best suited for this approach due to the flexibility in the interconnection network, adaptability to algorithmic reduction and silicon implementation [4].

In this paper, an algorithm for fault tolerance of VLSI/WSI arrays through reconfiguration is presented. Reconfiguration of VLSI/WSI arrays is constrained by many limitations. These are related both to the physical and logical aspects of array processing. These limitations can be identified as follows.

- Communication delays between cells due to reconfiguration should be minimized to maintain a synchronized flow of processing and avoid uneven delays. Either wait-states for unreconfigured fault-free cells, or complex switching circuits are required to avoid uneven delays [1,5]. In the first case cell utilization is low, in the second hardware overhead can be significant.
- Processing in a synchronous array shows significant dependency on the architectural/physical embedding of the algorithm [6]. In a synchronous array, a very stringent requirement is to achieve high throughput in fixed-size architectures;
- Data at inputs and outputs of the array along the directions of signal flow and the corresponding processing in the cells are closely related. This relationship should be preserved within the internal organization and dimensions of the reconfigured array [7].

A new approach to array reconfiguration is proposed to overcome the above limitations. This approach, referred as *orthogonal mapping*, exploits index mapping [8,3] and achieves

(+) This work was partially sponsored by NATO and AT&T.

reconfiguration of a two-dimensional array by minimizing the maximum and average interconnection delays between reconfigured cells in the reconfigured operational array. This reconfiguration approach has the following advantages.

- Uniformity in the average interconnection wire lengths between neighboring cells after reconfiguration. This allows to minimize uneven communication delays.
- High probability of survival in presence of large number of faults either randomly distributed or clustered.
- Possibility to define the maximum interconnection delay allowed in the reconfigured array.
- Reconfiguration by graceful degradation.
- To reconfigure arrays with different types of interconnection networks.

This paper is organized as follows. In section 2 the reconfiguration problem is defined. A set of benchmark criteria for evaluation of a reconfiguration algorithm is identified. The fault model is then discussed. Random fault distribution and cluster distribution are considered. The model and the technique are presented. The reconfiguration problem is formally introduced. Particular attention is given to the problem of communication delays among cells of the reconfigured array. The reconfiguration algorithm is described in detail in section 3. Finally, section 4 presents the characteristics of the algorithm implementation, the simulation results and the architectural overhead.

2. THE RECONFIGURATION PROBLEM

This paper deals with two-dimensional arrays made of equal processing cells connected by a regular interconnection network along rows and columns. This type of array is referred as an *orthogonal array*. The array is made of two cell banks: the *operational array*, constituted by n cells organized into M rows and N columns, and a set of s spare cells. Spare cells are kept inactive and are switched into an operative state when substituting a faulty processing cell. Spare cells are organized into R_s rows and C_s columns. Therefore, there are $s = R_s(N+C_s) + C_s(M+R_S) - R_sC_s$ spare cells in the array.

Fault tolerance through reconfiguration of VLSI/WSI arrays can be used to improve the probability of survival, and therefore to enhance the yield. The numerous and different approaches to reconfiguration can be distinguished into two classes: techniques that aim at overcoming production faults [9,1] and techniques that deal with run-time faults [2,10,11]. Both approaches, however, share many commonalities.

In both cases, off-line reconfiguration uses the outcome of testing to maximize the number of fault free cells that can be kept operational in the reconfigured array. During testing, the positions of faulty cells are logged into a map. This map is used to generate a reconfiguration-solution, whenever one exists. An ideal off-line reconfiguration solution should also satisfy the following benchmark criteria.

1. Execution time of the reconfiguration algorithm must be kept within reasonable bounds to meet application-time constraints (especially when dynamic host-driven reconfiguration is applicable).
2. The execution of the algorithm should be "aborted" at the earliest detection of impossibility of reconfiguration. Graceful degradation should be started (if foreseen as an alternative).
3. Adaptability of the algorithm to different interconnection configurations, as allowed by the nominal (original) array.
4. The number of fault-free cells in the reconfigured array must be maximized within a given execution-time of the algorithm.
5. Switching and routing circuits must be simple. This is required to preserve locality and structure regularity in the reconfigured array [10].
6. Reconfiguration algorithms must lead to reasonably low increase in area due to additional interconnections. In particular, global interconnections may lead to total system failure.

7. Reconfiguration must be implemented using interconnection links of bounded length between processing cells that become logical neighbors.
8. Area increase due to various sources of redundancy must be kept within bounds. Otherwise the apparent "gain" which could be attained by the addition of spare cells to the nominal array, will be offset by the actual loss in yield and reliability.

2.1. The Fault Model

The assumed fault model deals with both random and clustered faults to provide a unique framework for production and run time reconfiguration [16].

The assumed fault model is applicable to both VLSI and WSI devices: for VLSI a random fault occurrence is more appropriate as the cells are complex processors; for WSI it has been proved that clustering is the main source of yield degradation due to the smaller size of each cell [16].

2.2 Index mapping: a formalization

A technique referred as index mapping has been extensively advocated for the use of reconfiguration modeling, especially in the case of two-dimensional orthogonally interconnected arrays [8,3]. Each cell in the array is identified by two pairs of indices. Physical indices correspond to the physical position of the cell in the array. Logical indices are associated with the task performed by the given cell into the operating array. Cells with no task associated, have logical indices set to (0,0). If no faults at production time are present, then the logical and physical indices are equal. Each cell of the operating array ($n = MN$) has associated a row index i, (varying between 1 and M) and a column index j (ranging from 1 to N). Cells which belong to the spare rows R_s and to the spare columns C_s have logical indices set to (0,0), because no task is associated with them. Logical indices are associated to a logical matrix L, while physical indices are arranged into a physical matrix P. Both matrices have dimensions $(M+R_s)\times(N + C_S)$.

Reconfiguration consists of defining an algorithm that globally maps the array functions, given by the logical indices, onto fault free cells.

Primary requirements of the reconfiguration algorithm are:

- A correct logical matrix [8].
- All faulty cells must have null logical indices after reconfiguration is performed [3].
- Cells that are logical neighbors, must also be physical neighbors [3].

The first condition implies that for the MN operational array with R_s rows and C_s columns of spares, all logical pairs of indices (i,j) $(i = 1,M, j = 1,N)$ appear in the logical matrix. Each distinct index pair appears only once and s indices are null. This corresponds to a correct reconfiguration by definition [10]. A reconfiguration is correct if the generated logical matrix is correct and all faulty and unused spare cells are represented by null logical indices [10].

2.3. Cell distance and locality

The *distance* between two logically adjacent cells characterizes the concept of locality. This is pivotal in the proposed reconfiguration strategy. The array is analyzed using a graph approach. Every cell is represented by a vertex. Two physically adjacent cells are connected by a path of length 1. Therefore, the distance between two cells of the same physical column with physical indices (i_1,j_1) and (i_2,j_1) is given by $|i_1 - i_2|$. This applies also to cells which belongs to the same physical row. Let a cluster of faulty cells be present (i.e. a defective area of silicon). The distance between two cells which are located on opposite sides of the defective region must be computed differently. No path that passes through the cluster can be used. Therefore, the distance between the two cells is given by the path that follows the perimeter of the cluster. This implies that reconfiguration in presence of a cluster of faulty cells in the wafer requires a bypassing capability which is independent of the shape of the cluster [3].

Locality is a very important condition when considering reconfiguration algorithms for array structures [10]. It allows to reduce interconnection delays between logically (but not physically) adjacent cells and to avoid uneven delays.

This property can be formally defined. Let's consider two physically adjacent cells, say (i_1,j_1) and (i_1+1,j_1). Their logical indices (i'_1,j'_1) and (i'_1+1,j'_1), can be transferred by the reconfiguration algorithm to two cells such that their distance is limited by the requirements set in the array. This is a parameter of the reconfiguration algorithm presented in this paper. The concept of locality is related to the correctness of the logical matrix. Let (i_1,j_1), (i_2,j_2), (i_3,j_3) be the physical indices of three cells whose logical indices are respectively (i',j'), $(i',j'-1)$ and $(i-1,j')$. The distances between (i',j') and its logically adjacent neighbors (i.e. a cell on the same row and the other cell along the same column), are:

$$D^r(i',j') = |i_1-i_2| + |j_1-j_2| - 1 \qquad (01)$$

and

$$D^c(i',j') = |i_1-i_3|+ + |j_1-j_3|-1 \qquad (02)$$

The maximum distances are defined by a correct logical matrix L [8] which has been obtained by the correct execution of a reconfiguration algorithm. These are given by

$$D'_{\max} = \max\{D^r(i',j')\} \qquad (03)$$

and

$$D^{c_{\max}} = \max\{D^c(i',j')\} \qquad (04)$$

The maximum distances computed by the reconfiguration algorithm, represent the maximum physical lengths of the horizontal and vertical connections between cells. Hence, it is possible to state that a reconfiguration algorithm satisfies the locality condition if and only if the generated distances are within the bound, independently of the dimensions of the array.

The objective of the proposed reconfiguration algorithm is the minimization of the maximum distances $D^r_{\max}$ and $D^c_{\max}$ together with the minimization of the average distances along the rows D^r and the columns D^c (denoted by D^r_{av} and D^c_{av}, respectively). However, it is also desirable to allow the algorithm to choose the maximum distances (L_r,L_c) admissible during reconfiguration. This implies that reconfiguration of the array is performed under such constraint.

Therefore, the reconfiguration goals consists of the mapping of the physical matrix P onto a correct logical matrix L (as defined above) given the following constraints.

$$\min\{D^r_{\max};D^c_{\max}\} \text{ and } \min\{D^r_{av};D^c_{av}\} \qquad (05)$$

or given L_r and L_c such that

$$D^r_{\max} \le L_r \text{ and/or } D^c_{\max} \le L_c \qquad (06)$$

and

$$\min\{D^r_{av};D^c_{av}\} \qquad (07)$$

3. THE RECONFIGURATION ALGORITHM

The reconfiguration approach deals with a square array with one spare row and one spare column, i.e. $R_s = 1$ and $C_s = 1$ and an operational array of $n = N \times N$ cells. Hence, the physical matrix P and the logical matrix L will have dimensions $(N+1)\times(N+1)$. These assumptions will not restrict the application of the algorithm to higher redundancies, but it will simplify the description of the reconfiguration strategy without any loss of generality.

The following assumptions are made in the design of the reconfiguration algorithm.

1. Faults are associated with cells. Interconnections are assumed to be fault-free in case of a random fault distribution. This is a widely used assumption [5], because interconnection links require fewer fabrication steps than the more complex circuitry that composes the processing cells.

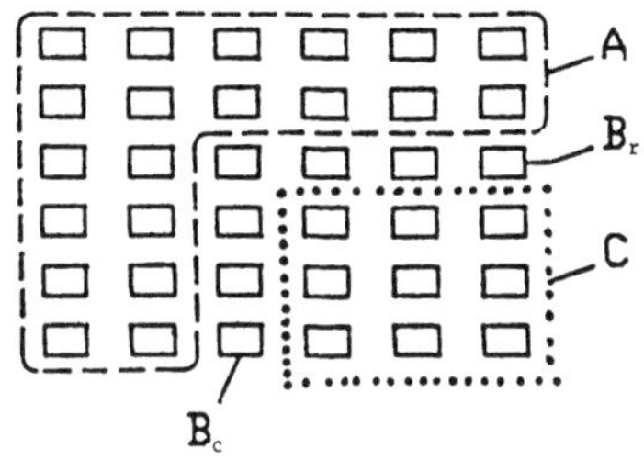

A - PREVIOUSLY RECONFIGURED ROWS AND COLUMNS

B - ROW AND COLUMN UNDER RECONFIGURATION

C - ELEMENTS STILL UNKNOWNS

Figure 1.

2. Reconfiguration-controlling and switching circuits are assumed to be either self-testing, or self-checking [3]. These circuits are not considered to be hard-core.
3. The reconfiguration algorithm is executed by a host computer that has full control over the array of processing cells.
4. Faults can occur either randomly (i.e. either independence of fault occurrence is upheld), or they can be clustered as for production defects.
5. Only permanent faults are considered. Transient faulty cells can be masked using appropriate techniques.
6. Processing proceeds synchronously in a strict lock-step fashion [1].
7. The number of simultaneous faulty cells at real-time is small. However, they can accumulate over a period of time. Ultimately, a growing number of faulty cells can be present in the array.
8. The array has already been tested prior to execution of the reconfiguration algorithm and fault sites have been stored into a fault map. This map is stored in the memory of the host-computer that executes the reconfiguration algorithm.

The proposed algorithm adopts a strategy of successive reductions of the overall reconfiguration problem. This is achieved by considering simultaneously, or alternatively (in case of different interconnection network) the reconfiguration of the lowest indexed row and column of the physical matrix. This technique is referred as *orthogonal mapping* due to the orthogonal nature of the set of indices which is modified at the same time in the construction of the logical matrix L. Minimization of distances between logically adjacent cells is also a primary requirement for both directions of reconfiguration. Orthogonal mapping considers all options for interconnecting cells along the row and column under reconfiguration. The distance set D (made of D^r and D^c for each distinct cell) is evaluated for each of these cases. The case with the minimum interconnection distance is chosen for reconfiguring either the row, or the column. The orthogonal process is executed iteratively $(N-1)$ times. A smaller orthogonal mapping is executed every time, until the entire logical matrix is fully generated. This is shown in Figure 1.

The conditions by which an orthogonal mapping is generated to fit the previous one, can be defined. The total interconnection distance (denoted as D) is computed by considering the distances between all logically adjacent cells along the prescribed direction and the distance of these cells with respect to the reconfigured cells. These latter cells belong to the previously reconfigured row or column.

Assume that testing of the array has been performed and that a fault map has been stored in the memory of the host computer. Hereafter, the steps in which the algorithm can be logically subdivided are given.

Algorithm 1: Orthogonal Mapping

Step 1: Create the physical matrix P.

Step 2: If (number of faulty cells > number of spares s)
then array unreconfigurable,
end.

Step 3: for (i=1,N; j=1,N) do

Step 4: Define the actual array to be reconfigured ($I = r,N$; $J = c,N$);

Step 5: Consider row r (column c);

Step 6: Compute in parallel all rows (columns) which are candidates to reconfigure row r (column c);
for each alternative compute the total interconnection distances D;
Choose as row r (column c) the row (column) with minimum total interconnection distance;
Map the chosen values into matrix L assigning correctly the entries.

Step 7: $r = i+1$; $c = j+1$;
goto step 3;

Step 8: end.

The first five steps are self explanatory and therefore they will not discussed any further. Step 6 however requires a more detailed explanation. The array under reconfiguration can be modeled as a graph. A weight is associated to each arc. This weight represents the distance between the cells it connects (as defined in the previous section). Let V denote the vertex set of all (fault-free and faulty) cells. This set includes also the I/O cells which are considered as fixed vertices. Let V' be the set of cells which have not been reconfigured yet.

The following strategy finds all cells which can be thought of as candidates for the construction of a given row r (column c). A theoretical reconfigured row r (column c) is built by extracting one cell at a time from the set V'. This process is repeated until the number of cells required to compose a row in the reconfigured array has been met, i.e. all N entries in row r (column c) of the logical matrix L are set to a value different from zero. This is in place of the $r-1$ ($c-1$) already defined in the previous iterations. This row (column) is referred as a *theoretical line T*. Construction of a theoretical line T characterizes also the choice of each cell from the set V' which must be inserted into the theoretical line. These conditions must satisfy the constraint of adjacency with the lines reconfigured during the previous iteration. A theoretical line T is defined to be adjacent to the previously reconfigured line T' if the following formula is satisfied:

$$\min\left\{\sum_{j=i}^{N+C_s} D^c\{t(i,j),t'(i-1,jj)\} + \sum_{j=1}^{N+C_s} D^r\{t(i,j),t(i,j+1)\}\right\} \qquad \textbf{(08)}$$

where the line T has been considered here as a row of index i made by $t(i,j)$ a cell of the theoretical row i and column j and by $t'(i-1,j)$ a cell of the previously reconfigured row $i-1$ and column j.

The theoretical line allows to specify all possible candidates of a reconfigured line. This is achieved by defining a reconfiguration *function*. The function of a theoretical line can be distinguished according to the following three cases.

1. The theoretical line has a number of faulty-free cells equal to the number of cells which is necessary to complete row r (or column c) in the reconfigured array. All cells (i,j)

for $1 \leq i \leq r-1$ and $1 \leq j \leq c-1$ have already been reconfigured in the previous iterations. Only the logical cell with the current indices must be added.

2. The theoretical line has a number of fault-free cells greater than the number of cells which are necessary to complete row r (column c).

3. The theoretical line is composed of fault-free cells whose number is less than the number of operating cells required to construct the entire row r (column c).

If the theoretical line satisfies the first condition, the reconfiguration function returns it as the new reconfigured line (row or column). If cases 2 or 3 are satisfied instead, the evaluation of the reconfiguration line is iterated on a new line. This new line is obtained differently, depending on which case has to be considered. If case 2 applies, a cell is deleted from the theoretical line, because there are a number of exceeding fault-free cells. In general, the deleted cell is the one which contributes to a maximum interconnection length between the logically adjacent cell in row $(r-1)$ and column $(j-2)$. This cell often corresponds to the cell which lies on the boundary, i.e. a spare cell of the physical array. Thus, the cell eliminated will be the one with the highest D^r_{max} and/or D^c_{max}. This new theoretical line must be evaluated by the reconfiguration function. If case 3 applies, a new cell set must be considered. Consider a faulty cell on the theoretical line T. A new theoretical line has to be built by excluding this faulty cell from the set V'. These operations are performed for each faulty cell included in the theoretical line. A new theoretical line is built and must also be evaluated at each iteration.

After a few iterations of this process, the set of theoretical lines can be represented as a nonbinary tree (assuming that relabeling of the nodes is performed). Each path of this tree constitutes a possible reconfigured line. All paths have length either equal to $(N-c)$ in the case of a row, or to $(N-r)$ in case of a column.

Selection of a reconfigured line is accomplished by considering each path from the root of the tree (given by cell (r,c)), to the leaves and its total interconnection length. This length is computed as a function of the following parameters in the construction of the theoretical line.

a) The sum of the interconnection lengths between every pair of cells, which belong to the theoretical line, must be minimized; this corresponds to the sum of the weights of the edges connecting the vertices extracted from V'.

b) The sum of the interconnection lengths of the cells along the theoretical line T and the logically adjacent cells on the line reconfigured during the previous iteration of the reconfiguration algorithm, must also be minimized.

These two parameters are computed during the construction of either each path, or possible theoretical line. The length is computed as a weighted sum of the two terms above. The weights associated with these two terms can be varied such that the type of reconfiguration can also be modified. This procedure alters the figure of merit by which the choice of the path of the tree is performed (therefore, the choice of the actual reconfigured line).

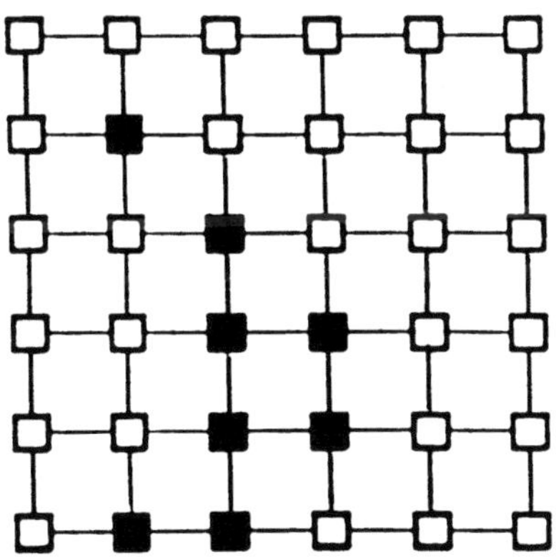

Figure 2: Example; dark cells are faulty.

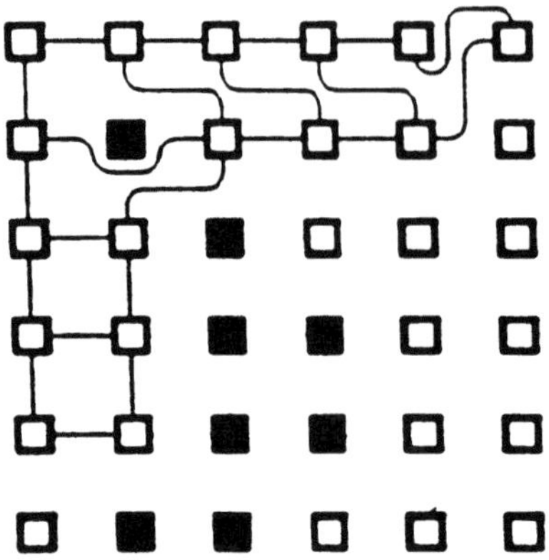

Figure 3: Example; scenario before column 3 is reconfigured.

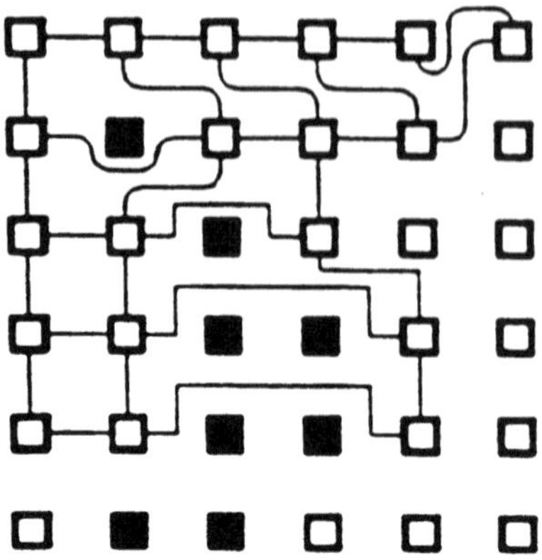

Figure 4: Example; scenario after column 3 is reconfigured.

In the simulated algorithm, the weight associated with the second term has been set to a greater value than the weight associated with the first term. Experimental results have shown that a higher weight for the second term allows to achieve a greater locality among the cells of the set V'. If the number of faulty cells is very close to its upper bound, (i.e. almost equal to the number of spare cells), a theoretical line cannot be found in certain cases. A fatal failure is said to occur. A smaller array can still be obtained using the previously reconfigured rows and columns. This is either a $(r-1)\times(r-1)$ square array, or a $I\times J$ rectangular array (with a value I ranging from 1 to $(r-1)$ and J from 1 to N). As an example consider the fault pattern shown in Figure 2. Figures 3 and 4 show the scenarios before and after the reconfiguration of the third column; Figure 5 shows the reconfigured array.

The complexity of the proposed reconfiguration algorithm can be computed. Orthogonal mapping involves each cell in the array only once along the two required theoretical lines. This is executed in $O(N^2)$ time by the nature of orthogonal mapping and algorithm (1). Hence, in the general case if the target array is a $N\times M$ rectangular array, the execution complexity of the algorithm is $O(N\times M)$.

4. IMPLEMENTATION AND SIMULATION RESULTS

The proposed reconfiguration algorithm has been extensively simulated. All possible candidates for a reconfigured line are generated by means of a recursive function. The recursive

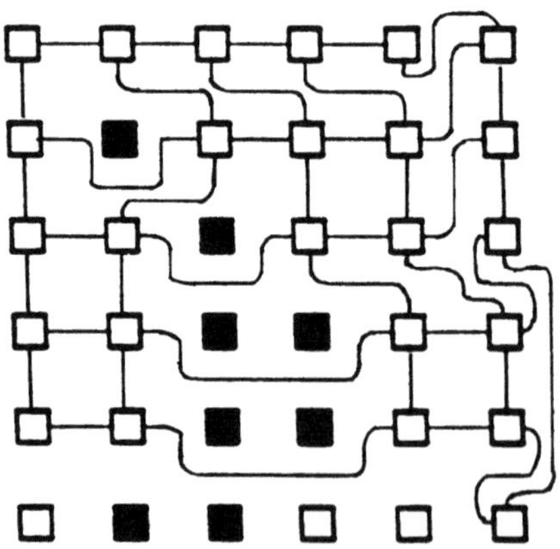

Figure 5: Example; reconfigured array.

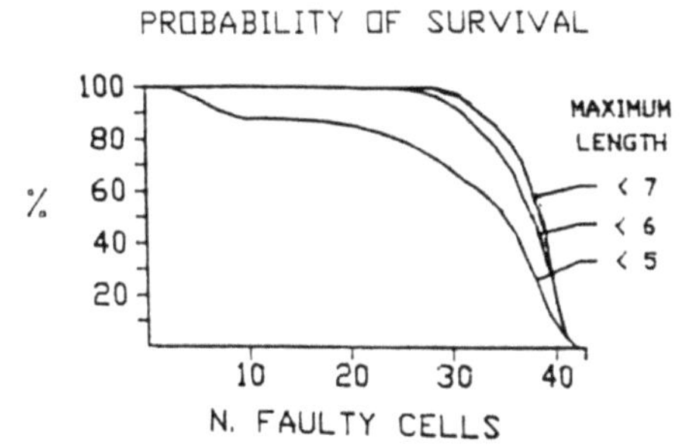

Figure 6.

function generates the non-binary tree corresponding to all possible alternatives of a theoretical line. Simulation of the construction and evaluation of the tree are performed in parallel by different processes. A square array composed of 21×21 cells has been simulated. The target is given by a 20×20 array. Hence, the redundancy is given by one spare row and one spare column. The maximum number of faulty cells is given by 41. Simulation has been performed with respect to an upper bound in the maximum interconnection lengths allowed between two logically adjacent cells, i.e. $D^r_{max} \leq L_r$ and $D^c_{max} \leq L_c$. Figure 6 shows the number of faulty cells that can be overcome with a given maximum length and probability. A good uniformity in the average interconnection delays between logically adjacent cells has been achieved. Simulation results show that in a 20×20 target array, the average delays range from 1 to 1.4 as the number of faults increases from 0 to 41. A comparison with other known reconfiguration algorithms is shown in Table 1; only randomly distributed faults are considered.

When reconfiguration of a rectangular array is considered, an additional algorithm must be executed with orthogonal mapping. This is due to the fact that in this case, it is not possible to alternate one row and one column regularly at each iteration. If for example the number of rows is twice the number of columns, the sequence of iterations should follow the sequence row, row, column to maintain a symmetry in the reconfiguration. This symmetry allows to achieve better results of the reconfiguration strategy. An algorithm can be developed to determine, in case of $M = N$ if at the ith iteration, a row or a column must be considered.

Consider now the reconfiguration of hexagonal arrays. A few modifications are necessary in the orthogonal mapping algorithm. In particular, it is necessary to evaluate the other connection between adjacent lines and cells. Hence, it is necessary to modify the function that chooses the theoretical line. This is required to take into account the distance between the element of logical indices (i,j) and $(i-1,j-1)$ (the immediately previous element on the diagonal). Note that in case of an hexagonally connected array the maximum length of the other connection from a cell whose logical indices in the matrix L are (i,j), to its logical neighbor $(i-1,j-1)$

Table 1

Reconfiguration Algorithm	Probability of survival 90%	75%	50%
Orthogonal Mapping	34	36	38
Complex Fault Stealing [3]	31.5	34.4	36.8
MORA [10]	32	34.5	37
Spanning Tree Based Rec. [14]	31.9	36.5	38

is fixed as the square root of the sum of the square of the two bounds (given in the vertical and horizontal directions L_c and L_r). The implementation for octagonal arrays can be deduced in a straightforward manner from the case of the hexagonal arrays. The requirements for the implementation of an array which allows this type of reconfiguration strategy are very straightforward. The CHiP architecture [4] consists of a two-dimensional matrix of identical processing cells, each pair of rows and columns being separated by a wiring channel. Connection tracks among processing cells run in a multiple bus organization.

For WSI arrays, laser-programmable connections at the intersection of a horizontal and vertical channel, can make a horizontal and vertical wire electrically equivalent. Connections between two cells can be made from the wires in the channel to the inputs and outputs of the two cells. This allows laser cutting configuration of the array at production-time [15,9]. Simulation results have proven that 3 vertical and 3 horizontal data tracks are sufficient to allow reconfiguration. An alternative solution, more suitable for host-driven dynamic reconfiguration during the operational life of the array, is the use of switched busses. The implications of switched bus structures have been extensively analyzed in [3]. It has been proven that they require a low silicon area overhead because they allow bus multiplexing and shared use of busses both for horizontal and for vertical tracks.

5. REFERENCES

[1] Leighton, T., and C.E. Leiseron, "Wafer-Scale Integration of Systolic Arrays," *IEEE Trans. on Computers,* Vol. C34, No. 5, pp. 448-461,1985.

[2] Rosenberg, A.L., "The Diogenes Approach to Testable Fault Tolerant VLSI Processor Arrays," *IEEE Trans. on Computers,* Vol. C-32, No. 10, pp. 902-910, 1983.

[3] Negrini, R., M.G. Sami and R. Stefanelli, "Fault Tolerance Technique for Array Structures used in Supercomputing," *IEEE Computer Magazine,* Vol. 19, No. 2, pp. 78-87, 1986.

[4] Hedlund, K. and L. Snyder, "Systolic Architecture - A Wafer-Scale Approach," *Proc. ICCD,* pp. 604-610, 1984.

[5] Greene, J.W. and A. El Gamal, "Configuration of VLSI Arrays in the Presence of Defects," *J. ACM,* Vol. 31, No. 4, pp. 694-717, 1984.

[6] Snyder, L., "Introduction to the Configurable Highly Parallel Computer," *IEEE Computer Magazine,* Vol.15, No. 1, pp. 47-56, 1982.

[7] Fortes, J.A.B. and C.S. Raghavendra, "Gracefully Degradable Processor Arrays," *IEEE Trans. on Computer,* Vol. C-34, No. 11, pp. 1033-1044, 1985.

[8] Lombardi, F., M.G., Sami and R. Stefanelli, "Reconfiguration of VLSI Arrays: An Index Mapping Approach," *Proc. IEEE Compeuro,* pp. 60-65, 1987.

[9] Aubusson, R.C. and I. Catt, "Wafer-Scale Integration: A Fault Tolerant Procedure," *IEEE J. Solid-State Circuits,* Vol. SC3, No. 3, pp. , June 1978.

[10] Lombardi, F., R. Negrini, M.G. Sami and R. Stefanelli, "Reconfiguration of VLSI Arrays: A Covering Approach," *Proc. FTCS17,* pp. 251-256, 1987.

[11] Evans, R.A., "A Self-organizing Fault-Tolerant 2-D Array," *Proc. VLSI,* Tokyo, Japan, 1985.

[12] Chandramouli, R. and M. Sucar, "Defect Analysis and Fault Modeling in MOS Technology," *Proc. Int. Test Conf.,* pp. 313-321, 1985.

[13] Stapper, C.H., "On Yield, Fault Distributions, and Clustering of Particles," *J. Res. Develop.,* Vol. 30, No. 3, 1986.

[14] Lombardi, F and D. Sciuto, "Reconfiguration in WSI Arrays Using Minimum Spanning Trees," *Proc. IEEE Compeuro,* pp. 547-550, 1987.

[15] Mangir, T.E., "Sources of Failure and Yield Improvement for VLSI and Restructurable Interconnect for R-VLSI and WSI: Part II", *Proc. IEEE,* No. 12, pp. 1687-1694, 1984.

[16] Howell, M.C. and V.K. Agarwal, "Yield and Reliability Enhancement of Large Area Binary Tree Architectures," *Proc. FTCS 17,* pp. 290-295, 1987.

A GENERAL MODEL FOR FAULT COVERING PROBLEMS IN RECONFIGURABLE ARRAYS

N. Hasan, J. Cong, and C.L. Liu

University of Illinois at Urbana-Champaign
Department of Computer Science
Urbana, IL 61801

Abstract. An increase in chip density leads to a reduction in the yield of chip production. [Schu78] showed that in most cases, only a small number of elements in defective chips are actually defective. Thus the idea of repairing a chip after fabrication becomes very appealing. Reconfigurable chips are often used for this purpose. These chips contain redundant elements that can be used to repair the defective elements. There are many different ways to reconfigure a chip using redundant elements. The *fault covering problem* is to assign redundant elements to replace the defective elements such that the chip will function properly. In this paper we introduce a general model to represent the relationships between redundant elements and defective elements in a uniform way. This model generalizes the models discussed in previous approaches. We also give a complete characterization of the complexity of the fault covering problems for all the subcases of our model, most of which have not been studied before.

1. Introduction

Reconfigurable chips are chips in which redundant elements are available for replacement of defective elements resulted from imperfection in the fabrication process. A *covering* is an assignment of redundant elements to replace the defective elements such that the chip will function properly. Configuration that has been studied extensively ([Evan81], [HaLi88], [KuFu87], [TaBo84], [WeLo87]) consists of rectangular arrays in which redundant elements are arranged in the form of spare rows and spare columns. Both heuristic and exhaustive search algorithms for finding a covering of a rectangular array using no more than a given number of spare rows and a given number of spare columns were proposed. In fact, the problem of determining the existence of such a covering was proved to be NP-hard in [KuFu87]. [Sing88] studied configurations in which redundant elements are placed between elements in an array. However, in general, redundant elements can be placed anywhere in a chip. Also, replacement of defective elements might not necessarily be restricted to one-to-one replacement. See Figure 1, for

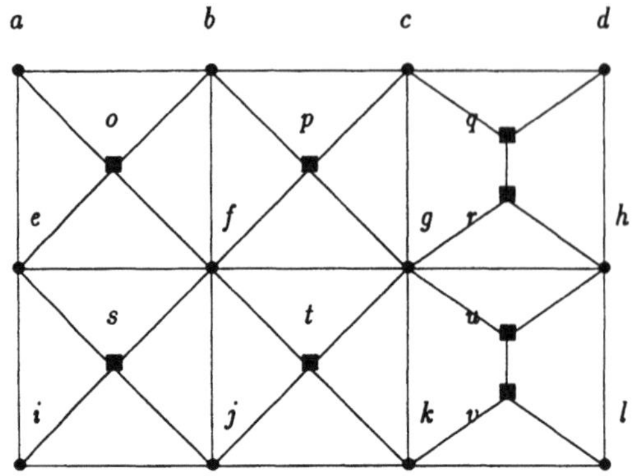

Figure 1

example. The •'s represent the regular elements of the chip and the ■'s represent the redundant elements. Redundant element o can be used to replace any one of elements a, b, e, f. Element f can be replaced by any one of the redundant elements o, p, s, t. We can also have the conditions that both s and t must be used to replace element j, and that both q and r must be used to replace either both c and g or both d and h.

In this paper, we present a general model to express many possible relationships between the redundant elements and the defective elements. For some subcases, we give simple linear time algorithms to solve the corresponding fault covering problems. For another subcase, we reduce the corresponding fault covering problem to a bipartite matching problem, which can be solved efficiently in $O(m\sqrt{n})$ time using an algorithm presented in [HoKa73], where m is the number of edges and n is the number of vertices in the graph. Furthermore, we show that the fault covering problems for the remaining subcases are NP-hard.

2. General Formulation of the Fault Covering Problem

The relationship between defective elements and redundant elements can be represented by a generalized bipartite graph which is a bipartite graph together with an assignment of Boolean expressions to the vertices in the bipartite graph. Let $G = \langle X \cup Y, E, W \rangle$ denote a generalized bipartite graph in which X and Y are two disjoint subsets of vertices, E is the set of edges connecting the vertices in X and the vertices in Y, and W is a function which maps each vertex in G to a Boolean expression. A vertex in X represents a single or a subset of defective elements, and a vertex in Y represents a single or a subset of redundant elements in the chip.

For each vertex v in X, $W(v)$ is a Boolean expression in which the variables correspond to the vertices in Y. For each vertex v in Y, $W(v)$ is a Boolean expression in which the variables correspond to the vertices in X. The Boolean operations in these Boolean expressions are restricted to *AND*'s and *OR*'s. In general, $W(v)$ is written in the form of a *sum-of-products*. For example, $W(v) = abc + cd$. The Boolean expression $W(v)$ captures the situation that a defective element can be replaced by one of several redundant elements, or that a defective element can only be replaced by several redundant elements jointly, or that a redundant element can be used to replace a number of defective elements at the same time. For example, let v be a vertex in X and

$W(v) = ab+cd$. It means that vertex v can be replaced either by both vertices a and b in Y, or by both vertices c and d in Y. If only vertices a and c are used to replace v then the values of a and c in $W(v)$ are true, and the values of b and d are false. Therefore, the truth value of $W(v)$ is equal to $ab + cd = \mathit{true} \cdot \mathit{false} + \mathit{true} \cdot \mathit{false} = \mathit{false}$ which means that vertex v can not be replaced. On the other hand, for the vertex a in Y, $W(a) = pq+s$ means that a can be used to replace either one or both of p and q or to replace s. If vertex a in Y is used in the replacements of vertex p, then the value of p in $W(a)$ is true and the values of q and s in $W(a)$ are false.

If x appears in $W(y)$ but y does not appear in $W(x)$ then we can delete x from $W(y)$ without changing the problem. This is so because although y is available to replace x, but x can not be replaced by y. If, on the other hand, y appears in $W(x)$ but x does not appear in $W(y)$ then we can also delete y from $W(x)$. This is so because although x can be replaced by y, but y is not available to be used to replace x. After all these deletion, we have *consistent* Boolean expressions, i.e. x appears in $W(y)$ iff y appears in $W(x)$. There is an edge between vertex x in X and vertex y in Y in G if and only if x appears in $W(y)$ and y appears in $W(x)$, which means that vertex y can be used in the replacement of vertex x. Notice that there is no "negation" in the Boolean expressions since negation corresponds to "vertex x cannot be used to replace vertex y" or "vertex u cannot be replaced by vertex v". These conditions have already been captured by not having any edge between vertex x and y and between vertex u and v.

Let $C = \{(x_1,y_1),(x_2,y_2), \cdots ,(x_m,y_m) \mid x_i \in X \, , \, y_i \in Y \, , \, 1 \leq i \leq m \}$ be a set of edges. C is said to be a *covering* if by setting the value of x_i in $W(y_i)$ and the value of y_i in $W(x_i)$ to *true* for $(x_i,y_i) \in C$, and the value of the rest of the variables to *false*, we have

1. for each vertex x in X, the value of $W(x)$ is *true*,
2. for each vertex y in Y, at most one product term in $W(y)$ assumes the value *true* and the values of all variables belong to the other product terms in $W(y)$ are *false*.

Intuitively, a covering C means that redundant element y_i is used to replace x_i, for $1 \leq i \leq m$, such that all defective elements are replaced ($W(x)$ is *true* for all x in X) and each redundant element is used to replace only one subset of defective elements (at most one term in $W(y)$ is allowed to be *true* and the variables belonging to other terms must be *false*).

The *fault covering problem* for a reconfigurable chip is to find a covering in the corresponding generalized bipartite graph.

We now show how a large variety of fault covering problems can be formulated in terms of our model. A classical problem that has been studied extensively is that of finding a covering of defective elements in a rectangular array (e.g. RAMs). Redundant elements are included in the form of spare rows and spare columns, and each spare row can replace a row of the array and each spare column can replace a column of the array. Figure 2a shows a rectangular array consisting of 4 rows and 5 columns of elements together with 2 spare rows and 3 spare columns. Each □ represents a defective element in the array. Figure 2b shows a representation of this array using our model. Each vertex in X represents a defective element. The two spare rows are represented by vertices r_1 and r_2 in Y and the three spare columns are represented by vertices c_1, c_2, and c_3 in Y.

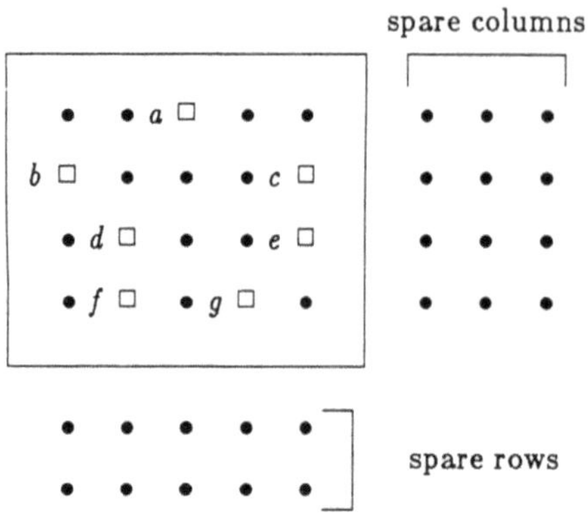

Figure 2a

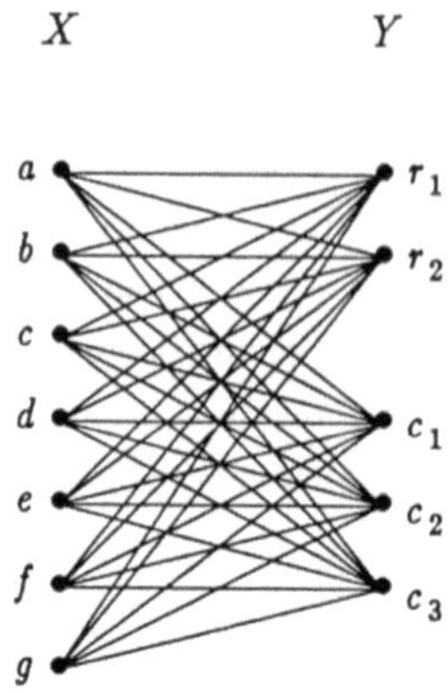

Figure 2b

$W(a) = W(b) = W(c) = W(d) = W(e) = W(f) = W(g) = r_1 + r_2 + c_1 + c_2 + c_3$, because each defective element can be replaced by any one of the spare rows or columns. $W(r_1) = W(r_2) = a + bc + de + fg$, because each spare row can be used to replace any one of the rows of the array. $W(c_1) = W(c_2) = W(c_3) = b + df + a + g + ce$, because each spare column can be used to replace any one of the columns of the array. It is evident that the fault covering problem formulated in our model is equivalent to the original problem.

An extension of this classical example is a case in which we have many arrays sharing spare rows and spare columns as shown in Figure 3 ([HaLi88]). Spare rows can be used to replace rows in the two arrays adjacent to them, and spare columns can be used to replace columns in the two arrays adjacent to them. Such a case can also be represented by our model. For details, see [HaCL88a].

In [Sing88], a number of configurations in which the defective elements are placed between regular elements were studied. Again, such configurations can be represented by our generalized bipartite graphs. Figure 4a shows an example of one of the configurations studied in [Sing88] where the •'s represent the regular elements of the array and the ■'s represent the redundant elements. In this configuration, a redundant element can be used to replace any one of the adjacent defective elements. For example, redundant element s can be used to replace any one of elements a, b, d, and e. Also, each defective element can be replaced by any one of the adjacent redundant elements. For example, element b

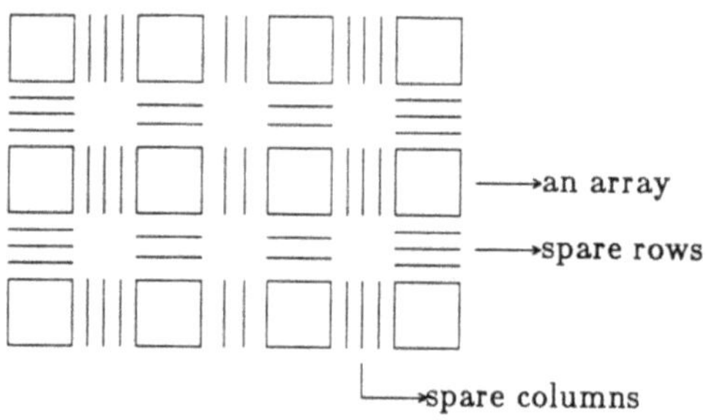

Figure 3

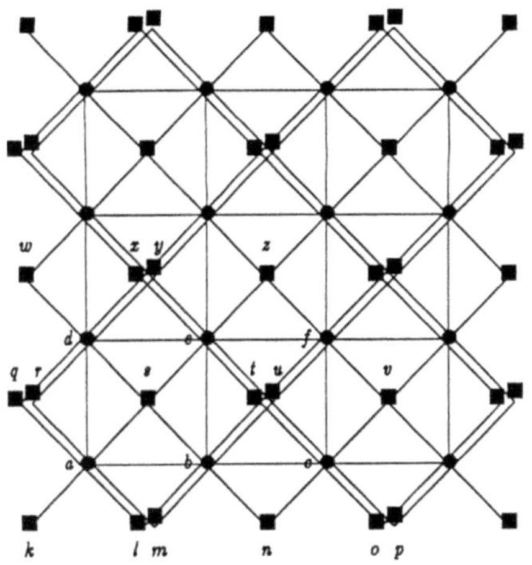

Figure 4a

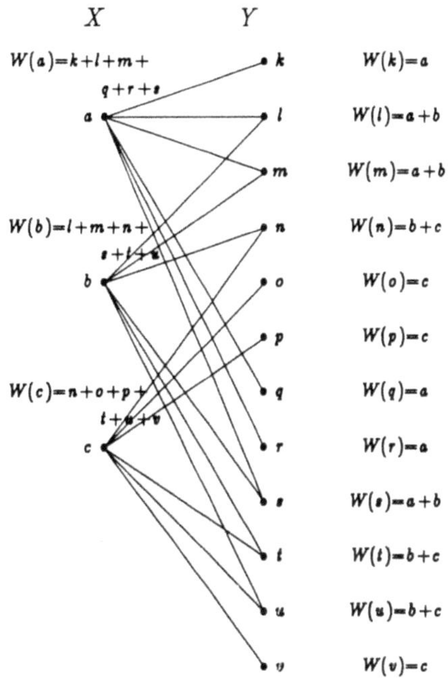

Figure 4b

can be replaced by any one of the redundant elements l, m, n, s, t, and u. These can be represented by a generalized bipartite graph with the Boolean expression $W(s)$ being $a + b + d + e$ and the Boolean expression $W(b)$ being $l + m + n + s + t + u$. Figure 4b shows the generalized bipartite graph assuming that elements a, b, and c are defective.

3. Complete Characterization of Subcases

We obtain the following results for the formulation presented in Section 2. Depending on the form of the Boolean expression associated with each vertex, we define nine different subcases. Table 1 shows the complexity results for the fault covering problems for the subcases. The OR–AND subcase in the table refers to the situation in which the Boolean expression associated with each vertex in X is a *sum* of individual variable in Y, and the Boolean expression accociated with each vertex in Y is a *product* of individual variable in X. The subcases OR–OR, AND–OR, and AND–AND are defined in a similar way.

We now discuss polynomial time algorithms and NP–hardness results for the fault covering problems that belong to the subcases described in Table 1.

3.1. Polynomial Time Algorithms

When we say that the complexity of a subcase is polynomial, we mean that every problem in that subcase is polynomial–time solvable. First of all, problems that belong to subcases 2, 3, 4, 7, and 8 can be solved easily in linear time in term of the number of vertices and edges in the graph G. For problems that belong to subcases 2, 4, and 7 we can use all the vertices in Y to replace the vertices in X. If all of the vertices in Y can cover all the vertices in X, then the chip can be repaired, and it is unrepairable otherwise.

Table 1

OR - OR Subcase 1	polynomial time
OR - AND Subcase 2	linear time
AND - OR Subcase 3	linear time
AND - AND Subcase 4	linear time
Sum of Products - OR Subcase 5	NP–hard
OR - Sum of Products Subcase 6	NP–hard
Sum of Products - AND Subcase 7	linear time
AND - Sum of Products Subcase 8	linear time
Sum of Products - Sum of Products Subcase 9	NP–hard

For problems that belong to subcases 3, and 8 we need to repair all defective elements. Thus, we can simply scan each vertex of X and see if it can be replaced by one or more of the vertices in Y. If the vertices in Y that are connected to the vertices in X form a covering, then the chip is repairable, and not otherwise.

Problems that belong to subcase 1 can be solved using a bipartite matching algorithm. Each defective element x can be replaced by any one of the redundant elements that appear in $W(x)$ and each redundant element y can be used to replace any one of the defective elements that appear in $W(y)$. There is a known $O(m\sqrt{n})$ algorithm [HoKa73] for bipartite matching, where n is the number of vertices and m is the number of edges in the graph.

3.2. NP–hardness Results

When we say that the complexity of a subcase is NP–hard, we mean that at least one of the problems in that subcase is NP–hard. We note that subcases 5,6, and 9 are NP–hard. Finding a covering in a rectangular array using no more than a given number of spare rows and no more than a given number of spare columns can be shown easily to be a problem that belong to subcase 6. The former problem was shown in [KuFu87] to be NP–hard, which implies that subcase 6 is NP–hard.

We proved that subcase 5 is also NP–hard by a reduction from the SAT–problem. The proof can be found in [HaCL88a]. This implies that subcase 9 is also NP–hard.

4. Conclusions and Future Research

We presented in this paper a model for fault covering problems that generalized the models studied in earlier works. We also showed the complexity of the subcases of the model.

In addition to the Boolean expression associated with every vertex we can also have partitions on subsets of vertices. These partitions can capture the situation in which only a certain number of defective elements in each block of defective elements need to be replace or in which only a certain number of redundant elements in each block of redundant elements can be used to replace defective elements. The complexity of the corresponding fault covering problems are studied in [HaCL88b].

We are currently looking into another extension to the model when the kind of constraints associated with the subset of the edges are of the form if edge (u,v) is used in the covering, meaning that u is used to replaced v, then edge (a,b) must also be included in and edge (c,d) must be excluded from the covering.

ACKNOWLEDGEMENTS

N. Hasan is supported by an International Business Machine fellowship, J. Cong is supported by a Digial Equipment Corporation fellowship, C. L. Liu's research is supported by the National Science Foundation under grant MIP 87–03273, and by the Semiconductor Research Corporation under contract 87–DP–109.

REFERENCES

[Evan81] R. C. Evans, "Testing Repairable RAMs and Mostly Good Memories," *Proc. IEEE Int. Test Conference*, pp. 49–55, 1981.

[HaCL88a] N. Hasan, J. Cong and C. L. Liu, "A New Formulation for Fault Covering Problems in Reconfigurable Chips," manuscript, 1988.

[HaCL88b] N. Hasan, J. Cong and C. L. Liu, "A New Formulation of Yield Enhancement Problems for Reconfigurable Chips," *Proc. Intl. Conf. on Computer-Aided Design,* pp. 520–523, Nov. 1988.

[HaLi88] N. Hasan and C. L. Liu, "Minimum Fault Coverage in Reconfigurable Arrays," *Proc. 18th Int. Symp. on Fault-Tolerant Computing,* pp. 348–353, June 1988.

[HoKa73] J. E. Hopcroft and R. M. Karp, "An $n^{5/2}$ Algorithm for Maximum Matchings in Bipartite Graphs," *SIAM J. Comput.*, Vol. 2, No. 4., pp. 225–231, Dec. 1973.

[KuFu87] S. Y. Kuo and W. K. Fuchs, "Efficient Spare Allocation For Reconfigurable Arrays," *IEEE Design and Test,* Vol. 4, No. 1, pp. 24–31, Feb. 1987.

[Liu68] C. L. Liu, *Introduction to Combinatorial Mathematics,* New York, McGraw Hill, 1968.

[Schu78] S. E. Schuster, "Multiple Word/Bit Line Redundancy for Semiconductor Memories," *IEEE J. Solid-State Circuits,* Vol. SC–13, No. 5, pp. 698–703, Oct.1978.

[Sing88] A. D. Singh, "Interstitial Redundancy: An Area Efficient Fault Tolerance Scheme for Large Area VLSI Processor Arrays," to appear in *IEEE Trans. on Computers, Nov. 1988.*

[TaBo84] M. Tarr, D. Boudreau, and R. Murphy, "Defect Analysis System Speeds Test and Repair of Redundant Memories," *Electronics, pp. 175–179, Jan. 1984.*

[WeLo87] C. L. Wey and F. Lombardi, "On the Repair of Redundant RAM's," *IEEE Trans. on Computer-Aided Design,* Vol. Cad-6, No. 2, pp. 222–231, March 1987.

10 FAULT-TOLERANT ARRAYS

DEFECT TOLERANCE IN A WAFER SCALE ARRAY FOR IMAGE PROCESSING

G. Saucier, J-L Patry, E-F Kouka, T. Midwinter*, P. Ivey*, M. Huch, and M. Glesner****

Institut National Polytechnique de Grenoble
46, Avenue Felix Viallet
38031 Grenoble Cedex, France

* **British Telecom Research Laboratories**
U.K.

** **Technische Hochschule Darmstadt**
Schloßgartenstrasse, 8
6100 Darmstadt, F.R.G.

Abstract

This paper describes the defect tolerance strategy used in the design of a full wafer 2D-array for image processing manufactured within an European ESPRIT project. The paper focuses on the global architecture which is partially based on yield consideration. The soft switch network allows the construction of the 2D-array even in the presence of defective processing elements (PEs). The configuration algorithm creates the target array with a very good "harvest" of functional PEs.

1 INTRODUCTION

A Wafer Scale array for image processing called ELSA (European Large SIMD Array) has been developed within the European WSI-824 ESPRIT project.

It is a SIMD architecture which implements a 2D-array of elementary Processing Elements (PEs) on a full wafer.

The Processing Element is simple but as general purpose as possible. The ALU operates on single bit data and processes multi-bit data in a bit serial fashion. There is no limit to the precision of the data when processed in this way although data can only be stored locally at the PE with a precision of 64 bits.

The PE has been optimized to perform adds and multiplies in the minimun number of clock cycles. In order to achieve this, the local storage (128 bits of RAM per PE) has a Read-Modify-Write architecture which is configured as independently addressable modules each of 64 bits. Therefore, in each clock cycle, a single bit of the output word can be generated so that, for example, an 8 bit add takes 9 clock cycles.

Each PE has a "flag" register which considerably enhances the functionality of the processor. At each "flagged" instruction each PE can perform either the given instruction or a null operation depending on the state of its flag register. By this means, the array can operate in a MSIMD fashion, executing in sequence different programs in different parts of the array.

Communication of the ELSA is very important. In many applications it is necessary to have high bandwidths to external image storage. ELSA has a communication register in each PE which is serially connected so that data can be scanned in from the edge of the array in background; the foreground process is only interrupted to load a complete plane of data into the PEs.

The target performance of the wafer level processor is in excess of 10,000 Mips with an array size of 128 by 128 PEs. This size of array can only be achieved with 6 inches wafers but our initial devices will be fabricated on 4 inches wafers during 1988. Such a computation density would enable many very complex image processing applications to be carried out in real time.

The goal of this paper is to present the yield enhancemend strategy used to reach an acceptable yield for the product. For this purpose defect tolerance has been implemented at two levels: At the "chip" level and at the "wafer level". The first section will present the global approach and the yield enhancement evaluation leading to the definition of the chip size.

The second section focuses on a switching network separating the chips in the array and a last section reports on the configuration algorithm allowing the creation of the 2D-array even in the presence of faulty chips.

2 General Defect Tolerance strategy and expected yield enhancement

2.1 Defect tolerance strategy

The defect tolerance is implemented in a hierarchical fashion. First, the PEs are grouped into basic arrays of 12 rows by 9 columns. So, at an architectural point of view, the chip is a 12*8 array of PEs with one spare column. This size was chosen as a result of yield calculations presented below and because of floor planning consideration. The chip is 6*6 square mm in a HCMOS 1.2µ technology. In a row of 9 PEs, a faulty PE can be bypassed by transmission gates. 8 PEs are used per row. Therefore, at most, one PE can be repaired and the transmission gates allow the West-East and North-South connections.

Secondly at a global level, a network of switches allows faulty chips to be bypassed and an operational 2D-array to be constructed with the good chips. Several switching networks have been considered and compared with respect to electrical parameters (size of the network, speed fixed to 30Mhz, number of vias, size of aluminium lines) and sensitivity to patterns of defects.

In order to evaluate the sensitivity to patterns of defects, a catalog of elementary patterns of defects was created and the length of the paths and the number of good chips which were unuseable, after reconfiguration was taken as the evaluation criteria. A double rail network with soft programmable switches (figure 1) was chosen and leads to the final architecture shown in (figure 2)

The switching network will be exhaustively studied in the next section. The irregular shape of the architecture is due to the fact that the masks are manufactured using step and repeat, with the result that, the reticle dimensions define the elementary block which may be repeated on the wafer. In this case, the reticle has a maximum size of 14*14 square mm, and therefore 4 chips fit in the reticle.

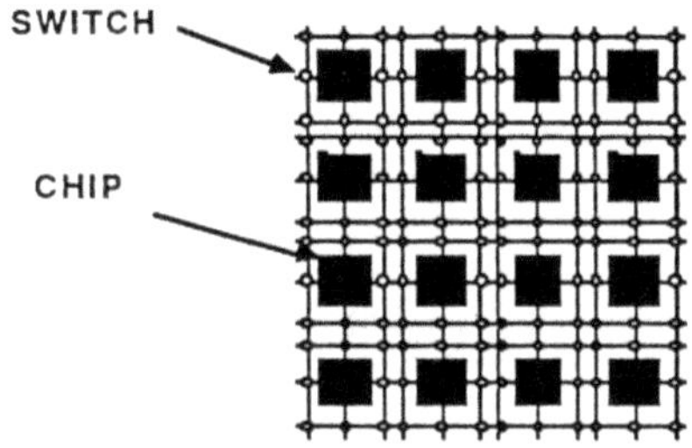

figure 1. Double rail network

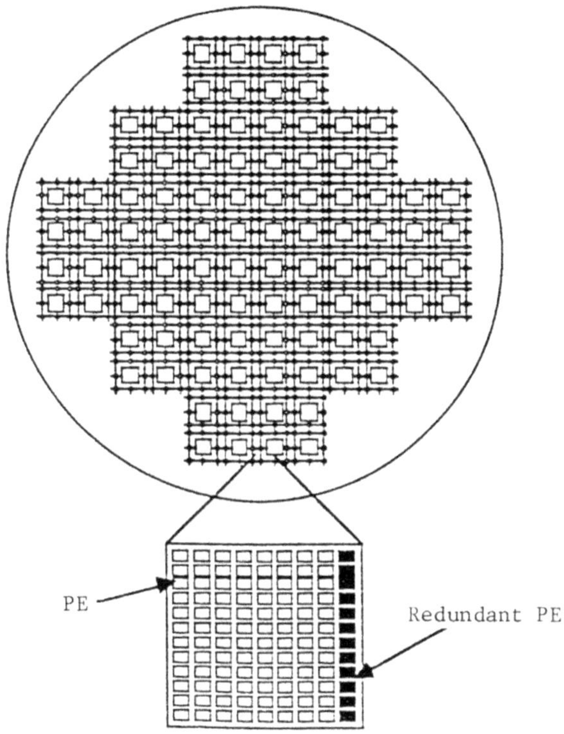

figure 2. Wafer architecture

2.2 Expected yield enhancement

The evaluation of the yield of the whole wafer is carried out in a hierarchical fashion. First we compute the yield at the chip level as a combination of the yields of the PEs included in each chip . Secondly the yield at the wafer level is computed as a combination of the yield of the chips implemented on the wafer. The analysis of the expected global yield has led to the final architecture. We explain first this yield calculation and then we show how it has led to the final architecture .

2.2.1 Yield calculation

a Chip Level

At first, it is crucial to choose an accurate probabilistic model for the yield prediction. With HCMOS3 technology, we chose different models depending on the value of D,S which are respectively the expected defect density and the area of the chip. To compute the probability of getting a valid chip We use the Murphy and Seeds model for low values of D.S, and the Bose-Einstein model for larger values of D S. Hence we got the following formula :

$$P(pe) = Min(P_{B\text{-}E}(pe) ; P_{M\text{-}S}(pe))$$

with: $P_{Bose\text{-}Einstein}(pe)= 1 / (1+S.D)\ M$

$$P_{Murphy\text{-}Seeds}(pe)=(((1-\exp(-M.D.S)\ )/\ M.D.S)\ 2 + \exp(-\sqrt{M.D.S}))/2$$

where **M** is the number of critical masks
S is the area of a chip
D is the defect density per square cm

A chip consists of rows of PEs. Using this probabilistic model, the yield of the row can be computed. Each row has at most one redundant PE. If (n+1) is the number of PE in a row then the yield depends on the number of combinations of n good PEs from n+1 PEs. Consequently:

$$Y(row)= C^{n}_{n+1}\ P(pe)^{n}(\ 1-P(pe)\ )\ C +\ P(pe)^{n+1}$$
$$= (n+1)\ P(pe)^{n}\ \ (1-P(pe)\ )\ C + P(pe)^{n+1}$$

C is a coefficient which takes into account the possibility to repair one defective PE. Here, when a PE is faulty we bypass it by a transmission gate. As a transmission gate is a small device, C may be taken equal to 1.

Hence: $Y(row)= [\ (n+1).P(PE)^{n}.(1-P(PE)) + P(PE)^{n+1}\ ]$ (yield of a row)

According to the architecture of the chip the four switches directly connected to the chip must be operational to get a valid chip. The yield of a chip depends on the yield of each row and of the switches.

The yield of a chip is given by: $Y(CHIP) = Y(row)^{P}.\ Y(switch)^{4}$

where **Y(switch)** is a coefficient which depends on the area of the switch.

b Wafer level

At the wafer level, we compute the yield with the following formula :

$$P = \sum_{j=0}^{j=Ni-Nu} \left(C^{Nu+j}_{Ni}\ P(chip)^{Nu+j}\ \ (1-P(chip))^{Ni-Nu-j} \right)$$

where **Nu** is the number of chips required to get the target array.
Ni is the number of chips integrated on the wafer.
P is the probability to have at least Nu good chips among Ni ones.

The influence of the reconfiguration algorithm on the yield evaluation has been taken into account by increasing this number Nu to a number Nu' > Nu; the difference being the result of the harvest loss of the reconfiguration algorithm. For a wafer with **Ni** chips implemented on it, we want to connect **Nu** chips to get a given target array (**Nu<Ni**). We have a reconfiguration algorithm with a harvesting (percentage of defect-free chips interconnected by the algorithm) equals to **R** (**R** in]0,1]).

We must have at least **Nu'= Nu/R** good chips for the algorithm to work in good conditions. Therefore the final wafer yield will be computed as the probability of choosing **Nu'** good chips among **Ni** chips.

Consider for example a target array with 128 X 128 PEs so **Nu**=128 X 128/(N.P) (N.P is the number of PEs in a chip). If R=0.9 the algorithm needs more than **Nu** chips to build the target array; it needs **Nu'=Nu**/0.9

Therefore, the calculation of the yield at wafer level is shown below:

$$Y(wafer) = \sum_{j=0}^{j=Ni-(Nu/R)} \left(C_{Ni}^{(Nu/R)+j} \; P(chip)^{(Nu/R)+j} \; (1-P(chip))^{Ni-(Nu/R)-j} \right)$$

where : $\mathbf{P(chip)=Y(row)^{P}.\ Y(switch)^{4}}$

$\mathbf{Y(row)=\ (n+1)\ P(pe)^{n}\ (1\text{-}P(pe))\ C + P(pe)^{n+1}}$

3 The switching network

The switching network consists of a double rail of programmable soft switches surrounding the chips. The goal of this network is first to ease the end of manufacturing test. A progressive exploration of the wafer associated with some self-test features of the PEs leads to the cartography of the wafer. This cartography points out the good chips, the good switches, the good connections. The test strategy defined by British Telecom Research Laboratory is outside the scope of this paper.

A reconfiguration program descibed in section IV defines the reconstruction of the 2D-array on the wafer. This leads to the programming of associated paths on the wafer. In a final step, the switches are set in their definitive position by floating-gate FETs or laser fuses. Both are implemented presently and the most reliable devices will be chosen at the end of the project . Concurrent researches within this ESPRIT project deal with the investigation of the reliability of these devices which appear to be much better than the already published papers.

3.1 The basic switch

Eight switches surround a chip. A switch can be programmed in four situations according to the values of two steering variables A and B (figure 3).

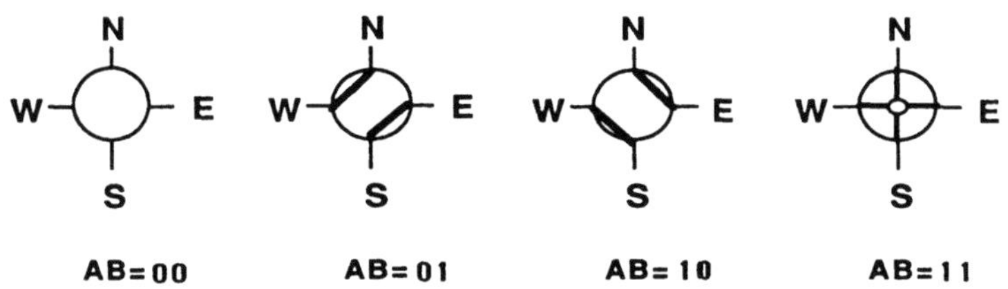

figure 3. The four possible configurations

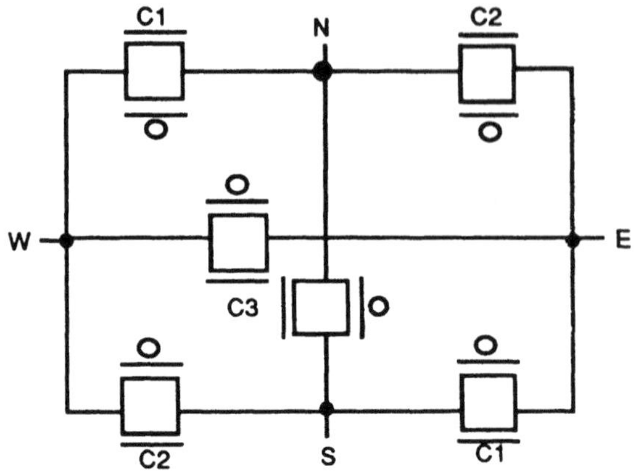

figure 4. The set of transmission gates

The basic switch is made up of two parts. The first part (figure 4) is a set of transmission gates realizing the set of paths. The second part (figure 5) is the control part of the switch in which two steering flip-flops A and B control the transmission gates signals through a random logic.

3.2 Programming paths in the switching network from the I/O pads

To program a path in the switching network, from the I/O pads, a shift register is implemented. For every switch two bits of the shift register (X,Y) contain steering information. This information is sent serially all over the wafer. If at a given instant, the steering information concerns a switch, this information will be sampled in its steering flip-flops; otherwise it will circulate further on. A control signal PROG is used to indicate wether steering information is just circulating (PROG=1) or wether this information has now reached the target switches and must be sampled in the steering flip-flops (Falling edge of PROG). Let us illustrate this and show the creation of a path between two successive switches (SW1,SW2) illustrated by figures 6 and 7.

3.3 Architecture of the switch

The figure 8 shows the general architecture of the switch.

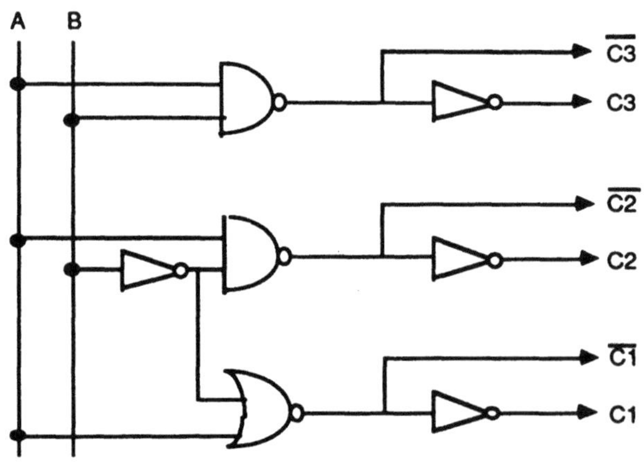

figure 5. The control logic

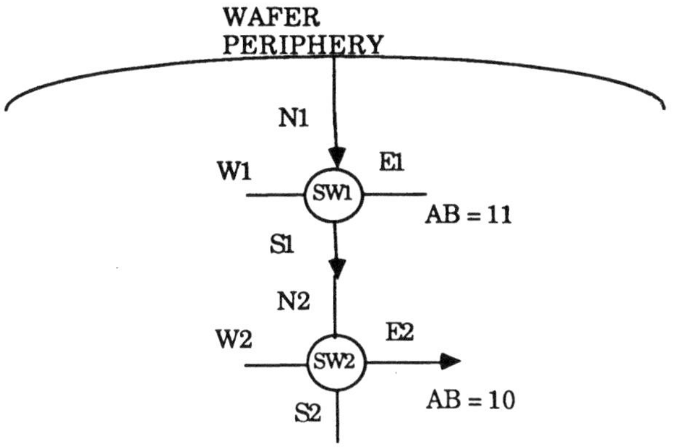

figure 6. Path programming informations

i) The TRANSFER block

The transfer block is made up of (n+2) basic transfer blocks (cf figure 4), where n is the number of data wires which cross a switch. One basic transfer block carries the CTRPROGx signal and one basic transfer block carries the PROGDATAx signal. The n others carry the data (figure 8).

ii) The CONTROL block

A block called "SELECT" is made up of a XOR gate, a block called "FLAG" and a block called "TFG". "FLAG" is made up of 4 flags (flagn, flags, flage, flagw) and "TFG" is made up of transfer gates. The block "SELECT" selects the inputs/outputs of the "CONTROL" block. It receives the programming control signal CTRPROG (ctrprogn, ctrprogs, ctrproge, ctrprogw) and the programming data signal PROGDATA (progdatan, progdatas, progdatae, progdataw) from all its four neighbours. The programming is detected when one and only one CTRPROG is high (see previous remark).

Therefore, the flags store the source of the programming by storing the CTRPROG wires status into the four flags (flagn, flage, flags, flags) when XOR(progn, progs, proge, progw)=1. At the beginning, all the basic transfer block are "open" or cut and all the flip-flops are at 0.

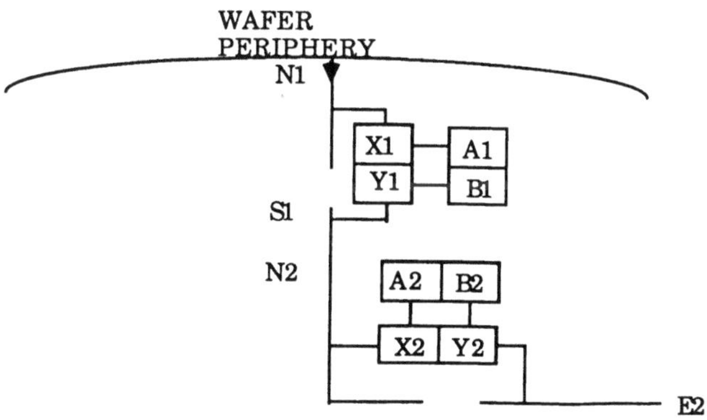

figure 7. Shift register and steering flip-flop values
associated with the two switches

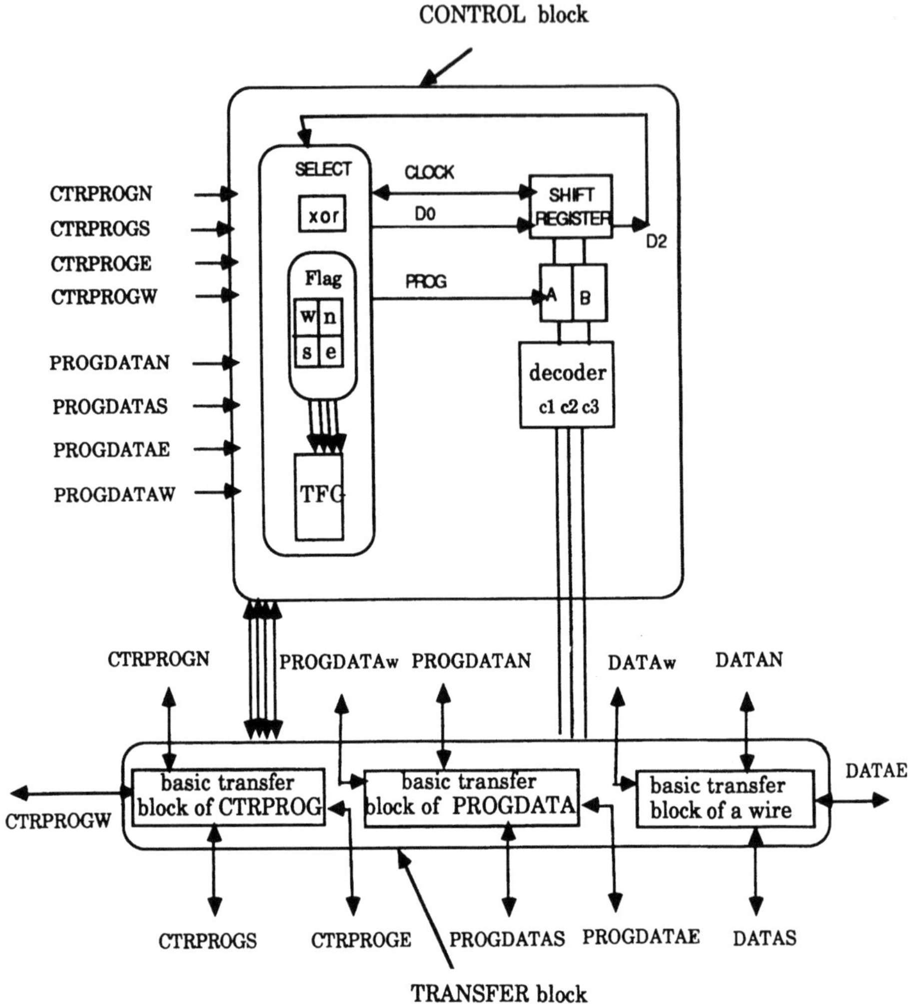

figure 8. Architecture of the switch

The load control signal of the flag is synchronized with the clock signal, to avoid intempestive loading. The flags control the "TFG", which selects the PROGDATA wires from which the programming data flow arrives. The status 1 of a flagn opens a transfer gate which connects the datan to the shift register etc ...

After the programming of the switches, the paths are fix with fuses and floating gates. The fuses and floating gates are implemented in the steering flip-flop (A,B).

4 The reconfiguration approach

In the presence of defects, the reconfiguration purpose is to interconnect good resources among defective ones in order to reproduce the target array, hence to emulate the function of a perfect wafer. This can be viewed as a mapping of the target array on the good resources of the wafer. We present here a placement and routing based approach to achieve such a mapping. The mapping is carried out slice by slice. Depending on the chosen procedure, a slice can be defined as a complete row of the target array, or a complete column, or a sector defined as two orthogonal partial row and partial column which intersect at the main diagonal of the target array.

In a general sense, the target is a rectangular array with R rows and C columns. The mapping must preserve the systolic properties of the target array. Due to communication delays, a connection length between two neighbouring chips must not exceed a prescribed limit, say D. The algorithm operates from a wafer cartography which gives the location of **defective resources (switches, segments, chips, I/O pads**).

The main algorithm consists of three routines. Each routine performs the mapping according to the given type of slice.

We will first briefly present the placement and routing procedure which is the kernel of the algorithm. Then we shall present the three procedures which are recursively called according to the nature of the chosen slice.

4.1 The placement and routing procedure

This is the basis of the presented approach. Placement and routing are carried out concurrently in order to achieve the following restrictions :

- the placement procedure must determine neighbouring chips in order that the routing of connections between chips does not exceed the D limit. So the chips placement is directed by the routing algorithm,
- the routing algorithm must find the peripheral I/O pads to be connected to bordering chips. This is the I/O pads assignment which is carried out as the bordering slices are pointed out,
- placement of chips must be locally reconsidered in order to reduce the deformations introduced by the routing process,
- a segment can be crossed by one path at most,
- earlier routed connections must not "freeze" good resources or block for latter routing.

We use a Lee based maze routing algorithm to achieve the above goals. When looking for a path between two chips, or when determining the best neighbour of a chip, we scan step by step the neighbourhood of all the visited chips and switches. The visited resources are labelled. A path development is aborted as soon as its length exceeds the D limit. If more than one path reach the goal with the same length, we keep the one which involves the least deformations for the routing of the corresponding connection line. Paths which have not been kept are released for further investigations.

4.2 The row by row mapping

This procedure tries to achieve the configuration by a progressive mapping of the target array on the wafer. One row is mapped at a time.

The first step determines the chip which will receive the virtual index (1,1), that means the embedding of the element located at the first row and first column of the target array. That is chosen at the left top of the wafer. The placement and routing algorithm is then invoked to connect this chip to an I/O pad at the north and to another I/O pad at the west of the wafer.

The second step completes the mapping of the first row of the target array. This is achieved by C - 1 recursive calls to the placement and routing procedure. Then we get C successive neighbouring chips connected by their "East to West" links to the previous chips. Connections to I/O pads are appropriately settled and the visited switches are configured according to the way they have been crossed.

The inner loop of the algorithm iterates on the mapping of the remaining rows until the target array is completely embedded in the wafer. During the mapping of successive rows, "South to North" connections are routed between vertically neighbouring chips. The bordering chips are connected to the appropriate I/O pads.

The mapping of a row is locally reconsidered if the routing of connections between the chips fails.

4.3 The column by column mapping

This is the symmetrical approach of the previously presented method. The target array is mapped column after column.

After the embedding of the chip which receives the virtual index (1, 1), R - 1 recursive calls to the placement and routing procedure will complete the embedding of the first column of the target array in the wafer. Then, as presented for the previous method, C - 1 calls to the kernel procedure will create C - 1 additionnal columns and provide a R*C array of interconnected chips with bordering chips connected to appropriate I/O pads.

These two mapping methods lead to $O(R*C*D^2)$ algorithms in run time-complexity. The reconfiguration is carried out in a reasonable time. One disadvantage is the sensitivity to the presence of clustered defects. In that case, the routing procedure introduces deformations of physical rows or columns that drastically reduce the harvesting. The following method which mixes the mapping of partial rows and partial columns tries to overcome this drawback.

4.4 The sector by sector mapping

The idea is to embed partial rows and partial columns at a time. A sector is defined as a partial row and a partial column which intersect at their end point, forming an elbow.

Each step of the inner loop tries to complete the embedding of a square portion of the target array. If the target array is rectangular in shape then the embedding is completed with one of the two previous procedures.

The first step consists in the embedding of a 1*1 array, that is the chip to be associated with the virtual index (1, 1). Let m = min (R, C) and M = max(R,C). Then m-1 recursive calls to the embedding of successive sectors will lead to the mapping of a m*m square subarray. Finally M - m calls to one of the previous methods ("Row-by -row" if M = R, otherwise "Column-by-column") will complete the embedding of the whole target array.

Unlike the two previous methods, the sector oriented mapping method better overcomes the presence of clustered defects. Deformations are limited to neighbouring sectors. The propagation of a deformation is hopefully slowed down after two iterations.

This method leads to an algorithm with a run time-complexity of the same order of magnitude as the two first ones however it seems to take more cpu time to deliver a configuration.

The cartography was randomly generated from a probabilistic yield model. An example of a configuration resulting from the sector by sector algorithm is presented in figure 9.

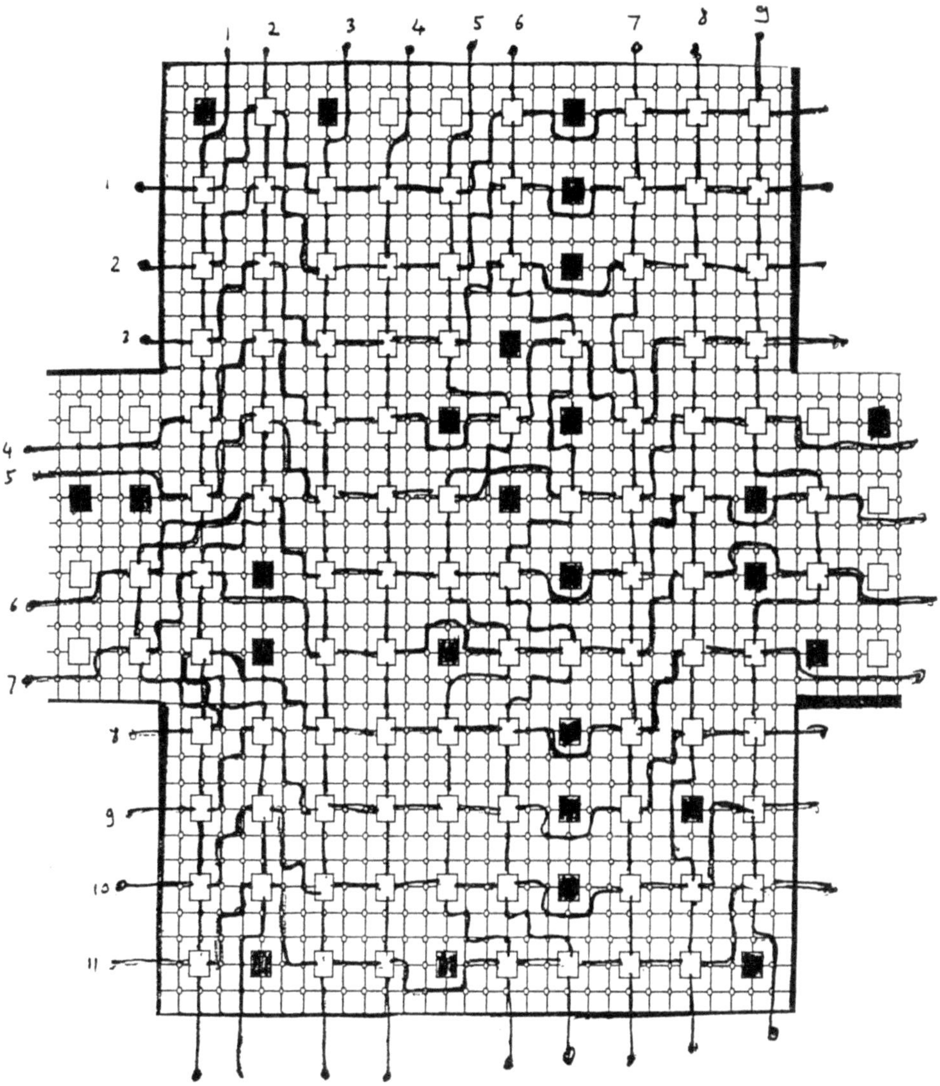

Figure 9. A result of reconfiguration using the Sector-by-Sector method on a wafer with 136 physical chips. There are 23 defective chips. The target is a 11 x 9 array.

Experiments on the basis of several simulations of wafer cartographies show that all the presented methods suffer the same drawbacks : early routing decisions are made blindly and adversely affect latter routing. Further investigations will try to perform a more intelligent routing. This needs the implementation of a domain specific knowledge for monitoring the routing process.

These ideas have been embodied in the **CRAWL/Systolic** system, an interactive system which performs **C**onfiguration and **R**econfiguration **A**t **W**afer **L**evel especially dedicated to the presented systolic array network. Programs are written in the C language under the Vax/VMS operating system. To give an idea of the CRAWL/systolic performances, let notice that the mapping of a 11*9 array takes less than 1sec cpu on a micro Vax II computer.

5. Conclusion

Defect tolerance techniques have been applied for the design of a wafer scale 2D-array for image processing. Such a study is complex as it is based upon good architectural choices relying on yield prediction, on adequate switching devices and on an efficient configuration algorithm. All these three aspects have to be optimized to succeed in such a design. Experimental results will be the final proof that all these factors have been successfully addressed and show the feasibility of wafer Scale Integration.

Acknowledgement

This project has been developped under EEC ESPRIT I 824 grant with the leadership of Dr J.TRILHE of SGS-Thomson microelectronics.

References

(1) Jim C. Harden and Noel L R.Strader II,"Architectural Yield Optimization for WSI",IEEE trans. on computers,vol.37,pp. 88-110,1988

(2) K.Hedlung,"Wafer Scale Integration of Parallel Processors" PHD Thesis, Purdue University, Xest Lafayette,Nov 82.

(3) Duncan Moore Henry Walker,"Yield Simulation for Integrated Circuits", Kluwer Academic Publishers.

(4) F.R.K Chung, F.T.Leighton,and A.L.Rosenberg,"Diogenes: a methodology for designing fault-tolerant VLSI processor arrays,"Proceeding of the IEEE Symposium on fault-tolerant Computing, June 1983, pp.26-31.

(5) D.Fusell and P.Varman, "Fault-tolerant wafer-scale architectures for VLSI," Proceeding of the 9th annual IEEE/ACM Symposium on Computer Architecture, April1982, pp.190-198.

(6) K.Hedlung and L.Snyder, "Wafer-scale integration of configurable, higly parallel (CHIP) processors," Proceeding of the IEEE International Conference on parallel Processing, 1982 , pp.262.264

(7) I.Koren and M.Breuer, "On area and yield considerations for fault tolerant VLSI processor arrays ," IEEE Transactions on computers, Vol.C33, No 1,1984, pp. 21-27

(8) H.T.Kung," Why systolic architectures," Computer magazines, IEEE , January 1982,pp.21-27.

(9) F.T Leighton and C.E Leiserson,"Wafer-scale integration of systolic arrays," IEEE Transactions on computers, vol. c-34, No.5 , May 1985, pp.448-461

(10) R.Negrini and R.Stefanelli,"Comparative evaluation of space and time redundancy approaches for WSI processing arrays", IFIP Workshop on Wafer-scale integration, Grenoble , 1986.

(11) M.Sami and R.Stefanelli,"Fault tolerance approaches for VLSI/WSI arrays", Proc.IEEE Pheonix Conf. on Comp and Comm; 1985, pp.460-468.

(12) G.Saucier and J.Trilhe, editors, proc. IFIP Workshop on Wafer-scale Integration, Grenoble,1986

(13) Roberto Negrini and Renato Stefanelli, "Comparative Evaluation and Time Redundancy Approaches for WSI Processing Arrays", Wafer Scale Integration, G.Saucier And J.Trilhe (Editors),p. 207-221.

(14) Manolis G.H Katevenis and Miriam G.Blatt, " Switch Design for Soft-Configurable WSI System", Wafer Scale Integration, G.Saucier And J.Trilhe (Editors),p. 255-270.

DISTRIBUTED FAULT TOLERANT EMBEDDING OF BINARY TREES AND RINGS IN HYPERCUBES*

Foster J. Provost and **Rami Melhem**

University of Pittsburgh
Department of Computer Science
Pittsburgh, PA 15260

1. INTRODUCTION

In this paper we first present fault tolerance techniques based on distributed algorithms for embedding binary trees in hypercubes. Starting with the root (invoked in some cube node by a host), each node is responsible for determining the addresses of its children, and for invoking the embedding algorithm for the subtree rooted at each child in the proper cube node. This distributed embedding, along with the wealth of communication links in the hypercube, leads to a high potential for fault tolerance. We demonstrate the fault tolerance capability by introducing restructuring techniques which may be used to tolerate faults during the initial embedding, as well as to remap nodes that fail at run-time. The distributed nature of the embeddings eliminates the need for global knowledge of faulty nodes; each node must only know the status of its neighbors.

Fault tolerance in tree connected multiprocessors has been considered by many researchers (see for example [3] and [7]). However in the previous research, fault tolerance has been achieved by adding redundant hardware to a structure such that the topological connection is preserved even in the presence of faults. Our approach, on the other hand, uses a hypercube-configured multiprocessor and takes advantage of the inherent redundancy of communication paths to remap faulty processors and avoid faulty links, thus producing a fault tolerant embedding of a binary tree structure. This approach is similar to that taken by Sami and Stefanelli [8] and Kung[5] for reconfiguration in meshes except that the reconfiguration is very local.

Several researchers have studied the embedding of balanced binary trees in hypercubes. In doing so it is important to discover embeddings which preserve the adjacency of nodes so that communication can take place efficiently. It has been proven that an n-tree cannot be embedded in an n-cube with adjacency preserved[9]. Different methods have been suggested to bypass this difficulty. Bhatt and Ipsen [1] and Deshpande and Jenevein[2] embed a "two-rooted" n-tree in an n-cube, preserving adjacency; they use the extra root as a communications processor which routes information from the root to one of its logical children. This provides communication with a maximum of one hop between processors. Johnsson[4] and Wu[9] take a different approach; they embed an $(n-1)$-tree in an n-cube preserving adjacency. Wu gives a recursively defined, bottom-up algorithm for determining a proper embedding; Johnsson, in contrast, gives a top-down algorithm. Both algorithms rely on a "bird's eye" picture of the entire embedding and hence the embedding has to be determined by a host and tree nodes have to be statically assigned to processors.

* This research is, in part, supported under ONR Contract N00014-80-C-04555.

The question of fault tolerance arises naturally at this point. If a tree node processor or communication link is faulty, the entire embedding has to be recalculated and tree nodes reassigned to processors. Even worse, if the fault occurs during processing, the entire structure has to be remapped before the processing can continue. This is time consuming and costly, especially if each tree node has to transport its program and data to the node's new location in the hypercube. In a time critical application, this expense would not be feasible.

We present, in Section 2, a distributed algorithm for embedding a binary $(n-1)$-tree in an n-cube in which each node determines the addresses of its children and invokes in these children the algorithm to embed the subtree rooted there. In order to overcome the problem of reassigning the entire embedding in the presence of faults, we introduce, in Section 3, a tree restructuring technique which may tolerate node or edge faults. The technique is extremely local and entirely distributed; in all but a single case, only the faulty node (or a node connected to a failed edge) is remapped and only its parent and children are aware of the change. The only case where more than just the faulty node is remapped is if a height 2 node N (height 1 nodes are leaves, height 2 nodes are leaves' parents, and so on) is faulty. In this case N and its two leaf children are remapped within the 2-cube that originally contained them. Any single fault can be tolerated as well as many multiple fault configurations. The tree is initially embedded with adjacency preserved; an embedding which bypasses a faulty processor (or a set of faulty processors) may contain *one-hop* communication between the remapped tree node and its logical neighbors, where *one-hop* denotes the passage of a message through one node, en route to another node (distance 2 communication). If the communications part of the processor is suitably designed, this one-hop degradation due to faulty processors will incur only a minor increase in communication time between nodes. The same applies if a faulty node may act as a switch (with trivial delay). We then enhance the algorithm so that it tolerates any configuration of two faults.

In Section 4, we briefly discuss a similar treatment for a distributed, fault tolerant embedding of rings (or linear arrays) in hypercubes. Finally, in Section 5, we give simulation results which show the robustness of the fault tolerant algorithms. Because of space constraints the proofs of correctness have been omitted; for a more comprehensive coverage of the topic we refer to[6].

2. A TOP-DOWN, DISTRIBUTED TREE EMBEDDING ALGORITHM

We have developed a top-down, distributed algorithm for embedding an $(n-1)$-tree in an n-cube, preserving adjacency. In this algorithm, each node receives a small packet of information from its parent, determines which nodes will be its children and sends them the information they need to continue the configuration. The information needed to determine the addresses of a node's children consists of: (i) the height of the subtree rooted at the node; and (ii) an "interchange word" which specifies some axes along which the subcube has to be reflected in order to yield a proper embedding.

The addresses of the node's children are calculated based on the bit positions which differ between the node's address and its child's. Since adjacent nodes in a hypercube differ in only a single bit position, a single integer will suffice to differentiate between the neighbors of a node. In the subsequent discussion, neighbors of a given node will be referred to by the bit position differentiating them from this node. The interchange word, which in our algorithm is passed from a parent to its children, consists of a string of bits. Numbering the bits with zero corresponding to the rightmost (least significant) bit, a "1" in the jth position specifies an interchange of the form (j/j+1), where an interchange of the form (i/k) indicates that subsequent children that are to be assigned to neighboring nodes across the ith dimension (if any) have to be reassigned to the neighboring nodes across the kth dimension and vice versa.

More specifically, a node receives the height h of the subtree rooted there, and initially takes its left child to be its neighbor in the h dimension, and its right child the neighbor in the $(h-2)$ dimension. These bit positions are then modified by applying the interchanges specified in the node's interchange word. The word is read right to left, applying a given interchange to the bit position addresses of both children if the corresponding bit in the interchange word is set. To the right child, whose dimension has just been determined, the node sends height $(h-1)$ and its interchange word unmodified. To the left child the node sends height $(h-1)$ and its interchange word, modified by setting bit $(h-2)$. The embedding can be rooted at any node in the cube by invoking procedure TREE at that node with the height $h=n-1$, where n is the size of the cube and the interchange word $rot=0$. When invoked at a particular node, the procedure TREE computes l and r, the dimensions across which its child nodes should be located, and then invokes TREE at these child nodes. The procedure TREE is described formally as follows.

```
procedure TREE(h,rot)
begin
    if h=1 then stop /* this is a leaf node */
    else /* h>1 */
        r:=h;
        l:=h-2;
        for i:= 0 to (length of rot) loop
            if (bit i of rot=1) then
                if (r=i) then        r:=i+1;
                    else if (r=i+1) then r:=i;
                if (l=i) then l:=i+1;
                    else if (l=i+1) then l:=i;
            end if;
        end for;
    end if;
    rot_l=rot
    rot_r=rot+2^(h-2)
    invoke TREE(h-1,rot_r) at the node across dimension r  /* right child */
    invoke TREE(h-1,rot_l) at the node across dimension l  /* left child */
end TREE
```

2.1. Modifications for Fault Tolerance

We make few assumptions regarding the architecture of the hypercube machine. Each node must have processing as well as communications capabilities. It is desirable, although not necessary, that these be separated into a computation processor and a communications processor, with message routing considerably faster than processing. Thus the one-hop incurred by the remapping of a failed processor will not significantly delay the network's (otherwise adjacent node) communications.

Our fault model takes into account both faulty processors and faulty communications links. A node is considered faulty if:

(i) The node itself does not function properly (i.e. the computation and/or the communication part fails).

(ii) A communications link used in the embedded tree structure for communications with the node's parent does not function properly.

We assume that a node can detect faults in it's immediate neighbors.

Our fault tolerance scheme is based on the fact that in embedding an $(n-1)$-tree into an n-cube (using our scheme), the n-cube can be divided into two $(n-1)$-cubes, one of which contains three-fourths of the nodes, the other contains only one-fourth. This can be seen by studying the embedding algorithm. Specifically, bit zero of the exchange word is never used; it is set by the height 2 nodes, but the height one nodes (the leaves) do not use it. This implies that any embeddings across dimension zero come from the initial assignment of l or r. Moreover note that in TREE, only l can ever be zero, and it will be zero for every height 2 node (and only for these nodes). Thus, dimension zero separates the n-cube into two subcubes, one of which contains one-half of the leaves of the tree (about one fourth of the total tree nodes), the other contains the rest of the tree (about three-fourths of the total tree nodes). This separation of the tree nodes yields a structure which is amenable to a very local and distributed fault tolerance scheme. If we look at a *mirror image* of the tree structure across dimension zero (those nodes whose addresses differ from those of the tree only in dimension zero), we can notice that the only nodes which do not have a free *image* are the height 2 nodes and their left children. Hence the image subcube may be used to remap faulty processors for any node at a height higher than 2. For the leaf nodes and their parents, a different strategy must be employed. In other words, the scheme is divided into two parts. One applies to the leaves and their parents; the other applies to the rest of the nodes.

The algorithm that we will describe for restructuring the tree due to faulty processors or links can be applied both at the initial embedding (to bypass already faulty nodes) or during execution (to locally and distributedly remap failing nodes, without disturbing processing elsewhere in the tree). Any single fault is guaranteed to be tolerated. This guarantee, by itself, downplays the strength of the algorithm. The extreme "localness" of the restructuring along with the relatively empty image subcube allows the system to tolerate many faults simultaneously, including groups and strings of faults in the original tree.

For nodes of height greater than two, if a child is detected to be faulty, the node will route all information (for that child) to the child's image across dimension zero. This will incur a one-hop delay. Once in the image subcube, the embedding algorithm of Section 2 will be used to determine each node's children. After the initial

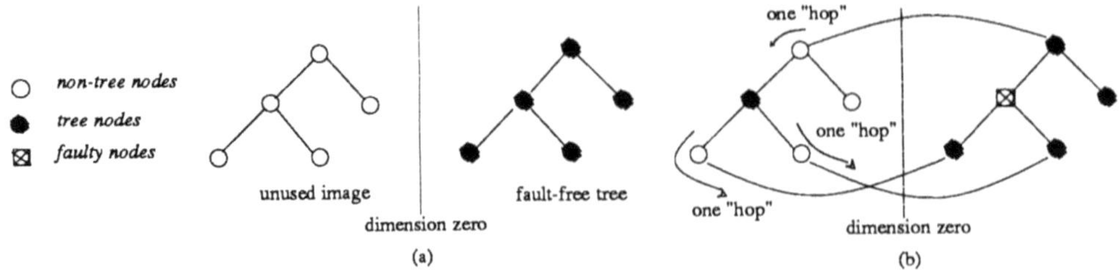

Figure 1
(a) a fault-free tree and its image
(b) local remapping for a single fault

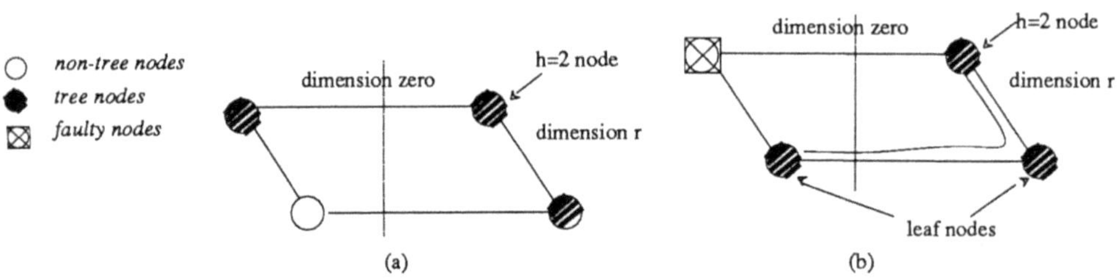

Figure 2
(a) fault-free leaf embedding
(b) remapping of a faulty leaf node

node remapping, a node in the image subcube will always attempt to *bounce back* (i.e. return processing to the original subcube) before taking over processing duties. Thus, for a single fault, the faulty processor's image will take over the processing duties, determine its children (the images of the faulty node's children), and route information to them. These image children will bounce back to the original children -- meaning that they will act as one-hop communications elements between the remapped node and its children. The net effect is a distributed shift of the node into the image subtree with the rest of the tree structure remaining unchanged (see Figure 1). If one of the original children is also faulty, its image will not be able to bounce back and hence will automatically take over the processing; the logical tree structure in each branch will return to the original subcube as soon as a non-faulty node is available. In this way sections or branches of the original tree can be faulty, as long as their images are not (see Figure 3).

For nodes on the bottom two levels (leaves and their parents) a special remapping must be employed. The parent will bypass a faulty node by routing the information for one leaf through the other (see Figure 2). At this point we must note that for each height 2 node (leaves' parent), the left child is originally connected across dimension zero and the right child is originally connected across some other dimension (we refer to this dimension as dimension r). If a child of a height 2 node is faulty, the algorithm will map it to the node diagonally

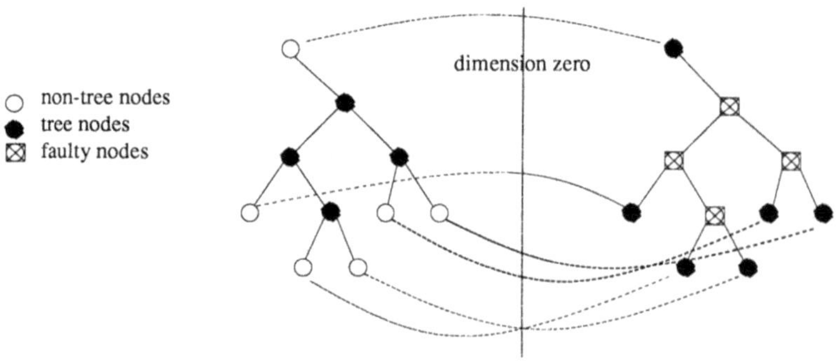

Figure 3 - remapping of a faulty section of the tree

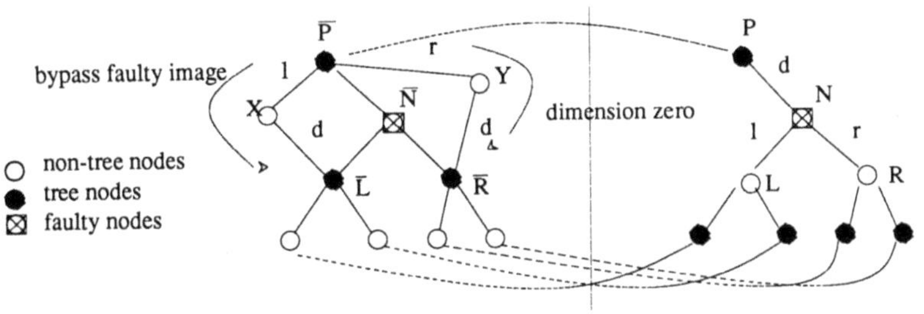

Figure 4 - a node and its image are both faulty

across dimensions r and zero. Whether the parent node is in the image subcube or the original subcube, its children occupy the same two cube nodes, mirrored about dimension zero. We consider the two-dimensional subcube containing the parent and its two children; if one of these four nodes is faulty, we will use the other three -- never incurring more than a one-hop delay in communications between logically adjacent nodes.

It can be seen that communications between any two logically adjacent nodes will incur at most a one-hop delay. This is the cost of the fault tolerant scheme. Notice also that the remapping of an interior ($h>2$) node will never interfere with the remapping of another node, and that no remapping of a node in one parent/leaf triad will interfere with the remapping of any node outside that triad.

2.2. An Enhanced Fault Tolerant Tree Algorithm

Although TREE_FT may tolerate multiple faults, certain double fault configurations cannot be tolerated. When a node is faulty, its image across dimension zero should not be faulty -- neither should the images of its parent or its children. In this section, we modify our algorithm so that any configuration of two faults will be tolerated. For this, we assume distance 2 fault detection. This can be implemented by having a node request a status report from some distance 2 neighbor via both one-hop paths, if it fails to receive an acknowledgement, it assumes (for the purpose of the algorithm) that this neighbor is faulty. Distance 2 fault detection is not needed -- an extra fault detection state can be added to the algorithm -- but for clarity in the description of the algorithm it is assumed.

The modifications stem from the fact that if a node can communicate with another node across dimension x and then y (one-hop), it can also communicate across dimension y and then x (also one-hop). Thus if we encounter a second fault when attempting to remap a node, we can initiate the embedding at a neighbor and one-hop to the children bypassing the faulty node. We also use backtracking if the embedding gets stuck due to faults. If a node discovers that it cannot embed its children, it declares itself faulty and lets its parent worry about reconfiguration.

Let us consider a tree node N', its parent P', and its children L' and R' which are initially mapped (in a fault-free embedding) to cube nodes N, P, L, and R, respectively; in addition, let the images across dimension zero of these nodes be denoted by $\bar{N}$, $\bar{P}$, $\bar{L}$ and $\bar{R}$. In our modified algorithm, if N is faulty N' is remapped to $\bar{P}$. L' and R' are then remapped to $\bar{L}$ and $\bar{R}$ (see Figure 4). Thus, there are two one-hop paths from N' to its children, and a second fault, in $\bar{N}$, can be bypassed. For example, if N is adjacent to P, L, and R across dimensions d, l and r, respectively, and $\bar{N}$ is faulty there is still a fault-free path from $\bar{P}$ to, say, $\bar{L}$ across dimensions l and then d. We refer to this method of avoiding faults in the image tree as *bypass* embedding. In our example if, say, $\bar{L}$ is faulty, L' can be mapped to node X, the node across dimension l from $\bar{P}$, and the

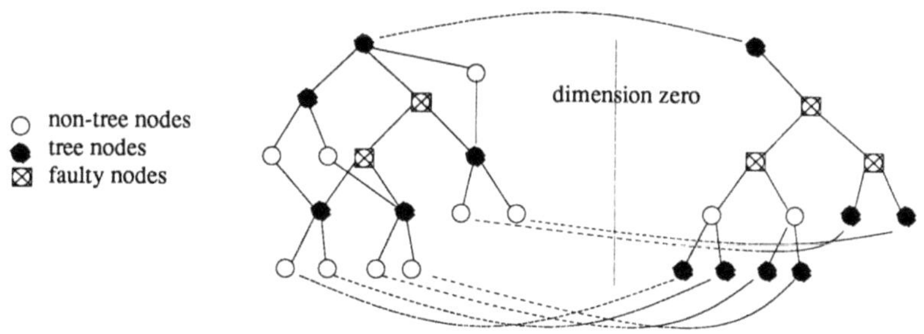

Figure 5 - a section of the tree and a section of its image are faulty

children of L' can be remapped using a similar bypass embedding. The embedding will bounce back as soon as it can do so (to a non-faulty node) and stay within the one-hop communication limit. Thus, if a second faulty node (N being the first) is L or R it will not affect the embedding. Likewise, if a second faulty node is one to which one of L's (or R's) children is mapped (in a fault-free embedding) the embedding will not bounce back to that node but continue in the image tree. In general, if a large subtree and its corresponding image are faulty, an embedding can still be realized with only one-hop communication; an example is given in figure 5. Finally, if both N and $\bar{P}$ (or N and the node to which the other child of P' is mapped) are faulty, P declares itself faulty and the node to which the parent of P' is mapped, call it PP, takes the responsibility for remapping P'; this is what we mean by *backtracking*. For PP, the node to which a child is mapped P and it's image $\bar{P}$ (or a section of a subtree rooted at P) are faulty. We have shown that such a case may be handled by the reconfiguration algorithm. The boundary cases of the root and the leaves take special treatment.

The algorithm utilizes the relative emptiness of the image subcube to embed leaf nodes which, because of certain fault configurations, would otherwise not be successfully embedded. A leaf node's parent (h=2 node) that has been remapped into the image subcube and cannot embed the leaf in the node specified by TREE attempts to embed the leaf in any unoccupied, fault-free node (at most distance 2) whose address's bit parity is different from that of a leaf node's. This will never interfere with the fault-free embedding of another part of the tree, since the nodes to which the leaves are assigned all have the same parity, and these are the only nodes occupied in the image subcube (in a fault-free embedding).

3. DISTRIBUTED FAULT TOLERANT EMBEDDING OF RINGS IN HYPERCUBES

We have also developed distributed algorithms for embedding rings (or linear arrays) into hypercubes in a fault tolerant manner. The embedding of a ring of size 2^n in an n-cube may be given in terms of a sequence $\{v_1, \cdots, v_{2^n}\}$ of nodes, where each adjacent pair of nodes v_i and v_{i+1} are neighbors across a given dimension. Alternatively the embedding may be specified in terms of a sequence $\{d_1, \ldots, d_{2^n-1}\}$ of dimensions such that nodes v_i and v_{i+1} are neighbors across dimension d_i. The latter specification sequence, called the *embedding sequence*, is more general than the first because it leaves the starting node v_1 unspecified.

In general, given the dimension n of the cube, an embedding sequence for a 2^n node ring may be found from the following algorithm (the notation $x \mid y$ denotes the sequence resulting from the concatenation of the two sequences x and y):

```
procedure Ring(n)
    x= {1}
    FOR i=2,...,n DO
        x=x|{i}|x
    return x
```

The resulting embedding sequence produces the sequence of nodes corresponding to a reflected Gray code; the closing dimension of the ring is n [4]. By using an incomplete Gray code assignment which is 75 percent dense, we can embed a ring of $3*2^{n-2}$ nodes in an n-cube such that any single fault (and many multiple fault configurations) can be tolerated. The following algorithm produces such a fault tolerant embedding sequence:

```
procedure Ring_FT(n)
    x= {1,2}
    FOR i=3, . . . ,n DO
        x=x|{i}|x
    return x
```

The spare nodes are distributed such that each node N is covered by a spare N' such that N' is adjacent to each neighbor of N. The method is similar to the binary tree technique in that the fault tolerance is very local -- only a faulty node is remapped and only its immediate neighbors are aware of the restructuring of the ring embedding (see Figure 6). This technique is also enhanced to tolerate any two faults and many more multiple fault configurations.

4. PERFORMANCE EVALUATION AND CONCLUDING REMARKS

In order to test the robustness of our fault tolerant algorithms, we performed simulation studies with various size hypercubes and varying numbers of faults. We always invoked the algorithms at a non-faulty node, and simulated the fault tolerant embeddings on many different random fault configurations. Attempts were made to embed a tree of a given height (or a ring of a given size) for different distributions of a specific number of faults, and results were collected as to the number of successful attempts. For the first fault tolerant

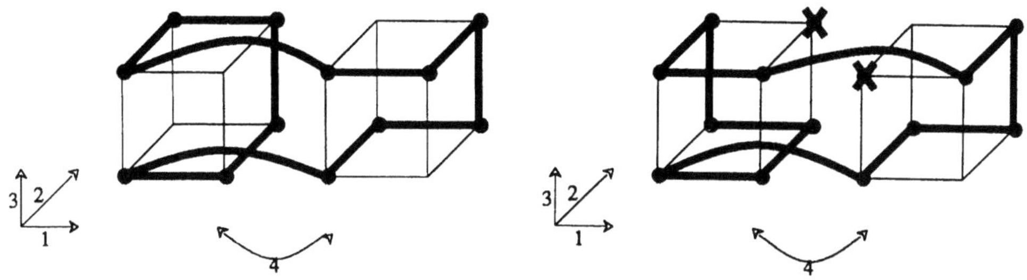

(a) a 12 node ring (b) reconfiguration due to faulty nodes

Figure 6 - Fault Tolerant Embedding of Rings in Hypercubes

algorithms the simulation results showed, as expected, that embeddings with zero or one fault were always successful (100 percent of the attempts succeeded). For the enhanced algorithm embeddings with up to two faults were always successful. The performance of the embedding algorithms degraded nicely as the number of faults increased. The results for trees of height 8 embedded in 9 cubes and the results for rings embedded in 9-cubes are shown in Figure 7. From these figures it is clear that the additional complexity in the enhanced fault tolerant algorithms provides more robust fault tolerant embedding algorithms.

The above results show the average number of faults that can be tolerated given an initial non-faulty starting node. This corresponds to the case of run-time fault-tolerance. At run-time, the basic structure of the tree or ring embedding has already been established and, as faults occur, this basic structure should not change.

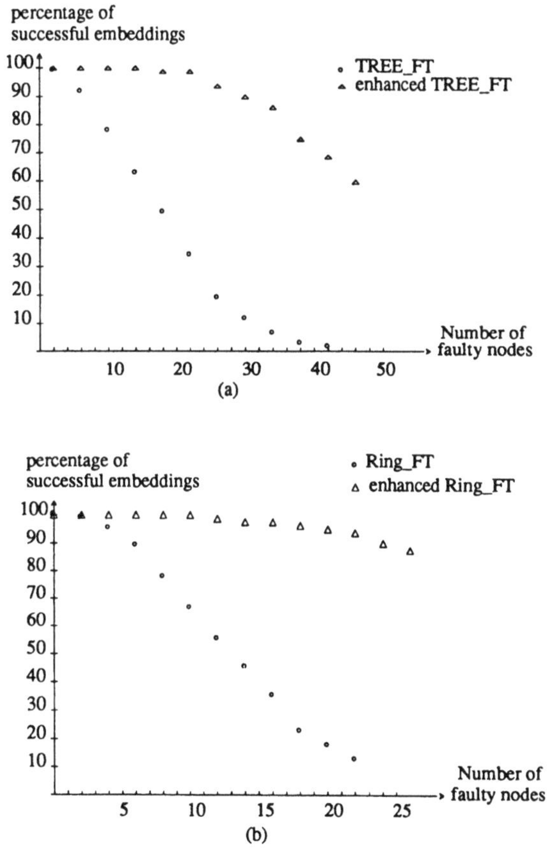

Figure 7 - Performance of the Fault Tolerant Algorithms
for (a) a tree and (b) a ring in a 512 node hypercube.

The problem of fault tolerance at the initial embedding of the tree or ring in the hypercube is slightly different. If the algorithm fails to embed the structure starting at the given node, there is no reason why it might not succeed from another starting node. Also, if the algorithm fails to embed the structure for some set of faults, it is possible that choosing a different assignment of logical dimensions to physical dimensions may produce a successful embedding.

We tried to make as few assumptions as possible regarding the hardware model. The fault model includes both node and link failures and assumes that the communications/routing facilities associated with a node fail along with the processing facilities. Certain hardware models would simplify the fault tolerant algorithms and make them more robust. For instance, in the tree case if the communications part of a node remains operable when the computation part fails (e.g. there is a separate routing part), this node would not have to be bypassed for communications purposes and more fault configurations could be tolerated; also, if upon failure a node short circuits all communication across a certain dimension (zero), a technique can be developed where only the relocated node is aware of any change in the embedded structure -- the parent and children send messages to the original location and these messages are transmitted across the short circuit to the node's image.

References

1. S. Bhatt and I. Ipsen, "How to Embed Trees in Hypercubes," Research Report YALEU/DCS/RR-443, December 1985.
2. S. Deshpande and R. Jenevein, "Scalability of a Binary Tree on a Hypercube," *Proceedings of the ICPP*, pp. 661-668, 1986 .
3. S. Hosseini, J. Kuhl, and S. Reddy, "Distributed Fault Tolerance of Tree Structures," *IEEE Transactions on Computers*, vol. C-36, no. 11, pp. 1378-1382, November 1987.
4. S.L. Johnsson, "Communication Efficient Basic Linear Algebra Computations on Hypercube Architectures," *Journal of Parallel and Distributed Computing*, vol. 4, pp. 133-172, 1987.
5. S.Y. Kung, C.W. Chang, and C.W. Jen, "Real-Time Configuration for Fault-Tolerant VLSI Array Processors," *Proceedings of Real-Time Systems Symposium*, pp. 46-54, December 1986.
6. F.J. Provost and R. Melhem, "Fault Tolerant Embedding of Binary Trees in Hypercubes," Technical Report 88-3, Department of Computer Science, University of Pittsburgh, Pittsburgh, PA, 1988.
7. C. Raghavendra, A. Avizienis, and M. Ercegovac, "Fault-Tolerance in Binary Tree Architectures," *IEEE Transactions on Computers*, vol. C-33, no. 6, pp. 566-572, June 1984.
8. M. Sami and R. Stefanelli, "Reconfigurable Architectures for VLSI Implementation," *Proc. Nat'l Computer Conf.*, pp. 565-577, May 1983.
9. A. Wu, "Embedding of Tree Networks into Hypercubes," *Journal of Parallel and Distributed Computing*, vol. 2, pp. 238-249, 1985.

ON THE ANALYSIS AND DESIGN OF HIERARCHICAL FAULT-TOLERANT PROCESSOR ARRAYS*

Y-X. Wang and **Jose A.B. Fortes**

Purdue University
School of Electrical Engineering
West Lafayette, IN 47907

1. INTRODUCTION

The relevance of Processor Arrays (PA's) stems not only from their ability to meet the computational demands of many real-time applications but also from their suitability for efficient implementations using Very Large Scale Integration/Wafer Scale Integration (VLSI/WSI) technology [1]. Fault-tolerant PA's (FTPA's) contain hardware redundancy for some of their components (globally referred to as "redundant hardware") as well as hardware for which spares are not provided (called "non-redundant hardware"). Typically, redundant hardware includes processor elements (PE's) and, sometimes, communication links, whereas non-redundant hardware may include circuitry for control, fault-detection, hardware reconfiguration, etc.. Low reliability in a bad FTPA design may result from, among other reasons, the unreliability of non-redundant hardware. This is particularly true for very large arrays and/or long operation times. This problem can be avoided in well designed hierarchical FTPA's, i.e., processor arrays organized as small fault-tolerant arrays of small FTPA's. This paper addresses the problems of analytically estimating and optimizing the reliability of hierarchical FTPA's.

Extensive work has been done towards devising FTPA's and a non-exhaustive list of references is [2–11]. There exist many fault-tolerant schemes such as the Column Redundancy (CR) ([3] [6] and others), Diogenes (DI) [5], Complex Fault Stealing (CFS) [7] and Triplicated Modular Redundancy (TMR) [10]. Hierarchical FTPA's or closely related approaches have been proposed ([13–16], [20–23] and others) for the yield improvement purposes. Because closed form expressions for the reliability of FTPA's are not easy to derive, one of the motivations of this paper is to provide the general reliability estimation model based on the properties of the hazard function that describes the instantaneous failure rate of FTPA's.

The basic ideas and the approximation model of the hazard function are studied in Section 2. Section 3 introduces two general approximation models for the reliability of FTPA's: the "single-Weibull" and "double-Weibull" approximation models. The first one is used to estimate the reliability of FTPA's where the reliabilities of unit *non-redundant area* (where the non-redundant hardware is laid) and *redundant area* (where the redundant hardware is laid) have the same value. The second one can be used to estimate the reliability of FTPA's where the two reliabilities are independent variables. In Section 4, the motivations for the use of hierarchically structured FTPA's are discussed and the methodology for the design of bi-level hierarchical FTPA's is presented. A case study is briefly

* This research was supported in part by the National Science Foundation under Grant DC1-8419745 and in part by the Innovative Science and Technology Office of the Strategic Defense Initiative Organization and was administered through the Office of Naval Research under contract No. 00014-85-k-0588.

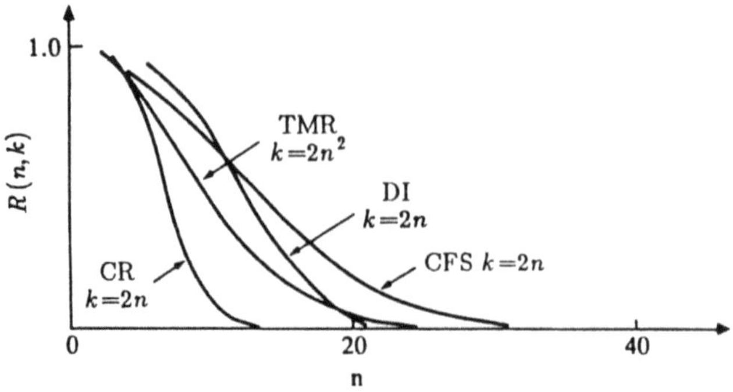

Figure 1. The reliability of different single-level FTPA schemes with (n×n)+k processors; different approaches are optimal for different values of n and low reliability results for large arrays.

described that illustrates the very high reliability improvement achievable by a bi-level structure in contrast with the poor reliability of single-level FTPA's with the same amount of redundancy. Section 5 is dedicated to conclusions.

2. THE OVERALL RELIABILITY AND HAZARD FUNCTION OF FTPA'S

The overall reliability of FTPA's depends on many factors, e.g., the reconfiguration scheme used, number of redundant components added, size of the array, complexity of PE's, reliability of switches and control circuits, etc. One reconfiguration scheme may have better performance than the others under some circumstances and have worse performance otherwise. For example, as shown in Figure 1, the DI scheme is better than CFS scheme if the size of FTPA's is small whereas for large size FTPA's, the opposite is true (see [13] for details and assumed design and technology parameters).

Without loss of generality, let's assume that an FTPA contains $(n \times n) + k$ PE's, where k is the number of spare PE's. Let $A_n(n,k)$ and $A_r(n,k)$ represent the area of non-redundant and redundant hardware in FTPA's, and $R_r(n,k,r_r)$ and $R_n(n,k,r_n)$ denote the reliabilities of the redundant and non-redundant hardware respectively, where r_r is the reliability of a unit of redundant area (in this paper, the area for a single PE is defined as the unit of area), and r_n is the reliability of a unit of non-redundant area. The reliability of an FTPA can be written as

$$R(n,k,r_n,r_r) = R_n(n,k,r_n) \cdot R_r(n,k,r_r) \quad . \tag{2.1}$$

Clearly, the overall reliability of an FTPA is at most as good as the lowest of the reliabilities of the redundant hardware and the non-redundant hardware.

In an FTPA, if some PE's fail, the reconfiguration scheme will try to replace these faulty PE's and make the whole system operational. For a given FTPA with $(n \times n)+k$ PE's, let r_r be the reliability of each PE, then the overall reliability can be written as

$$R(n,k,r_n,r_r) = r_n^{A_n(n,k)} \cdot \sum_{i=0}^{k} s_i \binom{n^2+k}{i} r_r^{(n^2+k-i)}(1-r_r)^i, \tag{2.2}$$

where s_i is the probability that the reconfiguration scheme is successful with i faulty PE's in the array. Expressions (2.1) and (2.2) are valid for any type of FTPA. However, in order to facilitate the presentation and discussion of the ideas and results, two assumptions are introduced next and apply to the remainder of this paper. For a given FTPA, it is assumed that, (1) all PE's are redundant hardware and the remaining circuitry corresponds to non-redundant hardware; (2) at the exception of the mechanisms and scheme used for reconfiguration, all other procedures and hardware required for fault tolerance are perfect (e.g., fault detection and location).

In general, the Reliability Distribution Function of any FTPA component is assumed to be exponential with operation time. Then the corresponding hazard function of each component is a constant [12]. In a simplex processor array (the array without any redundancy), a failure in any component causes the whole array to fail and the hazard function of the entire array, h(t), is also a constant, i.e., $h(t)=\lambda>0$. The corresponding reliability is $R(t)=e^{-\lambda t}$.

In FTPA's, a faulty redundant hardware component (e.g. a PE) may be replaced if there is a spare available in the array. Clearly, at time t=0, all components (including the spare ones) are assumed to be good, the probability of a successful reconfiguration is high, and the total number of mortal failures of the array per unit time is low. However, as time goes on, i.e., when the value of t becomes large, some of the spare components are exhausted. The possibility of recovery from additional failures becomes smaller, i.e., the probability of successful reconfiguration is lower and the value of h(t) will increase. Eventually, when most of the spare components are exhausted, the hazard function h(t) will tend to the constant which is the hazard rate of the array with redundancy exhausted. The value of h(t) depends on many factors and, as for the reliability R(t), it is very difficult to obtain the exact expression of h(t). However, the following simple properties of h(t) can be inferred from the discussions above: i) It should be a monotonic increasing function of time t and tend to a constant. ii) It is a continuous function and changes its value smoothly in range $(0, +\infty)$. The well known Weibull estimation model [12 17-18] will be studied and used to approximate the h(t) based on these properties. In the rest of this paper, $R^*(t)$ and $h^*(t)$ denote approximations of $R(t)$ and $h(t)$, respectively.

The two-parameter Weibull estimation model has the form

$$h^*(t)=a\cdot t^b, \quad a,\ b\in IR,\ a>0 \text{ and } b>0. \tag{2.3}$$

The parameters a and b are called the *scale parameter* and *shape parameter*, respectively. The corresponding formula for the reliability function is

$$R^*(t)=e^{-\int h(t)\,dt}=e^{-\frac{a}{b+1}\cdot t^{b+1}} \tag{2.4}$$

The expressions for both $h^*(t)$ and $R^*(t)$ are very simple compared to other possible approximations, such as Gamma, Beta and Lognormal distributions [12 19]. On the other hand, according to our experience and previous work [17-18], the Weibull distribution model is very flexible and capable of providing a rich family of approximations for increasing (and decreasing) hazard functions h(t) by adjusting the scale parameter a and the shape parameter b. Because, for FTPA's, $h(t)$ is a positive monotonic increasing function of t, parameters a and b are always positive numbers.

3. GENERAL MODELS FOR FTPA RELIABILITY

The reliability of FTPA's is not only a function of time but also a function of the size of FTPA's. A generalized reliability evaluation model that captures this dependence on size is studied in this section.

For a simplex PA of size N (i.e., N= number of PE's), the hazard function, in terms of the size, is a constant, i.e., $h(N)=\alpha=\ln\frac{1}{r_r}$. The corresponding reliability is $R(N)=r_r^N=e^{-\alpha\cdot N}$. For FTPA's, instead of a constant, the hazard function is a monotonic increasing function of N. This can be explained by the following two facts. First, *the redundancy factor* $\eta=\frac{N}{N-k}$, decreases significantly with the increasing of N. This may cause a shortage of the spares to replace faulty PE's in a very large array and implies that the probability of the failure of the whole system increases with the increase of N. Secondly, because the size of the non-redundant area A_n, usually, is proportional to the size of the array, the reliability of the non-redundant hardware $R_n=r_n^{A_n}$ decreases with the increase of N.

As mentioned in Section 2, the Weibull model can provide a simple and good approximation for those non-linear monotonic increasing (or decreasing) hazard functions. So, h(N) can be modeled by another Weibull formula, i.e.,

$$h^*(N)=a'\cdot N^{b'},\quad a',\ b'\in IR, \text{ and } a'>0,\ b'>0. \tag{3.1}$$

Without loss of generality, let's assume that the variable N is a continuous variable; then the corresponding reliability in terms of the size N can be written as

$$R^*(N) = e^{-\int h(N)\,dN} = e^{-\alpha N^{\beta}}. \tag{3.2}$$

where $\alpha = \frac{a'}{b'+1}$ and $\beta = b'+1$.

There is no time variable t in Equation 3.1 which implies the time is fixed in the equation, i.e., $t=t_0$. Assume that, for a given reconfiguration scheme, the parameter b', which describes the sensitivity of $h^*(N)$ to the size N, is independent of the time t; in other words, assumed to be the increasing rate of the hazard rate, $\frac{h^*(N+1)}{h^*(N)}$, for a given fault-tolerant scheme, is independent of time (however, parameter a' should be a function of t). If parameter b in Equation 2.3 is independent of the size N, then the hazard rate can be approximated by a model with two variables.

$$h^*(N,t) = c \cdot t^{b} \cdot N^{b'}, \quad c>0,\ b>0,\ b'>0,\ c,\ b,\ b' \in IR \tag{3.3}$$

where c is a new parameter. (The error of the two variable approximation model may increase if parameters b and b' are not independent of N and t, respectively.) The corresponding two variable model for the reliability function is

$$R^*(N,t) = e^{-\int\int h^*(N,t)\,dN\,dt} = e^{-\frac{c}{(b+1)(b'+1)} \cdot t^{b+1} \cdot N^{b'+1}} = e^{-\alpha \cdot t^{\beta} \cdot N^{\gamma}} \tag{3.4}$$

where parameters $\alpha = \frac{c}{(b+1)(b'+1)}$, $\beta = b+1$ and $\gamma = b'+1$ are positive real numbers.

Besides the time t and the size N, the reliability of FTPA also depends strongly on r_r, the reliability of single PE which is determined by the complexity of PE, technology etc.. Because $r_r(t) = e^{-\lambda t}$, where λ is the failure rate of a single PE, it follows that $t = \frac{-1}{\lambda} \ln r_r(t)$. Let $\delta = \frac{1}{\lambda} > 0$. By bringing t into Equation 3.4, the reliability then can be written in terms of N, and r_r as

$$R^*(N, r_r(t)) = e^{-\alpha \cdot \delta^{\beta} \cdot (-\ln r_r(t))^{\beta} \cdot N^{\gamma}}. \tag{3.5}$$

Instead of a function of time t, r_r itself can be an independent variable. Letting $\alpha' = \alpha \cdot \delta^{\beta}$, Equation 3.5 can be rewritten as

$$R^*(N, r_r) = e^{-\alpha' \cdot (-\ln r_r)^{\beta} \cdot N^{\gamma}} \tag{3.6}$$

In general, if the reliability of unit non-redundant area always has the same value as the reliability of a single PE, i.e., $r_n = r_r$, then the reliability of the array can be described by a ***single-Weibull distribution formula*** in terms of N and r_r as in Equation 3.6. However, in some cases, e.g., in hierarchical fault-tolerant structures [14–16], r_n and r_r may vary independently (details are shown in Section 4). Therefore, the reliability of redundant hardware and non-redundant hardware should be described separately in terms of variables r_r and r_n. Because the overall reliability of the system is the product of the reliability of the non-redundant hardware and the redundant hardware, another model, ***double-Weibull distribution formula,*** is used to estimate the reliability,

$$R^*(N, r_n, r_r) = R_n^*(N, r_n) \cdot R_r^*(N, r_r) = e^{-\alpha_n (-\ln r_n)^{\beta_n} N^{\gamma_n}} \cdot e^{-\alpha_r (-\ln r_r)^{\beta_r} N^{\gamma_r}} \tag{3.7}$$

where α_n, β_n, γ_n, α_r, β_r and γ_r are the six parameters that must be estimated for different FTPA's. Equations 3.6 and 3.7 are used in next section to study the different reconfiguration schemes and to derive a methodology to find the optimal structure of a bi-level FTPA.

4. THE ANALYSIS AND DESIGN OF HIERARCHICAL FTPA'S

The basic idea behind the design of a hierarchical FTPA is best explained if the particular case of a ***bi-level FTPA*** is considered first. A bi-level FTPA consists of a set of fault-tolerant subarrays. In other words, the full array is partitioned into subarrays and can be thought of as an array of subarrays. Both the subarrays and the array of subarrays use some fault-tolerance scheme. The subarrays are hereon denoted as ***1st-level FTPA's*** and the array of subarrays is referred to as the *2nd-level FTPA*. A 2nd-level FTPA can be thought of as an FTPA where the basic modules are themselves FTPA's and, physically, it is the same as the bi-level FTPA. The extension to multi-level FTPA's can be easily made by realizing that an n-level FTPA consists of an FTPA whose basic modules are (n-1)-level FTPA's which are FTPA's composed of (n-2)-level FTPA's, etc. For convenience of presentation, bi-level arrays are assumed hereon and, unless stated otherwise, the basic ideas and results apply to multi-level arrays as well.

When faults occur in a bi-level array, reconfiguration is first attempted at the 1st-level FTPA's, and, when reconfiguration fails at the 1st-level, then reconfiguration at the 2nd-level FTPA is attempted. The total non-redundant hardware in a bi-level structure consists of the switches, links and control circuits at the 2nd-level. Therefore, bi-level FTPA's can be expected to have better reliability than single-level FTPA's for at least two reasons: (1) the area of non-redundant hardware in bi-level FTPA's can be smaller than that in single-level FTPA's and (2) the size of arrays and the reconfiguration approach used at each level can be chosen so that optimal reliability results, thus avoiding the inevitable reliability degradation that occurs when the size of single-level arrays grows too large.

Having realized the potential benefits of multi-level FTPA's, from an engineering point of view, it is essential to have a systematic and formal approach to the design of these systems so to optimize the reliability of the overall FTPA. This approach is described in the remainder of this paper.

4.1 The optimization of hierarchical structure

Without loss of generality, let the size of the 1st-level and 2nd-level FTPA's be $(n_1 \times n_1) + k_1$ and $(n_2 \times n_2) + k_2$, respectively. Here, k_1 and k_2 correspond to the number of spare PE's in 1st-level and the number of spare subarrays in 2nd-level, respectively. Also, the number of processors in the bi-level array is $(n \times n) + k$ where $n = n_1 \times n_2$ and $k = k_1\ n_2^2 + k_2\ (n_1^2 + k_1)$. The reliability of the bi-level FTPA is essentially the reliability of the 2nd-level FTPA, i.e., $R_2 = R_2\ (n_2, k_2, R_1)$ where R_1 is the reliability of one 1st-level FTPA, i.e., $R_1 = R_1\ (n_1, k_1, r_r)$. Assume that k_1 and k_2 can be expressed as the functions of n_1 and n_2 for the given reconfiguration schemes, R_1 and R_2 can be simplified as $R_1 = R_1(n_1, r_r)$ and $R_2 = (n_2, R_1)$. For a given array size n, different values of n_2 significantly affect both R_1 and R_2. A small value of n_1 may yield a higher value of R_1 but does not necessarily provide a high value on R_2 because of the increase of n_2 $(n_2 = n/n_1)$. The example of the reliability of a bi-level FTPA against the number of subarrays n_2 is shown in Figure 2. The value of n_2 for which R_2 is optimal is the solution of the equation

$$\frac{dR_2}{dn_2} = \frac{\partial R_2}{\partial n_2} + \frac{\partial R_2}{\partial R_1} \cdot \frac{\partial R_1}{\partial n_2} + \frac{\partial R_2}{\partial k_2} \cdot \frac{\partial k_2}{\partial n_2} = 0 \tag{4.1}$$

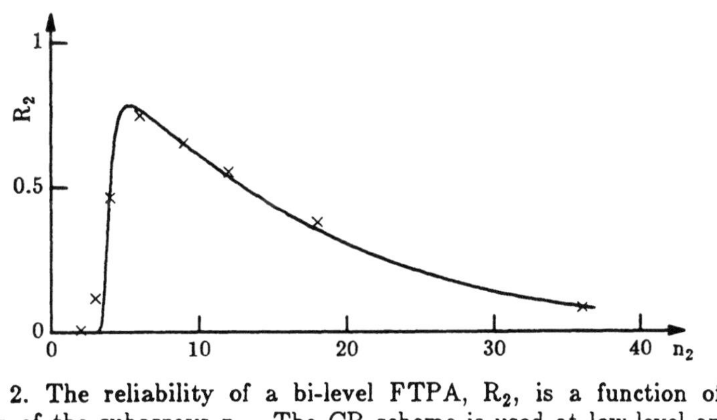

Figure 2. The reliability of a bi-level FTPA, R_2, is a function of the number of the subarrays n_2. The CR scheme is used at low-level and DI scheme is used at high-level with n=36. The values denoted by "×" are obtained by simulation and the solid line is according to the approximation model, i.e., Equation 4.10.

Once the solution n_2^* is obtained, n_1^* can be found because n is assumed to be known. Unfortunately, for a bi-level FTPA design, Equation 4.1 is hard to obtain and solve. In order to deal with this problem, a new methodology based on the general single- and double-Weibull approximation models discussed in Section 3 is proposed in which individual models are used to estimate the reliabilities of the first and the second levels.

Without loss of generality, assume that the area of hardware is normalized so that a single PE always occupies one unit of chip area. Let the reliability of hardware in any unit of area have the same value regardless the kind of circuitry in it. For the reconfiguration schemes used at the first level, because the reliability of a basic PE is given as a constant, the single variable for R_1 is the size of subarrays n_1. The single-Weibull approximation model can be used to estimate $R_1(n_1)$ as

$$R_1^*(n_1) = e^{-\alpha' \cdot n_1^\beta} \tag{4.2}$$

where α' and β are real positive constants.

The approximation form R_2^* of the reliability for the second-level as a function of n_2 is more involved than R_1^*. This is because the reliability of one module (subarray) in 2nd-level is $R_1(n_1)$ which is the function of n_1 (it is also a function of n_2 for a fixed n because $n_1 = n/n_2$) instead of a constant (as r_r in the 1st-level). However, the reliability of unit non-redundant area r_n is still a constant. The reliability of redundant and non-redundant area R_r and R_n in 2nd-level will not always vary congruously with n_2. For example, sometimes, a decrease of R_n due to a larger value of n_2 may cause R_r to increase. Because the goal here is to find the trade-off between n_2 and n_1 on which the overall reliability is optimized, the single-Weibull model can not be used here any more and, the reliability R_r and R_n should be estimated separately in order to represent their variations with n_2. According to the general expression of the reliability in Equation 2.1, the reliability in 2nd-level for some reconfiguration scheme can be written as

$$R_2(n_2, r_n, R_1(n_2)) = R_n(n_2, r_n) \cdot R_r(n_2, R_1(n_2)) \tag{4.3}$$

where

$$R_n(n_2, r_n) = r_n^{A_n(n_2)} \tag{4.4}$$

Because r_n is a constant, $R_2(n_2, r_n, R_1(n_2))$ can be written in terms of n_2 and R_1

$$R_r(n_2, R_1(n_2)) = \sum_{i=0}^{k_2} s_i \binom{n_2^2 + k_2}{i} R_1^{(n_2{}^2 + k_2 - i)} (1 - R_1)^i \quad . \tag{4.5}$$

A_n is the non-redundant area in 2nd-level and k_2 is the number of spare subarrays which is, usually, a known function of n_2 for a given reconfiguration scheme. The double-Weibull approximation model then should be used to express the reliability of the reconfiguration schemes used in 2nd-level of FTPA. Because r_n is a constant, R_n can be modeled as

$$R_n^*(n_2) = e^{-a \cdot n_2^b} \tag{4.6}$$

where a and b are positive real number parameters. From (4.4) and (4.6), it is clear that $A_n(n_2)$ is approximated by $\frac{-a}{\ln r_n} \cdot n_2^b$, i.e., the non-redundant area is assumed to grow proportionally with some power of the number of processors in the array.

The approximation form used for $R_r(n_2, R_1)$ has two variables, n_2 and R_1,

$$R_r^*(n_2, R_1) = e^{-(-\ln R_1)^u \cdot v \cdot n_2^w} \tag{4.7}$$

where u, v and w are positive real constants. In summary, the double-Weibull approximation form R_2 is

$$R_2^* = e^{-a \cdot n_2^b} \cdot e^{-(-\ln R_1)^u \cdot v \cdot n_2^w} \tag{4.8}$$

Because $n_1 = n/n_2$, letting $\alpha = \alpha' \cdot n^\beta$, (4.2) can be rewritten as

$$R_1^* = e^{-\alpha \cdot n_2^{-\beta}} . \tag{4.9}$$

Substituting R_1^* into (4.8) yields

$$R_2^* = e^{-(a\cdot n_2^b + \alpha^u\cdot n_2^{-\beta u}\cdot v\cdot n_2^w)}$$
$$= e^{-(a\cdot n_2^b + \alpha^u\cdot v\cdot n_2^{(w-\beta\cdot u)})}. \tag{4.10}$$

Equation (4.1) can now be solved, i.e.,

$$\frac{dR_2^*}{dn_2} = [a\cdot b\cdot n_2^{b-1} + \alpha^u\cdot v(w - \beta\cdot u)\cdot n_2^{(w-\beta\cdot u-1)}]\ e^{-(a\cdot n_2^b + \alpha^u\cdot v\cdot n_2^{(w-\beta\cdot u)})} = 0. \tag{4.11}$$

Because $R_2^*\neq 0$, $\frac{dR_2^*}{dn_2} = 0$ if and only if

$$a\cdot b\cdot n_2^{b-1} + \alpha^u\cdot v(w - \beta\cdot u)\cdot n_2^{(w-\beta\cdot u-1)} = 0. \tag{4.12}$$

Also let's assume $n_2\neq 0$, (4.12) can be simplified as

$$a\cdot b\cdot n_2^b + \alpha^u\cdot v(w - \beta\cdot u)\cdot n_2^{(w-\beta\cdot u)} = 0. \tag{4.13}$$

Because all parameters are positive real numbers, the necessary and sufficient condition for the existence of a real positive solution to (4.12) is

$$w - \beta\cdot u < 0 \tag{4.14}$$

There are many fault-tolerant schemes (with different values on the parameters u, w and β) which can be used to construct the bi-level structure. If (4.14) is not satisfied for two chosen schemes, then Equation 4.13 has no real number solution and R_2 gets its maximum at one of the boundary points, i.e., n_2=1 or n_2=$n\times n$. This means the bi-level structure can not provide better reliability with these two schemes than a single-level structure with one of these schemes. If this happens, other schemes should be considered and checked by (4.14). If (4.14) holds for two specific schemes, Equation (4.13) can be rewritten as

$$a\cdot b\cdot n_2^{b-(w-\beta\cdot u)} + \alpha^u\cdot v(w - \beta\cdot u) = 0. \tag{4.15}$$

Letting $\theta = w - \beta\cdot u$ and $\phi = -\alpha^u\cdot v\cdot\theta/(a\cdot b)$, then the optimal solution is

$$n_2^* = \phi^{1/(b-\theta)} \tag{4.16}$$

Substituting (4.16) in (4.10) yields the maximum reliability attainable with a bi-level FTPA using two given fault-tolerance schemes

$$R_2^* = e^{-[a\cdot\phi^{b/(b-\theta)} + \alpha^c\cdot d\cdot\phi^\theta/(b-\theta)]}. \tag{4.17}$$

In summary, the optimal hierarchical structure can be obtained by the following four steps: 1) Getting the parameters of Weibull distribution model (via simulation or estimation) for all previous fault-tolerant schemes that are possibly to be used on the hierarchical structure. 2) Choosing different combinations of the schemes in 1st and 2nd level and checking Equation 4.14 to find the possible candidates. 3) Substituting the parameters of the possible combinations of the schemes into (4.17), the one which maximizes R_2^* is the optimal combination of the schemes. 4) Substituting the chosen parameters into (4.16) to achieve the optimal partition n_2 of the hierarchical structure.

4.2 An illustrative example

As an example, the design of a bi-level FTPA with n=36 is considered where the simple CR approach is to be used in the 1st-level and the DI scheme is used in the 2nd-level. For both CR and DI schemes a column of spare processors is used. The area and reliability estimations were computed for both the CR and the DI methods and the reliabilities for each of the levels were approximated as

$$R_1 \approx e^{-14.88 n_2^{-2.957}} \tag{4.18}$$

$$R_2 \approx e^{-[0.027 n_2^{1.26} + 0.64\ (ln\frac{1}{R_1})^{4.742}\ n_2^{4.739}]} \tag{4.19}$$

Assume that $r_r = r_n = 0.99$. The values of the variables in (4.16) are $\theta \approx -11.24$, $b \approx 1.26$, $\phi \approx 1.144\times10^9$ and the optimal value of n_2 is $n_2^* = 5.3$. Because both n_1^* and n_2^* should be integer numbers with $n_1^* \times n_2^* = 36$, the number $n_2^* = 6$ is used. Then, there are total 1764 PE's in the array and the redundancy factor η is 1.36. The overall reliability is 0.75, where the reliabilities for a single-level array using the DI approach or the CR approach with the same redundancy factor are 0.08 and 0.31, respectively.

5. CONCLUSIONS

Several important related conclusions can be made from the work reported in this paper. First, according to the properties of the hazard rate of the FTPA's, the Weibull estimation model can give a good evaluation on the reliability of different FTPA's. The second conclusion is that, based on reliability estimations, it can be found that non-redundant hardware for fault-tolerant purposes does limit the usefulness of single-level FTPA's beyond a certain size. For different fault-tolerant schemes, different array sizes and different area and technology parameters, there does not exist a scheme for FTPA's which is universally optimal. In other words, different fault-tolerant schemes are optimal for different array sizes and different technologies of VLSI. The third conclusion is that the hierarchical FTPA's do not suffer from the disadvantage pointed out in the second conclusion for single-level FTPA's and can be made highly reliable.

To achieve the third conclusion mentioned above, the problem of designing optimal bi-level FTPA's was addressed and a methodology for its solution has been described. The key to this methodology is to avoid the complexity of the exact analytical expressions by using accurate functional approximations of the reliability of FTPA's at different levels. These approximations are based on Weibull reliability functions.

6. REFERENCES

[1] C. Mead and L. Conway, "Introduction to VLSI Systems," Addison-Wesley, Reading, Mass., 1980.

[2] J. A. Abraham, P. Banerjee, C-Y. Chen, W. K. Fuchs, S-Y. Kuo and A. L. N. Reddy, "Fault Tolerance Techniques for Systolic Arrays," Computer, July 1987, pp. 65-74.

[3] I. Koren and M. A. Breuer, "On Area and Yield Considerations for Fault-Tolerant Processor Arrays," IEEE Trans. on Computers, Jan. 1984, pp. 21-27.

[4] S. Y. Kung, "VLSI Array Processors," Prentice-Hall, Englewood Cliffs, N.J. 1987.

[5] A. L. Rosenberg, "The Diogenes Approach to Testable Fault Tolerant Arrays of Processors," IEEE Trans. on Computers, Oct. 1983, pp. 902-910.

[6] J. A. B. Fortes, C. S. Raghavendra, "Gracefully Degradable Processor Arrays," IEEE Trans. on Computers, Nov. 1985, pp., 1033-1044.

[7] M. Sami and R. Steffanelli, "Fault-Tolerance and Functional Reconfiguration in VLSI Arrays," International Conference on Circuits and Systems, 1986.

[8] H. T. Kung and M. S. Lam, "Fault-Tolerance and Two-Level Pipelining in VLSI Systolic Arrays," Proc. Conf. Advanced Research in VLSI, Jan. 1984, pp. 74-83.

[9] J-H. Kim and S.M. Reddy, "A Fault-Tolerant Systolic Array Design Using TMR Method," Proc. IEEE Int'l Conf. Comp. Design: VLSI in Computer, Oct. 1985, pp. 769-773.

[10] J. Von Neumann, "Probabilistic Logics and the Synthesis of Studies," C.E. Shannon and J. McCarthy (Ed.), Princeton University Press, Princeton, New Jersey, 1956.

[11] I. Koren, "Comments on "The Diogenes Approach to Testable Fault-Tolerant Arrays of Processors,"" IEEE Trans. on Comp. Vol. C-35, No.1, Jan. 1986, pp. 93.

[12] D.P. Siewiorek and R.S. Swarz, "The Theory and Practice of Reliability System Design," Digital Press, Educational Services, DEC, Bedford, MASS., 1982.

[13] Y. Wang and J.A.B. Fortes, "Hierarchical Approaches to Fault-Tolerance in Processor Arrays," SPIE's 1987 Tech. Symp. on Optics, Electro-Optics, and Sensors, 1987.

[14] T. Ishikawa, S. Momoi, S. Shimada and Y. Ogawa, "Hierarchical Array Processor (HAP) Featuring High Reliability and High System Performance," 1986 Int'l. Conf. on Parallel Processing

[15] M. Wang, M. Cutler and S.Y.H. Su, "On-Line Error Detection and Reconfiguration of Array Processors with Two-Level Redundancy," Proc. Comp. on VLSI and Computers, May, 1987.

[16] N. Tsuda and T. Satoh, "Hierarchical Redundancy for a Linear-Array Sorting Chip," I.F.I.P. international workshop on Wafer-Scale Integration, 23-25 September 1987.

[17] A. Adatia and L.K. Chan, "Robust Estimators of the 3-Parameter Weibull Distribution," IEEE Trans. Reliability, October 1985.

[18] H.L. Harter, A.H. Moore and R.P. Wiegand, "Sequential Tests of Hypotheses for System Reliability Modeled by a 2-Parameter Weibull Distribution," IEEE Trans. Reliability, October 1985.

[19] J.B. McDonald and D.O. Richards, "Hazard Rates and Generalized Beta Distributions," IEEE Trans. Reliability, October 1987.

[20] K.S. Hedlund and L.Snyder, "Systolic Architectures-A Wafer Scale Approach," IEEE 1984 Int'l. Conf. on Comp. Design: VLSI in Computers, 1984, pp. 604-610.

[21] S.Y. Kung, C.W. Chang and C.W. Jen, "On Fault-Tolerance in Array Processors," IEEE 1985 Int'l. Conf. on Comp. Design, pp.764-768 1985.

[22] C. Jesshope and L. Bentley, "Techniques for Implementing two-dimensional wafer-scale processor arrays," IEE Proceedings, Vol. 134, Pt.E, No.2, March 1987, pp. 87-92.

[23] J.H. Hwang and C.S.Raghavendra, "VLSI Implementation of Fault-Tolerant Systolic Arrays," Proc. IEEE Int'l. Conf. on Comp. Design: VLSI in Computers, 1986, pp. 110-113.

Contributors

Vinod K. Agarwal • Department of Electrical Engineering, McGill University, Montreal, Quebec H3A 2A7, Canada

Vishwani D. Agrawal • Computer Systems Research Department, AT&T Bell Laboratories, Murray Hill, NJ 07974

Robert Aitken • Department of Electrical Engineering, McGill University, Montreal, Quebec H3A 2A7, Canada

Brian Boerman • Intel Corporation, 3065 Bowers Avenue, Santa Clara, CA 95051

C.W. Chang • Department of Electrical Engineering, University of Southern California, Los Angeles, CA 90089

Ming-Feng, Chang • Coordinated Science Laboratory, University of Illinois, Urbana, IL 61801

Savio Chau • Department of Computer Science, University of California, Los Angeles, CA 90024

J. Cong • Department of Computer Science, University of Illinois, Urbana, IL 61801

R.J. Cosentino • The MITRE Corporation, Bedford, MA 01730

Michal Cutler • Department of Computer Science, State University of New York at Binghamton, Binghamton, NY 13901

A. Danor • National Semiconductor (Israel) Ltd., Herzlia, Israel

A.V. Ferris-Prabhu • IBM General Technology Division, Essex Junction, VT 05452

Jose A.B. Fortes • School of Electrical Engineering, Purdue University, West Lafayette, IN 47907

W. Kent Fuchs • Coordinated Science Laboratory, University of Illinois, Urbana, IL 61801

Donald Fussell • Department of Computer Sciences, The University of Texas at Austin, Austin, TX 78712

Reinhard Glang • Federal Systems, IBM Corporation, Manassas, VA 22110

M. Glesner • Technische Hochschule Darmstadt, Schloßgartenstrasse, 8, 6100 Darmstadt, F.R.G.

Gensuke Goto • Fujitsu Laboratories Ltd., Atsugi 10-1, Morinosato-Wakamiya, Atsugi 243-01, Japan

C.G. Guy • Department of Electronic Engineering, UCL, Torrington Place, London WC1E JE, UK

R. Hadaway • Northern Telecom Electronics, Ottawa, Ontario K1Y 4H7, Canada

James J. Hammond • Intel Corporation, 3065 Bowers Avenue, Santa Clara, CA 95051

Nany Hasan • Department of Computer Science, University of Illinois, Urbana, IL 61801

Shinpei Hijiya • Fujitsu Laboratories Ltd., Atsugi 10-1, Morinosato-Wakamiya, Atsugi 243-01, Japan

J.-L. Houle • Department of Electrical Engineering, Ecole Polytechnique de Montreal, Montreal, Quebec H3C 3A7, Canada

Michael C. Howells • Department of Electrical Engineering, McGill University, Montreal, Quebec H3A 2A7, Canada

M. Huch • Technische Hochschule Darmstadt, Schloßgartenstrasse, 8, 6100 Darmstadt, F.R.G.

P. Ivey • British Telecom Research Laboratories, UK

Ajai Jain • Department of Electrical Engineering, McGill University, Montreal, H3A 2A7, Canada

S.N. Jean • Department of Electrical Engineering, University of Southern California, Los Angeles, CA 90089

Leonardo Jervis • Department of Electronics, Politecnico di Milano, Milano, Italy

B.L. Johnson • The MITRE Corporation, Bedford, MA 01730

V. Karpati • National Semiconductor (Israel) Ltd., Herzlia, Israel

Israel Koren • Department of Electrical and Computer Engineering, University of Massachusetts, Amherst, MA 01003

E.-F. Kouka • Institut National Polytechnique de Grenoble, 46, Avenue Felix Viallet, 38031 Grenoble Cedex, France

S.Y. Kung • Department of Electrical Engineering, Princeton University, Princeton, NJ 08544

Regis Leveugle • Institut National Polytechnique de Grenoble, 46, Avenue Felix Viallet, 38031 Grenoble Cedex, France

C.L. Liu • Department of Computer Science, University of Illinois, Urbana, IL 61801

Fabrizio Lombardi • Department of Computer Science, Texas A&M University, College Station, TX 77843

Steven Magdo • IBM General Technology Division, East Fishkill, NY 12533

Miroslaw Malek • Department of Electrical and Computer Engineering, The University of Texas at Austin, Austin, TX 78712

D. Marron • SGS-Thomson Microelectronics, France

Nobutake Matsumura • Fujitsu Limited, 1015, Kamikodanaka, Nakahara-ku, Kawasaki 211, Japan

B. Maytal • National Semiconductor (Israel) Ltd., Herzlia, Israel

P. Mazumder • Department of Electrical Engineering and Computer Science, University of Michigan, Ann Arbor, MI 48109

Rami Melhem • Department of Computer Science, University of Pittsburgh, Pittsburgh, PA 15260

T. Midwinter • British Telecom Research Laboratories, UK

D.Y. Montuno • Bell Northern Research, Ottawa, Ontario K1Y 4H7, Canada

Will R. Moore • Department of Engineering Science, Oxford University, Oxford OX1 3PJ, UK

B. Nasreddine • Institut National Polytechnique de Grenoble, 46, Avenue Felix Viallet, 38031 Grenoble Cedex, France

R. Nassrallah • National Semiconductor (Israel) Ltd., Herzlia, Israel

C.A. Njinda • Department of Engineering Science, Oxford University, Oxford OX1 3PJ, UK

J.-L. Patry • Institut National Polytechnique de Grenoble, 46, Avenue Felix Viallet, 38031 Grenoble Cedex, France

Douglas L. Peltzer • Tactical Fabs, Inc., San Jose, CA 95112

Foster J. Provost • Department of Computer Science, University of Pittsburgh, Pittsburgh, PA 15260

Janusz Rajski • Department of Electrical Engineering, McGill University, Montreal H3A 2A7, Canada

Sampath Rangarajan • Department of Computer Science, The University of Texas at Austin, Austin, TX 78712

David Rennels • Department of Computer Science, University of California, Los Angeles, CA 90024

Gabriele Saucier • Institut National Polytechnique de Grenoble, 46, Avenue Felix Viallet, 38031 Grenoble Cedex, France

Yvon Savaria • Department of Electrical Engineering, Ecole Polytechnique de Montreal, Montreal, Quebec H3C 3A7, Canada

Peter Schvan • Northern Telecom Electronics, Ottawa, Ontario K1Y 4H7, Canada

Donatella Sciuto • Department of Computer Science, Texas A&M University, College Station, TX 77843

Sharad C. Seth • Department of Computer Science, University of Nebraska, Lincoln, NE 68588

E. Shihadeh • National Semiconductor (Israel) Ltd., Herzlia, Israel

Y. Sidi • National Semiconductor (Israel) Ltd., Herzlia, Israel

Adit D. Singh • Department of Electrical and Computer Engineering, University of Massachusetts, Amherst, MA 01003

G.S. Sohi • Department of Computer Sciences, University of Wisconsin, Madison, WI 53706

Mohammad Soueidan • Institut National Polytechnique de Grenoble, 46, Avenue Felix Viallet, 38031 Grenoble Cedex, France

Charles H. Stapper • IBM Corporation, General Technology Division, Essex Junction, VT 05452

Stephen Y.H. Su • Department of Computer Science, State University of New York at Binghamton, Binghamton, NY 13901

C. Thibeault • Department of Electrical Engineering, Ecole Polytechnique de Montreal, Montreal, Quebec H3C 3A7, Canada

J. Trilhe • SGS-Thomson Microelectronics, France

J.J. Vaccaro • The MITRE Corporation, Bedford, MA 01730

Fred W. Voltmer • Intel Corporation, 3065 Bowers Avenue, Santa Clara, CA 95051

Minghsien Wang • Department of Computer Science, State University of New York at Binghamton, Binghamton, NY 13901

Y. Wang • Institut National Polytechnique de Grenoble, 46, Avenue Felix Viallet, 38031 Grenoble Cedex, France

Y.-X. Wang • School of Electrical Engineering, Purdue University, West Lafayette, IN 47907

Norbert Wehn • Technische Hochschule Darmstadt, Schloßgartenstrasse, 8, 6100 Darmstadt, F.R.G.

Koichi Yamashita • Fujitsu Laboratories Ltd., Atsugi 10-1, Morinosato-Wakamiya, Atsugi 243-01, Japan

Hee Yong Youn • Department of Computer Science, University of North Texas, Denton, TX 76703

Index

MIX
Papier aus verantwortungsvollen Quellen
Paper from responsible sources
FSC® C105338

If you have any concerns about our products,
you can contact us on
ProductSafety@springernature.com

In case Publisher is established outside the EU,
the EU authorized representative is:
Springer Nature Customer Service Center GmbH
Europaplatz 3, 69115 Heidelberg, Germany

Printed by Libri Plureos GmbH
in Hamburg, Germany